PATTERN FORMATION IN THE PHYSICAL AND BIOLOGICAL SCIENCES

PATTERN FORMATION IN THE PHYSICAL AND BIOLOGICAL SCIENCES

Editors

H. F. Nijhout
Department of Zoology
Duke University

Lynn Nadel
Department of Psychology
University of Arizona

Daniel L. Stein
Department of Physics
University of Arizona

Lecture Notes Volume V

Santa Fe Institute
Studies in the Sciences of Complexity

The Advanced Book Program

CRC Press
Taylor & Francis Group
Boca Raton London New York

CRC Press is an imprint of the
Taylor & Francis Group, an **informa** business
A CHAPMAN & HALL BOOK

Director of Publications, Santa Fe Institute: Ronda K. Butler-Villa
Production Manager, Santa Fe Institute: Della L. Ulibarri
Publication Assistant, Santa Fe Institute: Marylee Thomson

First published 1997 by Addison-Wesley Publishing Company

Published 2018 by CRC Press
Taylor & Francis Group
6000 Broken Sound Parkway NW, Suite 300
Boca Raton, FL 33487-2742

CRC Press is an imprint of the Taylor & Francis Group, an informa business

Visit the Taylor & Francis Web site at
http://www.taylorandfrancis.com

and the CRC Press Web site at
http://www.crcpress.com

ISBN 13: 978-0-201-15691-1 (pbk)
ISBN 13: 978-0-201-40844-7 (hbk)

This volume was typeset using T$_E$Xtures on a Macintosh IIsi computer.

About the Santa Fe Institute

The *Santa Fe Institute* (SFI) is a private, independent, multidisciplinary research and education center, founded in 1986. Since its founding, SFI has devoted itself to creating a new kind of scientific research community, pursuing emerging science. Operating as a small, visiting institution, SFI seeks to catalyze new collaborative, multidisciplinary projects that break down the barriers between the traditional disciplines, to spread its ideas and methodologies to other individuals, and to encourage the practical applications of its results.

All titles from the *Santa Fe Institute Studies in the Sciences of Complexity* series will carry this imprint which is based on a Mimbres pottery design (circa A.D. 950–1150), drawn by Betsy Jones. The design was selected because the radiating feathers are evocative of the outreach of the Santa Fe Institute Program to many disciplines and institutions.

Santa Fe Institute
Studies in the Sciences of Complexity

Proceedings Volumes

Vol.	Editors	Title
I	D. Pines	Emerging Syntheses in Science, 1987
II	A. S. Perelson	Theoretical Immunology, Part One, 1988
III	A. S. Perelson	Theoretical Immunology, Part Two, 1988
IV	G. D. Doolen et al.	Lattice Gas Methods for Partial Differential Equations, 1989
V	P. W. Anderson, K. Arrow, & D. Pines	The Economy as an Evolving Complex System, 1988
VI	C. G. Langton	Artificial Life: Proceedings of an Interdisciplinary Workshop on the Synthesis and Simulation of Living Systems, 1988
VII	G. I. Bell & T. G. Marr	Computers and DNA, 1989
VIII	W. H. Zurek	Complexity, Entropy, and the Physics of Information, 1990
IX	A. S. Perelson & S. A. Kauffman	Molecular Evolution on Rugged Landscapes: Proteins, RNA and the Immune System, 1990
X	C. G. Langton et al.	Artificial Life II, 1991
XI	J. A. Hawkins & M. Gell-Mann	The Evolution of Human Languages, 1992
XII	M. Casdagli & S. Eubank	Nonlinear Modeling and Forecasting, 1992
XIII	J. E. Mittenthal & A. B. Baskin	Principles of Organization in Organisms, 1992
XIV	D. Friedman & J. Rust	The Double Auction Market: Institutions, Theories, and Evidence, 1993
XV	A. S. Weigend & N. A. Gershenfeld	Time Series Prediction: Forecasting the Future and Understanding the Past, 1994
XVI	G. Gumerman & M. Gell-Mann	Understanding Complexity in the Prehistoric Southwest, 1994
XVII	C. G. Langton	Artificial Life III, 1994
XVIII	G. Kramer	Auditory Display, 1994
XIX	G. Cowan, D. Pines, & D. Meltzer	Complexity: Metaphors, Models, and Reality, 1994
XX	D. H. Wolpert	The Mathematics of Generalization, 1995
XXI	P. E. Cladis & P. Palffy-Muhoray	Spatio-Temporal Patterns in Nonequilibrium Complex Systems, 1995
XXII	H. Morowitz & J. L. Singer	The Mind, The Brain, and Complex Adaptive Systems, 1995
XXIII	B. Julesz & I. Kovács	Maturational Windows and Adult Cortical Plasticity, 1995
XXIV	J. A. Tainter & B. B. Tainter	Economic Uncertainty and Human Behavior in the Prehistoric Southwest, 1995
XXV	J. Rundle, D. Turcotte, & W. Klein	Reduction and Predictability of Natural Disasters, 1996
XXVI	R. K. Belew & M. Mitchell	Adaptive Individuals in Evolving Populations: Models and Algorithms, 1996

Contributors to This Volume

Current contributor affiliations are shown here; those shown at the beginning of each chapter are the authors' affiliations at the time the chapter was first printed.

Bergman, Aviv
Interval Research Corporation, 1801 Page Mill Road, Building C, Palo Alto, CA 94304

Finkel, Leif H.
University of Pennsylvania, Department of Bioengineering, Philadelphia, PA 19104

Fontana, Walter
Universität Wien, Inst. für Theoretische Chemie, Währingerstraße 17, A-1090 Wien, AUSTRIA

Gershenfeld, Neil A.
Massachusetts Institute of Technology, Media Lab, 20 Ames Street, E15-495, Cambridge, MA 02139

Goldstein, Raymond E.
Princeton University, Physics Department, J. Henry Labs., Princeton, NJ 08544

Gray, Charles M.
University of California-Davis, The Center for Neuroscience, Oakland, CA 94602

Hogg, Tad
Xerox PARC, Palo Alto Research Center, 3333 Coyote Hill, Palo Alto, CA 94305

Huberman, Bernardo A.
Xerox PARC, Palo Alto Research Center, 3333 Coyote Hill, Palo Alto, CA 94305

Jen, Erica
Santa Fe Institute, Santa Fe, NM 87501

Lebowitz, Joel L.
Rutgers University, Math Department, New Brunswick, NJ 08903

Li, Wentian
Columbia University, Department of Psychiatry, Unit 58, 722 West 168th Street, New York, NY 10032

Maes, Christian
Inst. voor Theoretical Fysica, Department Natuurkunde, K. U. Leuven Celestijn 200D, B-3030 Leuven, BELGIUM

Newell, Alan C.
University of Arizona, Department of Mathematics, Tucson, AZ 84721

Nijhout, H. F.
Duke University, Department of Zoology, Durham, NC 27706

Shatz, Carla J.
    Stanford University Medical School, Department of Neurobiology, Stanford,
    CA 94305

Speer, E. R.
    Rutgers University, Math Department, New Brunswick, NJ 08903

Tam, Wing Yim
    University of Arizona, Department of Physics, Tucson, AZ 85721

Weigend, Andreas S.
    University of Colorado, Department of Computer Science, Box 430, Boulder,
    CO 80309-0430

Contents

Preface

This Lecture Notes Volume grew out of the recognition that it would be desirable to collect previously published Complex Systems Summer School lectures that dealt with a common theme. In previous years, Summer School volumes encompassed all of the lectures in a given summer school session, which typically covered a broad diversity of topics and problems, related by varying degrees of similarity or tenuousness. This volume is the first effort to collect a set of related lectures in a general category. Most of these are articles that first appeared in the volume corresponding to the Summer School in which the lecturer presented his or her topic. Exceptions to this are the introductory article by Nijhout, which was written especially for this volume; the article by Gray, and the article by Gershenfeld and Weigend, both of which are reprinted from other sources. In preparation of this collection the authors were asked to edit and update their previously published articles, and several availed themselves of this opportunity.

We thank the authors for seeing the virtues of collecting these lectures into one volume, and for their efforts in updating and improving their original articles. We also thank Ronda K. Butler-Villa, Della Ulibarri, and Marylee Thomson of the SFI for their hard work in putting this volume together.

We would also like to acknowledge past funding from the National Science Foundation, MacArthur Foundation, Center for Nonlinear Studies, Department of Energy, University of California system, National Institute of Mental Health, and North Atlantic Treaty Organization.

H. F. Nijhout, L. Nadel, and D. L. Stein

October 1996

H. F. Nijhout
Department of Zoology, Duke University, Durham, NC 27708-0325

Pattern and Process

MECHANISM

An early insight into the mechanisms responsible for order in some biological structures came from the work of D'Arcy Thompson,[31] who offered the first organized attempts to explain the growth and form of biological structures in terms of physical and mathematical principles. Thompson begins his book with a series of references to scientists and philosophers of previous centuries who held that no science is a true science unless it can be expressed in mathematical terms. He then proceeded, with inimitable style, to show that if biological structures obey the laws of physics, many aspects of biological form emerged as a simple consequence of the external and internal forces acting on cells, tissues, and organs during development. Furthermore, simple differences in the relative growth of different portions of a biological structure (a skull, or an appendage, or the carapace of a crab), could lead directly to the change in the shape of that structure observed during its ontogeny, and more importantly, could completely account for the species-specific differences in the shape of the structure. This work laid the foundation for subsequent studies on allometry,[13] and the mathematical analysis of form.[1,2]

Modern interest in pattern formation dates from the publication of the ground-breaking work of Alan Turing. Turing[32] provided the first demonstration of how an organized and dynamically changing pattern could emerge in an initially homogeneous and randomly perturbed system. Prior to the advent of molecular genetic technology, it was Turing's reaction-diffusion system and Wolpert's theory of positional information[34] that breathed new life into an area of developmental biology that was depressed by the failure, after decades of intensive search, to find the chemical basis of embryonic induction.[30] The subsequent explosion of molecular genetic technology, however, soon deflected attention from the mechanism behind emergent order. As biologists became uniquely able to study the spatial and temporal patterns of gene expression in ever greater detail, it soon became conventional to think that the control of form resided entirely within the genome.

Harrison[10] has drawn attention to the great conceptual gulf that has arisen in modern developmental biology between those involved in the analysis of genes and proteins and those interested in the macroscopic organization of organisms. To bridge this gulf it is necessary for both parties to agree on a common language, which has been absent until now. Molecular geneticists have been primarily interested in local events of gene expression and have not paid particular attention to description and analysis of the dynamic changes in the spatial patterns of that expression. Theoreticians, by contrast, usually focus on dynamics but have not always been interested in incorporating all the known details of a developmental system into their models. So each party habitually ignores what the other thinks is the most interesting or important feature of the system.

Taken together, the work of Thompson, Turing, Wolpert, and their successors provided a basis for the view that biological pattern and form are not specified by a detailed genetic program or blueprint, but are emergent properties of relatively simple processes occurring in particular physical or chemical contexts. On such a view, understanding the details of mechanism is imperative. And since genes play a crucial role in the process, understanding exactly how genes can be fit explicitly into the current theoretical models of pattern formation should be of prime interest.

CHEMICAL AND PHYSICAL PATTERNS

Chemical and physical pattern formation is better understood at present than pattern formation in biological systems. The fit between theory and experiment is also better in chemical and physical systems, probably because they are generally simpler and better defined than most biological systems. Their component parts can be fully specified and their understanding requires no new chemistry or physics. But the systems are typically nonlinear and most can not be solved with conventional analytical mathematics, which is why numerical simulation and cellular automata have played such a great role in the analysis of pattern formation. The results of such simulations are generally non-intuitive.

The systems that are at present best understood from both a theoretical and detailed mechanistic perspective are spatially patterned reactions in chemical solution, such as the Belousov-Zhabotinski reaction, and the orderly convection patterns that appear as scrolls or polygons in thin fluid layers, such as Rayleigh-Bérnard cells. The starting point in most models of physical pattern formation is a set of equations of motion for which a uniform solution is assumed to exist. The equations can be realistic descriptions of the system in question, or a mathematical model chosen because its solutions mimic the behavior of the system one wishes to study.[4] The best understood systems, clearly, are those for which a formal mathematical treatment is possible. Progress in understanding pattern formation in these systems is determined in large measure by the limitations of the available analytical or numerical methods.

The study of pattern formation in defined chemical and physical systems is an active area of innovative research as the papers in this volume and the recent reviews by Cross and Hohenberg[4] and Kapral and Showalter[14] illustrate. For a biologist these systems are interesting because they provide wonderful examples of the fact that self-organization is not a property restricted to living systems. But whether physical or chemical systems for pattern formation provide good models for understanding biological pattern formation is an open question, and the answer today depends in large part on the system in question. In most cases, biological pattern formation cannot be simply reduced to corresponding questions in physics or chemistry because biological systems are far more complex, and their components are not all known. Biological models therefore do not have the strong phenomenological base we find in physical models. Furthermore, biological systems can not be subjected to controlled quantitative experiments the way physical systems can,[4] and that inevitably limits the degrees to which they can be understood by methods of approach developed in physical systems. Probably the closest analogies between chemical/physical and biological systems are found between the spiral waves seen in the Belousov-Zhabotinski reaction, and those found in aggregating swarms of *Dictyostelium* and in fibrillation patterns in the heart.[33] The patterned distribution of free moving organisms in an ecological setting also probably behaves more like a chemical system than does pattern formation in development.

BIOLOGICAL PATTERNS

What distinguishes biological patterns from chemical and physical ones is that the former have both an ontogenetic and an evolutionary history. Not only have the patterns evolved, but the mechanisms that produce them are different today than they were in the past. Furthermore, biological patterns of any given type are diverse. Understanding biological pattern formation therefore requires explanation of the pattern, its ontogeny, and its diversity.

Held[12] has produced a formal taxonomy of pattern-forming models in development (Table 1). Unfortunately, the actual mechanisms of biological pattern formation remain, in large measure, a mystery. Attempts to extract chemical substances that are part of the normal mechanism of pattern development have been such spectacular failures that many biologists now doubt the existence of classically conceived morphogens. Accordingly, the fact that theoretical models for biological pattern formation generally rely on diffusion or reaction-diffusion of molecules has been criticized by experimental developmental biologists on the grounds that the existence of diffusion morphogens has never been proven.

Harrison[10] has pointed out that, in practice, diffusion means nothing more than communication. It is simply because reaction-diffusion theory is so well worked out that many theoretical models of pattern formation are phrased in a way that they use diffusion for long-distance communication. But many processes, such as action potentials in neural nets and stress fields in epithelia, can send long- and short-range signals whose dynamics can be similar to those of a reaction-diffusion system.[7,8,9] Diffusion is, therefore, best treated as a metaphor for long-range communication, and reaction-diffusion equations, as all good models, are best seen as descriptors of minimal conditions required for pattern formation in a given system. Reaction-diffusion models do not make strong predictions about the vehicle of communication nor about the form of the message.

The molecular-genetic mechanisms for patterning that have been elucidated to date appear to be relatively simple. In early embryonic development of *Drosophila* most of the pattern-forming interactions appear to be of the gradient-threshold type. Patterning of the segmental subdivision of the embryo is initiated by a quantity of *bicoid* RNA that is laid down in anterior pole of the egg by the mother during oogenesis, a quantity of *nanos* RNA laid down at the posterior pole, and *hunchback* RNA homogeneously distributed in the egg. The *bicoid* and *nanos* RNAs are translated into proteins which become distributed, presumably by diffusion, into anterior-posterior gradients. The *bicoid* protein controls transcription of the embryo's own *hunchback* gene in a concentration-dependent manner, while the *nanos* protein inactivates the maternally supplied *hunchback* RNA. The result is that *hunchback* protein becomes expressed in a graded fashion in the anterior half of the developing egg. The *hunchback* gradient appears to control the expression of various other genes in a concentration dependent manner, resulting in several broad bands of protein synthesis normal to the antero-posterior axis of the egg. These bands of proteins, in turn, interact in their areas of overlap, and control the expression of other genes.[17] The overall field thus becomes subdivided into ever narrower bands of gene expression, some of which come to correspond to the subsequent segmental pattern of the embryo. Segmental patterning in early embryonic development, thus, appears to be controlled by the subdivision of a simple initial gradient, which leads to a new set of gradient that are in turn subdivided, and so forth. This process takes advantage of the absence of cellular boundaries in the early embryo which allows macromolecules to diffuse over considerable distances and interact directly in the control of gene expression.

TABLE 1 Lewis Held's[12] taxonomy of model mechanisms that form periodic patterns in biological systems.

Category	Distinguishing Features	Model Types
Position Dependent Class	The relative position of a cell dictates it subsequent fate	
Positional Infor-mation Subclass	Cells know where they are relative to an external coordinate system, which they "interpret" as particular states of differentiation	Gradient Models Polar Coordinate Models Progress Zone Models
Prepattern Subclass	Cell's state is determined by mechanical or chemical signals from within the cell layer, or by induction from adjacent cell layers	Physical Force Models Reaction-Diffusion Models Induction Models
Determination Wave Subclass	Cell states are determined within a zone that traverses an array of cells	Chemical Wave Models Sequential Induction Models Clock and Wavefront Models Inhibitory Field Models
Darwinism Subclass	Each cell adopts a state and then examines the states of its neighbors; if it matches another cell some type of action is taken	Cell Death Models State Change Models
Rearrangement Class	Each cell adopts a state and then begins to move until it reaches a particular location	Adhesion Models Repulsion Models Interdigitation Models Chemotaxis Models
Cell Lineage Class	Cells divide asymmetric-ally, placing each daughter in a definite position and assigning it a particular state	Quantal Mitosis Models Stem Cell Models Cortical Inheritance Models

These experimental findings stand in sharp contrast to the assumptions of theoretical models that have addressed the problem of segmental patterning of the *Drosophila* embryo, which have generally used reaction-diffusion models.[6,11,15,16,21] These models have been successful to varying degrees in producing simple banding patterns as well as the progression of bands of protein synthesis that are observed in embryos. But the models clearly did not predict anything remotely like the actual processes that were eventually revealed by molecular genetic methods. Experimental investigations imply the operation of a simple sequential diffusion-threshold mechanism, and not a typical reaction-diffusion mechanism with autocatalysis and long-range inhibition. Insofar as diffusion-threshold mechanisms are simpler than reaction-diffusion mechanisms, it seems that the processes that occur in life are simpler than those required by theory.

This appears to be a reversal of the usual relationship between a model and that which is modeled, where the model is typically a simplified skeletal version of the actual process. Undoubtedly, part of the reason for theoreticians having "missed" the real mechanism is that diffusion-threshold processes are seldom looked at seriously in mathematical modeling. There are several reasons for this. On a philosophical level diffusion and other types of gradient models are not considered "interesting," because they can be used to generate whatever pattern one wishes by simply, and arbitrarily, increasing the number of diffusing and interacting species. On a practical level, diffusion models suffer from the drawback that when the gradients are shallow or noisy the threshold is spatially poorly defined. This results in fuzzy boundaries between regions that are above and below threshold. Real biological patterns usually have sharp boundaries, with different properties in adjacent cells, no cells with intermediate values, and, at the cellular level, long and sharp boundaries between one region and another. Patterning by diffusion gradients is, therefore, generally believed to require an additional boundary-sharpening mechanisms. Lewis et al.,[18] for instance, have described a simple positive feedback mechanism, based on realistic biochemical process, which can form thresholds of arbitrary sharpness on a simple continuous gradient in a single step. Reaction-diffusion models, by contrast, are attractive because they define the minimal conditions necessary to produce a fully developed pattern from a single continuous process, without requiring complex initial conditions. That nature does not do things in such a minimalist fashion comes as no surprise to biologists. After all, we are dealing with evolved systems that are seldom optimized and in which mechanism is inherited with modification and thus often highly constrained by history. From a biological perspective, Meinhardt's emphasis on the importance of boundaries, and on the hierarchical nature of biological pattern formation,[19,20,21,22] provides a more realistic and useful context for theoretical development than do theories that start with homogeneous and randomly perturbed systems.

But there is another side to the failure of theory to predict the correct mechanism. It is almost certainly true that what has been described so far by molecular techniques is only part of the mechanism for segmental patterning. The *Drosophila* embryonic system has been studied with the limited, though admittedly powerful, tools of molecular genetics. But these tools can only detect the macromolecular features of a process; that is, those that involve the synthesis of RNA and proteins. It can be argued that if you look with techniques that can only detect genetic mechanisms that use macromolecules, the only important thing you will find are genetic mechanisms that use macromolecules. The undesirable "noise" that appears at this level of analysis may be evidence of additional processes at work that the available techniques cannot detect. It is not unreasonable to assume that there are important but still covert processes at work in segmental patterning which current techniques cannot detect, and these may well be of the sort envisioned by various reaction-diffusion theories. But for the moment a simple sequential diffusion gradient-threshold mechanism appears to suffice, and theoreticians should pay heed.

Later in embryonic development, when the *Drosophila* embryo becomes cellular, long-distance communication by diffusion of macromolecules becomes difficult. Macromolecules either must be secreted and then diffuse in the extracellular environment, or must be attached to the surface of the cell, but that would restrict communication to direct cell-to-cell contact. Cell-to-cell communication through the intracellular medium would be restricted to small molecules that can pass through gap junctions. In eyespot determination on butterfly wings it appears that diffusion through gap junctions could operate over the required distances (about 0.5 mm or 20 to 50 cell diameters) without needing amplification of the signal,[24,29] so it is at least theoretically possible for simple diffusion of small molecules to act as a patterning mechanism. But the types of patterns one gets from a diffusion-threshold mechanism depend entirely on the shapes and distribution of sources and sinks for the diffusing substances. In order to get complex and "interesting" patterns you need an interesting distribution of sources and sinks, and that, of course, pushes the question of pattern formation back to one that asks what determines the positions of sources. In other words, the overt pattern must be preceded by prepattern of generating sources, much as the segmental pattern of the *Drosophila* embryo is preceded (and in a sense controlled) by a prepattern of *bicoid* and *nanos*. To generate sources and sinks (or any type of periodic pattern) from a single event, without a further prepattern, it is generally accepted that lateral inhibition (typically mediated by an appropriate reaction-diffusion process) is required.

There is suggestive evidence that lateral inhibition processes may be at work later in development, for instance during the determination of bristle spacing in the cellular epidermis.[17] Two genes, *Notch* and *Delta*, appear to be required for the development of normal bristle spacing. Both genes code for transmembrane proteins. The *Notch* protein appears to be a receptor for the inhibitory signal that affects bristle spacing, while the *Delta* protein is part of the inhibitory stimulus.[17] As neither gene product can diffuse, direct cell contact is required for the interaction of *Notch* and *Delta*.

What kinds of genes are likely to be involved in lateral inhibition and reaction-diffusion processes? The parameters of reaction-diffusion and lateral inhibition models are the reaction rate constants and the diffusion coefficients of the interacting species. The models therefore suggest that processes that are autocatalytic, that control reaction rates, and that can extend excitatory and inhibitory influences over distance are of critical importance in the formation of periodic patterns. These are not the properties of regulatory genes (which merely turn other genes on or off) but belong to the category we often call housekeeping genes that produce enzymes that catalyze biochemical reactions. Such reactions can produce small signaling molecules that can more easily travel from cell to cell than the gene products themselves. Such small signaling molecules may, in turn, act as regulators of gene expression and may be ultimately responsible for the spatially patterned expression of regulatory and structural genes that precedes local differentiation. Alternatively, and remembering that "diffusion" means nothing more than "communication," it is possible to send a signal over a long distance without using small diffusible molecules through the interaction of membrane-bound cell-surface molecules whose synthesis is stimulated or inhibited by a feedback mechanism.

PATTERN AND PROCESS

A model that faithfully reproduces a pattern from nature may not do so via the same mechanism tht produced the natural pattern. A nice illustration of how even a simple pattern gives no clue as to the process that gave rise to it is provided by Harris'[7] dilemma of the definition of a circle (Figure 1). A modeler must therefore ask the question whether it is sufficient to know in principle how a pattern *could be* made, or whether we want a model to capture the essence of how a pattern *is actually* made? Most of us would argue the latter, but in many cases we do not have sufficient information about process to structure the model appropriately, and modelers of biological systems have more often than not had to be satisfied with the former. It is in this regard that chemical/physical system and biological systems for pattern formation differ enormously. Chemical and physical systems for pattern formation can usually be fully defined and started from initially homogeneous conditions with random perturbation. If initial conditions are not homogeneous, they can at least be fully defined at the beginning of the experiment.

In biological systems, by contrast *there is always an antecedent pattern.* It is absolutely critical to recognize this when attempting to model any biological pattern. No matter how far back one looks, there is always an inhomogeneity or asymmetry. Often there is a relatively complex preexisting spatial pattern that influences the next step of pattern formation.

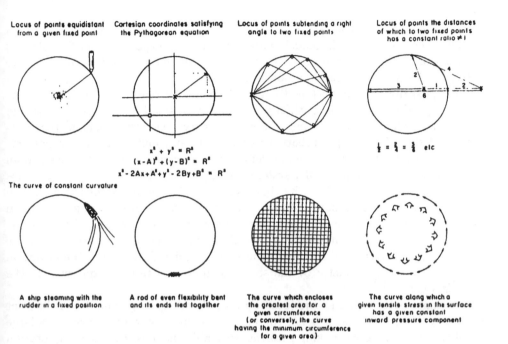

Locus of points equidistant from a given fixed point

Cartesian coordinates satisfying the Pythagorean equation

Locus of points subtending a right angle to two fixed points

Locus of points the distances of which to two fixed points has a constant ratio ≠ 1

$$x^2 + y^2 = R^2$$
$$(x-A)^2 + (y-B)^2 = R^2$$
$$x^2 - 2Ax + A^2 + y^2 - 2By + B^2 = R^2$$

$$\tfrac{1}{2} = \tfrac{2}{4} = \tfrac{3}{6} \text{ etc}$$

The curve of constant curvature

A ship steaming with the rudder in a fixed position

A rod of even flexibility bent and its ends tied together

The curve which encloses the greatest area for a given circumference (or conversely, the curve having the minimum circumference for a given area)

The curve along which a given tensile stress in the surface has a given constant inward pressure component

FIGURE 1   Eight different definitions of a circle. Fundamentally different processes can give rise to identical patterns, and how one approaches the pattern from a theoretical point is governed by what one knows (and, as importantly, what one does not know) about the process that produced the specimen that is the object of one's investigation.[7] Copyright ©Springer-Verlag, reprinted with permission.

While we can usually detect antecedent patterns, we generally do not know which aspects of those patterns are important for the subsequent development of pattern in the system, or indeed, whether or not we have detected all relevant prepatterns. Good modeling practice requires that under such conditions one tries to find the simplest model that *could* produce the desired final pattern, and that almost always leads to the assumption of initial homogeneity. This strong reliance on a *ceteris paribus* clause, while justified in simple systems, will almost certainly lead to incorrect inferences about process in biological systems. The development of periodic gene expression pattern in *Drosophila* embryos, outlined above, is a good example. So there is a real danger that models of biological pattern formation that do not take careful account of antecedent patterns, and make simplifying assumptions about boundary and initial conditions, may not be able to capture the essence of the process that leads to pattern in a given system.

How should we then proceed? Obviously, we would like to know which simplifying assumptions are safe to make without losing the essence of the process, and which simplifying assumptions lead to a model of little biological relevance. But discovering what those assumptions are is often the very purpose of constructing a model in the first place. The model should, of course, incorporate as much biological information as possible, and that is why modeling should be an active collaborative enterprise between biologists and mathematicians.

Perhaps the most important thing that should be required of a model is not that it reproduces a pattern faithfully, but that with small quantitative changes in parameters values, it can produce the evolutionary diversity present in that pattern, and the effects of perturbation experiments and mutations on the pattern. This requirement for diversity places very severe constraints on what is deemed an acceptable model for a particular system. Finally, a model should not only produce the final pattern but mimic in its time evolution the succession of patterns seen during ontogeny of the system. The two-stage model for color pattern formation on butterfly wings, for instance, meets these criteria.[25,26] It produces a broad diversity of natural patterns, it simulates the effect of surgery and mutations, and the dynamic spatial pattern of activator production predicted by the reaction-diffusion model is reproduced exactly by the spatio-temporal pattern of expression of the *distalless* genes.[3,27]

A single pattern can be produced by an indefinitely large number of different processes (Figure 1). But it is not common for a single process to produce a large number of very different-looking patterns. Probably the most extensive exploration of the pattern-generating potential of a single reaction-diffusion model is the work of Meinhardt.[19,23] This work makes it clear that in order to get a great diversity of spatial patterns it is not sufficient to merely permute parameter values of one model, but it is necessary to add to, or significantly alter, the reaction terms in the model. Nijhout[28] has also provided a demonstration of the sensitivity of the pattern to the reaction terms in a reaction-diffusion model. Meinhardt[19] provides an intuitive interpretation of why pattern morphology is sensitive to the type of reaction, but the matter has as yet received relatively little attention from a theoretical perspective.

REFERENCES

1. Bookstein, F. L. *The Measurement of Biological Shape and Shape Change.* Lecture Notes in Biomathematics, Vol. 24. Berlin: Springer-Verlag, 1978.
2. Bookstein, F. L. *Morphometric Tools for Landmark Data: Geometry and Biology.* Cambridge: Cambridge University Press, 1991.
3. Carroll, S. G., J. Gates, D. Keys, S. W. Paddock, G. F. Panganiban, J. Selegue, and J. A. Williams. "Pattern Formation and Eyespot Determination in Butterfly Wings." *Science* **265** (1994): 109–114.
4. Cross, M. C., and P. C. Hohenberg. "Pattern Formation Outside of Equilibrium." *Rev. Mod. Phys.* **65** (1993): 851–1112.
5. Edelstein-Keshet, L. *Mathematical Models in Biology.* New York: Random House, 1988.
6. Goodwin, B. C., and S. A. Kauffman. "Saptial Harmonics and Pattern Specification in Early *Drosophila* Development. I: Bifurcation Sequences and Gene Expression." *J. Theor. Biol.* **144** (1990): 303–319.
7. Harris, A. H. "Cell Traction and the Generation of Anatomical Structure." *Lectures Notes in Biomath.* **55** (1984): 103–122.
8. Harris, A. H. "Testing Cleavage Mechanisms by Comparing Computer Simulations to Actual Experimental Results." *Ann. NY Acad. Sci.* **582** (1990): 60–77. Also In *Cytokinesis: Mechanisms of Furrow Formation During Cell Division.*
9. Harris, A. H., D. Stopak, and P. Warner. "Generation of Spatially Periodic Patterns by a Mechanical Instability: A Mechanical Alternative to the Turing Model." *J. Embryol. Exp. Morphol.* **80** (1984): 1–20.
10. Harrison, L. G. *Kinetic Theory of Living Pattern.* Cambridge: Cambridge University Press, 1993.
11. Harrison, L. G., and K. Y. Tan. "Where may Reaction-Diffusion Mechanisms be Operating in Metameric Patterning of *Drosophila* Embryos?" *BioEssays* **8** (1988): 118–124.
12. Held, L. I. *Models for Embryonic Periodicity.* Basel: Karger, 1992.
13. Huxley, J. *Problems of Relative Growth.* New York: Dover, 1972.
14. Kapral, R., and K. Showalter. *Chemical Waves and Patterns.* Dordrecht: Kluwer, 1995.
15. Lacalli, T. C. "Modeling the *Drosophila* Pair-Rule Pattern by Reaction-Diffusion: Gap Input and Pattern Control in a 4-Morphogen System." *J. Theor. Biol.* **144** (1990): 171–194.
16. Lacalli, T. C., D. A. Wilkinson, and L. G. Harrison. "Theoretical Aspects of Stripe Formation in Relation to *Drosophila* Segmentation." *Development* **104** (1988): 105–113.
17. Lawrence, P. A. *The Making of a Fly.* Oxford: Blackwell, 1992.
18. Lewis, J., J. M. W. Slack, and L. Wolpert. "Thresholds in Development." *J. Theor. Biol.* **65** (1977): 579–590.

19. Meinhardt, H. *Models of Biological Pattern Formation.* London: Academic Press, 1982.
20. Meinhardt, H. "Hierarchical Induction of Cell States: A Model for Segmentation in *Drosophila.*" *J. Cell Sci. Suppl.* **4** (1986): 357–381.
21. Meinhardt, H. "Models for Maternally Supplied Positional Information and the Activation of Segmentation Genes in *Drosophila* Embryogenesis." *Development Suppl.* **104** (1988): 95–110.
22. Meinhardt, H. "Biological Pattern Formation: New Observations Provide Support for Theoretical Predictions." *BioEssays* **16** (1994): 627–632.
23. Meinhardt, H. *The Algorithmic Beauty of Sea Shells.* Berlin: Springer-Verlag, 1995.
24. Murray, J. D. *Mathematical Biology.* Berlin: Springer-Verlag, 1989.
25. Nijhout, H. F. "A Comprehensive Model for Color Pattern Formation in Butterflies." *Proc. Roy. Soc. London Ser. B* **239** (1990): 81–113.
26. Nijhout, H. F. *The Development and Evolution of Butterfly Wing Patterns.* Washintgon: Smithsonian Institution Press, 1991.
27. Nijhout, H. F. "Genes on the Wing." *Science* **265** (1994): 44–45.
28. Nijhout, H. F. "Pattern Formation in Biological Systems." This volume.
29. Safranyos, R. G. A., and S. Caveney. "Rates of Diffusion of Fluorescent Molecules via Cell-to-Cell Membrane Channels in a Developing Tissue." *J. Cell. Biol.* **100** (1985): 736–747.
30. Saxén, L., and S. Tovoinen. "Primary Embryonic Induction in Retrospect." In *A History of Embryology,* edited by T. J. Horder, J. A. Witkowski, and C. C. Wylie, 261–274. Cambridge: Cambridge University Press, 1986.
31. Thompson, D'A. W. *On Growth and Form.* Cambridge: Cambridge University Press, 1917.
32. Turing, A. M. "The Chemical Basis of Morphogenesis." *Phil. Trans. Roy. Soc. London, Ser. B* **237** (1952): 37–72.
33. Winfree, A. D. *The Geometry of Biological Time.* Berlin: Springer-Verlag, 1990.
34. Wolpert, L. "Positional Information and the Spatial Pattern of Cellular Differentiation." *J. Theor. Biol.* **25** (1969): 1–47.

Aviv Bergman
SFI International, 333 Ravenswood Ave., Menlo Park, CA 94025

Self-Organization by Simulated Evolution

This chapter originally appeared in *1989 Lectures in Complex Systems*, edited by E. Jen, 455–463. Santa Fe Institute Studies in the Sciences of Complexity, Lect. Vol. I. Reading, MA: Addison-Wesley, 1990. Reprinted by permission.

One of the most interesting properties of neural networks is their ability to learn appropriate behavior of being trained on examples. However, most neural network models suffer from a critical limitation: these systems must be "handcrafted" to deal with nontrivial problems. To have a chance of success, the designer of a neural network must embed much of his detailed knowledge about how to solve the task into the initial configuration. For problems of nontrivial complexity these systems must be handcrafted to a significant degree, but the distributed nature of neural network representations make this handcrafting difficult. This critical and difficult part of the design process is not formally specified, but rather depends on the intuition of the network designer. We present here an evolutionary approach to learning that avoids this problem. This approach simulates a variable population of networks which, through processes of mutation, selection, and differential reproduction, converges to a group of networks well suited to solving the task at hand. By providing a mechanism for evolving new structure (as opposed to just modifying weights in a fixed structural framework), we extend the range of problems that can be addressed successfully by network-based computational systems.

1. INTRODUCTION

Learning requires some level of *a priori* knowledge about the task to be learned; research on learning has oscillated between two poles, according to the amount of knowledge one is willing to assume is already present in the system before learning starts. Predominating today, in artificial intelligence circles,[12] is the idea that solving a particular problem entails repeated application, to a data set representing the starting condition, of some chosen *predefined* set of operations; the task is completed when the data set is found to be in "goal" state. This approach ascribes to the system, "from nature," the capabilities required for a successful solution. The other approach, quite popular in its early version,[13] has been favored recently by physicists,[4,6] and rests on the idea that "learning machines" are endowed, not with specific capabilities, but with some *general architecture* and a set of *rules*, used to modify the machines' internal states, so that progressively better performance is obtained upon presentation of successive sample tasks. The system is initially "ignorant," that is, begins from some "null" situation where all parameters to be modified by the rules are either set to zero or to some random values.[14]

Recent consideration of human or animal intelligence, however, has revealed that biological reality seems actually to be located somewhere between these two extremes. We are neither perfect *tabula rasa*, nor totally preprogrammed robots. A good illustration is provided by song learning in birds,[9] where initially almost random singing patterns develop slowly into the characteristic species' song, or into some modification thereof, depending on the auditory environment of the bird.

The question of whether "intelligent" behavior can be generated efficiently by using a proper balance of "nature" versus "nurture" thus remains open. Given that the information encoded into the brain at birth must mainly be of genetic origin, and that such information is conceived today as having emerged through a process of evolution, we are led naturally to think in terms of populations of individuals (see, e.g., Mayr[10]), each endowed with a structure it has inherited (for the most part) from its "parent(s)," and learning as a result of adaptation to its environment. Those individuals who are better learners leave more descendants than the others; statistically, the population cannot fail to improve its performance. Though the concept is a simple one,[2,5] its implementation can be difficult.

Our approach is to simulate a variable *population* of network automata. In a manner directly analogous to biological evolution, the population converges, under the influence of selective pressure, to a group of networks that are well suited for solving the task at hand. Normally the process does not start "from scratch" with completely unstructured networks: the designer still specifies an initial network that represents a best guess at the final solution. In fact, the process could start with a variety of networks representing distance prototypical designs. The evolutionary simulation then modifies these prototypes to produce one (or perhaps several) networks that more effectively learn to solve the task.

We visualize the approach as comprising two separate kinds of processes: a low-level "performance" process that operates on a population of "phenotypes," and a higher level "metaperformance" process that operates on a population of "genotypes." A genotype is a *description* of a class of neural networks in some specified language, while a phenotype is an *expression* of, or an *instantiation* of, a genotype—that is to say, an actual neural network that may be modified by learning. The performance process is essentially a standard learning algorithm that runs in parallel over all networks in the population of phenotypes. In addition, the performance process gathers information on the actual behavior of each of the networks, which is fed into a fitness function, yielding a grade for each. The metaperformance process examines the fitness of each network and permits only those that are most fit to *reproduce*, thereby increasing the frequency of the *genes* of the successful networks in the genetic pool of the population as a whole.

In what follows, we describe the "metaperformance" process and the initial structure of the machines we are considering; the basic scheme is general enough that a large class of tasks is in principle executable; however, at the outset, we do *not* attempt in any way to "program" a particular task into the system. We observe the emergence of structured networks and computational abilities through strictly genetic adaptation.

2. INITIAL ARCHITECTURE

In his "genetic algorithms," Holland has introduced the notion of genetically adaptable programs. These programs consist of modules that can be combined in various ways, and that contain a variety of parameters with specified values in each individual program.[5] The design of such modular programs calls for knowledge about the task to be executed; here, we would like to avoid the need to provide such advance knowledge, and propose to start from an architecture sufficiently general, so that a wide variety of tasks can in principle be performed. We expect the population of machines to adapt progressively, in such a way that one or several machines will ultimately meet our requirements: examining their structure should then inform us about how they are actually producing a solution. Presumably, since the procedure "found" by the successful machines is of evolutionary origin, it will exhibit various redundancies and useless complications: this has been referred to by Jacob as "evolutionary tinkering."[7] However, it should be possible to iron out these accidental features, leaving us with an efficient structure.

We consider cellular systems consisting of successive layers. Connections to and from cellular elements ("neurons") run exclusively between nearest and next-nearest neighbors on the square lattice (see Figure 1). All links have the same strength but may be excitatory (+) or inhibitory (−), and two elements may be connected by more than one link. Connections are established according to a set of probabilities of generating connections in the different directions: the machine's "genetic code."

Each unit is a simple processor with a sigmoidal transfer function:

$$O(u)\frac{1+\tanh(\beta u)}{2}$$

where u is the sum of its inputs (with their sign), and β is the "gain" of the unit.

This operation is carried out in parallel at each layer, at discrete time intervals: in the following, time will always be measured in units of the basic updating process. The top elements have their outputs forced to a set of integers $\{I_i\}$: this set constitutes the *input string* of the system. The neurons at the bottom of the array then generate a new set of integers $\{O_i\}$, the *output string*.

The network just defined certainly has the potential for carrying out complex computational tasks; in particular, the possibility of lateral and backward connections, which allow for feedback control, sets our model clearly apart from perceptron-like architectures.[11] However, carrying out a particular goal—that is, obtaining a given (possibly time-dependent) input-output relationship—requires setting up a very precise set of connections. We avoid having to specify this by running a population-dynamical simulation as described in Sections 3 and 4.

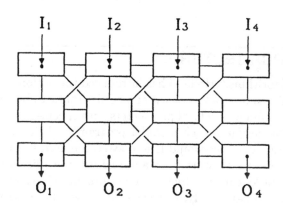

FIGURE 1   A three-layer, four-neuron-wide automaton. The links indicate which elements may be joined by one or more connections going either way. The state of each of the top elements is *forced* by an input integer I_j. The outputs are collected as the integers O_j delivered by the neurons in the last layer.

3. SIMULATED EVOLUTION

The basic idea is to *select*, as contributors to a new generation of automata, those machines that are performing best. Several tasks have been implemented. Here we choose the following problem as an illustration: three strings, comprising ten integers each, are presented successively to our machines for a certain time ("persistence time"); two of these can be matched by amplitude invariance, e.g., $S_1 = \{0, 0, 0, a, b, c, d, 0, 0, 0\}$ and $S_2 = \{0, 0, 0, na, nb, nc, nd, 0, 0, 0\}$ with n some integer constrained between 1 and 16; while the third is unrelated to these e.g., $S_3 = \{0, 0, 0, f, g, h, i, 0, 0, 0\}$. To a population of N ten-neuron-wide arrays, each of which typically contains six layers, we now present these strings, with a persistence time of at least five units, and in random order. We require that, upon successive presentation of $S_1 S_2$ (case AMP1) or $S_2 S_1$ (case AMP2), the output cells will assume some large activity, while after $S_1 S_2$ (i.e., NAMP1), $S_3 S_1$ (NAMP2), $S_2 S_3$ (NAMP3), or $S_3 S_2$ (NAMP4), these neurons will be as close to zero activity as possible. In a sense, we are thus asking for a "neuron assembly"[3] which fires specifically when detecting amplitude invariance (AMP). Programming the array *a priori* to perform this is not easy. All we have to do here is assign to each machine, upon completion of the test, a *grade*; the closer the machine comes to achieving the goal, the higher its grade: this can be, for instance, the sum of output values delivered after AMP minus the sum of outputs after NAMP, properly normalized. After testing, we set up a new population, to which each machine contributes in rough proportion to its grade. Each newborn automaton inherits the set of probabilities of generating its connections of its ancestor, but small perturbations (mutations) are added randomly. The new population is submitted to another test, and so on, until satisfactory performance is obtained.

At this stage the configuration of the best-performing machines is inspected. We look for a common structure shared by them, then impose this structure on future generations, improving their performance. This emergent structure reflects the environment's structure. A second important consequence from this operation is to produce a compact description of a machine in terms of its "genetic code."

Certainly, a solution will ultimately be found provided the task we have chosen can at all be carried out using the architecture that we use. However, the question remains as to whether this can be achieved in a reasonable computer-processing time, and as to how efficient and elegant the solution will be. Here, the answers are quite subtle.

4. RESULTS

The problem described above, as well as other related problems,[8] has been successfully solved; the quality of the results, however, depends strongly on how gradually we present problems to our machines and on how constrained the machine architecture is. Ideally, we should like not to supply the automata with any sort of initial information. Therefore, the populations which we begin with may typically consist of machines with a *random* set of probabilities of generating connections. Not surprisingly, random machines do not perform well; neither do they improve rapidly. Thus, we encounter the nature versus nurture problem: a stupid start leads you nowhere.

Much better results are obtained when rather unspecific bits of information are supplied. An example is the information that a successful machine probably possesses more forward than backward or lateral connections; another concerns the likely predominance of inhibition, that is, negative connections. If we build our random initial population such that these conditions are statistically satisfied, the first useful machines emerge. It is remarkable that populations where no biases of this type were imposed seemed to evolve them, but very slowly, and that so little information makes such a difference. Even then, however, before it can be learned in a reasonable amount of time (say, 30 minutes of CPU time, or roughly 1500 generations), the task must be presented in a rather restricted form. A "startup" procedure, where the strings are not modified at all but only the order of presentation is changed, is necessary; thereafter, the integers $a, b, c, \ldots, i$ (or *alphabet*) used in the definition of S_1, S_2, and S_3 are not varied, but only the *scaling factor* n is modified from one test (used to grade generation j) to the next (used with $j + 1$).

We now look at the structure of our automata. It should be stressed firstly that there are many machines, with rather different "innards," which perform satisfactorily a given task. Certain regularities are, however, apparent. Layer-wise periodicity, i.e., all neurons in a given layer are identically connected (LP-arrays), is the first regulatory pattern to emerge.

A striking difference appears at this stage when we impose this regulatory pattern on future generations.

The LP-machines are by far better than other architectures. Not only do they perform the limited test just described, but also the better ones among them are able to go on to the more difficult feat of performing well no matter what the alphabet. After they have learned with their first alphabet, they learn the second one much faster, and the learning time drops even further with the third. Note that, with each new alphabet learned, there is a loss in performance using the *previous* alphabets. This can be alleviated with a measure of "reinforcement," that is, by presenting instances of all alphabets previously learned even while teaching the new one. At any rate, machines that have gone through some ten alphabets seem to recognize amplitude invariance no matter what the alphabet or the scaling factor (success rates are around 95 percent). They still do well, even when persistence times are changed randomly: it would seem that they have learned the AMP *concept*.

By learning the concept of amplitude invariance, we mean that, embedded implicitly in the connectivity pattern, is an *algorithm* which embodies this concept. It is not difficult to imagine what an explicit algorithm would be; compared with this, it can be said that our arrays, performing an equivalent set of operations in ten ticks of their internal clock (i.e., two persistence times), appear to be fairly efficient!

Furthermore, the algorithm is naturally parallel, which allows us to implement it on modern processors built with this sort of application in mind. In addition, simulations of population dynamics, used to discover the procedure, are themselves genuinely parallel, in the sense that, at a given generation, each machine can be tested independently of the others: this is the time-consuming part of the calculation.

We now continue the evolutionary process, this time with LP-machines, and take a second look at the structure of our automata. Other regularities become apparent. Thus, sometimes excitatory and inhibitory connections alternate in a layer in a way very suggestive of multistable electronic circuits; this part of the automaton seems then to be in charge of *short-term memory*, storing temporarily the results of some computation on a first string. Another readily visible feature is the presence of strong lateral inhibition.

Thus, with no design effort beyond that of setting up the initial architecture, and designing a testing routine that progressively and faithfully represents the task at hand, we can claim to have reached a stage where efficient success was possible. The general procedure illustrated above with amplitude invariance has proved its flexibility by being used just as successively to solve a variety of other computational problems, with virtually no additional conceptual effort.

It seems clear to us that, by further developing the analysis of such circuitry, one should be able to abstract the essential elements of the algorithm involved from the "raw" structures obtained through evolution, thus paving the way for the design of actual VLSI chips. These chips would come in various "strains," each capable of certain predefined operations (*nurture*), and presumably able to learn a slightly more difficult set of tasks (*nature*). The non-uniqueness of structure among the "learned" automata certainly suggests that it must be worthwhile to maintain a whole population of chips: while all automata may do certain things, they are probably not equivalent in terms of their future learning potential.

It may well be that the major weakness of the present approach is that it is currently impossible to predict whether a given problem can be solved using some predefined set of architectural constraints. We cannot even say what depth (i.e., number of layers) is needed in general to solve problems like AMP. Perhaps one should let the automata themselves adjust some of the variables involved, such as the number of layers, but we have seen that increasing the dimensions of the space being searched is not always a good idea.

5. DISCUSSION

The initial structure of an organism can be viewed as its "nature," while the process that "fine-tunes" the organism's structure plays the role of "nurture." Clearly, intelligent behavior requires a balance of nature and nurture. One initial structure might be fine-tuned to solve one task and to generalize one set of examples, while a quite different configuration would be required to solve a different task. Currently, the initial configurations of any computational process (that is, their nature) must be handcrafted, and this is an art requiring knowledge and experience, both about the computational process and the task to be solved.

In the present work, we have tried to tread a path intermediate between the extremes of, on the one hand, a fixed operations repertoire acquired at birth, and on the other, random but adaptive architectures. We have discovered that the automata that we start with, which in general already possess computational capabilities, can refine themselves in many interesting ways in the process of competition. In a limited sense, we have thus demonstrated the complex interactions between nature and nurture in learning. We found that, from generation to generation, specific connectivity patterns develop; concomitantly, one can observe that certain groups of cells become active, while others lapse into disuse. This brings us in close contact with the group-selection theory of the brain.[3] In this respect, one should remark that singling out "grandmother neurons" (i.e., single cells instead of layers) for the completion of various tasks leads to an explosion in computation time, making the more difficult tasks impossible.

It is difficult to say *a priori* whether the introduction of recombinations—i.e., sexual relationships among our automata—is warranted.[1,5] The reason is that the tasks that we are attempting to teach them are *epistatic*: we need an entire set of connections to be established, and there are strong correlations between them. Therefore, it is not likely that combining some of the synapses of one machine with some of another's will lead to much improvement, quite the contrary. The choice of the *units* which undergo recombination is crucial and, most probably, recombination will, if ever, be of advantage only at a higher level of combining the capabilities of large assemblies of cells.

A clear shortcoming of the approach presented here is that adaptation is entirely *Mendelian*, that is, properties inherited from the ancestor are immediately translated into fixed properties of the descendent; the next natural step would consist of introducing a new level of adaptation, that of the child learning to behave in its environment. These later adaptations would *not* be transferred to the following generations, but would presumably interact profoundly with the heritable properties.

REFERENCES

1. Bergman, A., M. W. Feldman. "More on Selection For and Against Recombination." *J. Theor. Pop. Biol.*, 1990.
2. Braitenberg, V. *Vehicles, Experiments in Synthetic Psychology.* Cambridge, MA: MIT Press, 1984.
3. Edelman, G. M., and V. B. Mountcastle. *The Mindful Brain.* Cambridge, MA: MIT Press, 1978.
4. Hogg, T., and B. A. Huberman. "Parallel Computing Structures Capable of Flexible Associations and Recognition of Fuzzy Inputs." *J. Stat. Phys.* **41** (1985): 115–123.
5. Holland, J. *Adaptation in Natural and Artificial Systems.* Ann Arbor, MI: The University of Michigan Press, 1975.
6. Hopfield, J. J. "Neural Networks and Physical Systems with Emergent Collective Computational Abilities." *Proc. Nat. Acad. Sci. (USA)* **79** (1982): 2554–2558.
7. Jacob, F. *The Possible and the Actual.* New York, NY: Pantheon Books, 1982.
8. Kerszberg, M., and A. Bergman. "The Evolution of Data Processing Abilities in Competing Automata." In *Computer Simulation in Brain Science*, edited by R. M. J. Cotterill. Cambridge, MA: MIT Press, 1988.
9. Marler, P. "Song Learning: Innate Species Differences in the Learning Process." In *The Biology of Learning*, edited by P. Marler and H. S. Terrace. Berlin: Springer-Verlag, 1984.
10. Mayr, E. *The Growth of Biological Thought.* Cambridge, MA: Bellknap Press of Harvard University Press, 1982.
11. Minsky, M., and S. Papert. *Perceptrons.* Cambridge, MA: MIT Press, 1969.
12. Nilsson, N. J. *Principles of Artificial Intelligence.* Palo Alto, CA: Tioga, 1980.
13. Samuel, A. L. "Some Studies in Machine Learning Using the Game of Checkers." *IBM J. Res. & Dev.* **3** (1959): 211–229.
14. Toulouse, G., S. Dehaene, and J. P. Changeux. "Spin-Glass Model of Learning by Selection." *Proc. Nat. Acad. Sci. (USA)* **83** (1986): 1695–1698.

Leif H. Finkel
Department of Bioengineering, University of Pennsylvania, Philadelphia, PA 19104-6392

Limiting Assumptions in Models of Somatotopic Map Organization

This chapter originally appeared in *1990 Lectures in Complex Systems*, edited by L. Nadel and D. Stein, 269–284. Santa Fe Institute Studies in the Sciences of Complexity, Lect. Vol. III. Reading, MA: Addison-Wesley, 1991. Reprinted by permission.

Several recent computational models of somatotopic map organization are reviewed.[15,20,25,31,33] These models were originally developed to account for experimental findings from Merzenich, Kaas, and others on receptive field plasticity and dynamic map organization in somatosensory cortex. We examine the ability of these models to account for more recent experimental results on both rapid and long-term receptive field plasticity,[4,26,27] and discuss the assumptions underlying their successes and failures.

INTRODUCTION

One of the most striking features of cortical organization is the widespread occurrence of topographically organized neural maps. Maps of varying degree of topographic organization are found at all levels of the nervous system, from the spinal cord to the frontal cortex, and are a basic feature of most sensory and motor systems. Somatosensory cortex contains a number of topographic maps in which nearby sites on the body surface are represented at nearby sites in cortex. The mapping between the body surface and cortex is locally but not globally continuous as it contains differential magnifications of various body regions, in addition to fractionated and multiple representations.

Experimental observations have shown that in a number of brain regions, including somatosensory, auditory, visual, and motor cortex, map organization is dynamic and undergoes adaptive changes as a function of ongoing input stimulation. This involves changes in both "what" is represented "where" as well as in the functional receptive field properties of the participating neurons.

We will review several recent computational models of map organization. These models focus on one aspect of the mapping problem, namely, the local organization of the map. The major issues of concern are (1) the mechanism by which local continuity is established and maintained, (2) control of the size of receptive fields (how much of feature space does each neuron survey), and (3) control of the local magnification factor (how much map surface is devoted to a given area of feature space).

We will focus on results from areas 3b and 1 of monkey somatosensory cortex which each contain a topographic map of the body surface. In normal animals, there is a great deal of variability between the maps found in different individuals (for example, in the size of the cortical region representing the thumb). There are also significant changes in any one individual's map over time. These map changes involve shifts in the locations of representational borders (e.g., between thumb and index finger). Merzenich and his colleagues[17] made detailed mappings of areas 3b and 1 in monkey somatosensory cortex after a number of perturbations including nerve transections, nerve crush, digit amputation, syndactyly, and skin island pedicle transfer, and focal cortical lesions. They also studied the effects of repetitive tactile stimulation to a skin region under conditions in which the animal attended to the stimulus, or did not. The results of their extensive studies[1,2,6,17,19,22,23,29] are that, in general, within the limits allowed by the anatomy, the cortical map reflects spatial relationships in the stimulus domain. Regions that are costimulated in a temporally correlated manner tend to be represented in adjacent cortical locations. Map changes can be observed within minutes of a perturbation, but continue to reorganize over weeks and months to reflect new input conditions. After cortical lesions, skin sites formerly represented in the damaged region can come to be represented outside of the damaged area. These studies suggest that cortical representations are plastic, and may be alterable through some combination of the pattern of stimulation, control of synaptic plasticity, and the general excitability of cortical cells.

The process of reorganization appears to follow several general rules: (1) except at representational discontinuities, the receptive fields of nearby cells overlap considerably and the amount of overlap falls off monotonically with distance between the cells; (2) there is a distance limit of roughly 800 microns (probably determined by anatomical constraints) beyond which a de-innervated representation cannot spread; (3) there is an inverse relationship between the size of the representation and the size of the receptive fields of cells in that region; and (4) as map borders shift, they remain sharp.

A number of models have been developed to account for these findings—they differ in biological realism and mathematical tractability. Despite these differences, several common assumptions are made regarding network anatomy, neuronal processing properties, and synaptic modification rules. Some of these assumptions, which may not be necessary to account for the mapping data, are responsible for a failure of the models to account for more recent data that will be discussed, after a brief review of the computational models.

THE KOHONEN MODEL

Kohonen[20] developed the first general model of topographic map organization.[1] He was not specifically concerned with the somatosensory system, but rather with a general topographic map of features. In later papers he extended his algorithm to handle more cognitive tasks such as speech recognition. But for our purposes, the most important aspects of the paper have to do with the organization of an initially unstructured neural map.

Kohonen proposed a simple algorithm to carry out map organization. It consists of four steps:

1. An array of units (usually 2-D) receives inputs from a receptor sheet and each unit computes an output based on simple nonlinear summation processes.
2. There is a mechanism, whose exact implementation is left unstated which compares the outputs of all units in the array, and selects the unit with the maximum output.
3. Local interactions lead to the concurrent activation of the selected unit and its nearest neighbors in the array.
4. An adaptive process (involving synaptic plasticity) then operates to increase the output of the selected units to the present input.

[1]The related problem of map formation during development, e.g., the retino-tectal projection,[7] had been studied earlier by Willshaw and von der Malsburg,[36] Takeuchi and Amari,[34] and others.

The model uses a form of unsupervised learning that consists of normalized synaptic modifications. Input connections are modified in such a way that on each iteration, the synaptic weight vector is slightly rotated toward the input vector.

It is shown in several simulations of two-dimensional maps, that starting from a random initial condition, after several thousand iterations of the adaptive process, a global topographic organization emerges. The adaptive process is an intrinsically local process, affecting only a unit and its eight nearest neighbors. However, mutually incompatible changes produced at neighboring locations tend to cancel each other out; thus, only sets of changes with long-distance coherence survive. Kohonen observed that global organization usually propagated inward from the boundaries of the network. This was due to the fact that the boundaries have fewer neighbors to satisfy, and thus attain an appropriate global organization before the interior of a map can do so.

In an attempt to embed a degree of neural structure in the model, Kohonen considered a network composed of local excitatory interactions and long-distance inhibitory interactions. In such a network, there is a natural tendency of the network to develop clusters of small groups of activity of roughly the same dimensions as the spread of local excitatory interactions. In a series of simulations and analytical studies, he argued that once these groups have formed (by whatever means), they then undergo the same adaptive process as was discussed above for selected units.

Kohonen uses a synaptic sum rule in which the total strength of all inputs to a cell remains constant. Given this synaptic sum rule embedded in a network with activity clusters already formed, Kohonen makes the important discovery that the network obeys a magnification rule. Namely, as the frequency of activation of a particular input region is increased, the fraction of the map devoted to the stimulated region is proportionately increased. He did not comment on the related physiological property that as map representation increases, receptive field size decreases.

Kohonen's model does not incorporate realistic anatomical or physiological detail. He does make some speculations regarding the possible roles of various cortical cell types (chandelier cells, bipolar cells, etc.) which he proposed might play roles analogous to basis functions in determining the proper range of synaptic interactions. Overall, however, the argument is a mathematical one. Nonetheless, Kohonen clearly sets up the problem of how map order can be imposed by structured inputs, he anticipates the importance of local groups of cells, and he makes a series of remarkable observations about the major properties of topographically organized maps.

Schulten, Ritter, and coworkers have taken the basic Kohonen model and explicitly applied it to the case of the somatosensory system. Their results show that an organized map of the hand of a monkey can be formed using a formalized version of Kohonen's rules. They also show a simulation in which one of the "fingers" is amputated. Results correspond to the experimental observations in which the zone formerly devoted to representing the amputated digit comes to be occupied by a representation of the adjacent, remaining fingers.

Ritter and Schulten[31] use analytic techniques to examine the stationary state of the Kohonen mapping with regard to the problem of the magnification factor. In contrast to Kohonen's arguments, they find that in the general case of the two-dimensional map, the magnification factor (i.e., the amount of map devoted to a given peripheral region) cannot be expressed as any simple function of the frequency of stimulus input. Furthermore, in the case of one-dimensional maps, the magnification factor appears to depend on the stimulus frequency raised to the 2/3 power rather than on the linear relationship found by Kohonen.

These differences point out the importance of the exact assumptions. Moreover, they illustrate that analytical techniques, useful and powerful as they are, very quickly reach the limit of their applicability in dealing with even such simplified versions of complex systems.

THE MODEL OF FINKEL, EDELMAN, AND PEARSON

Finkel, Edelman, and Pearson[25] developed a model of topographic map organization in somatosensory cortex which explicitly attempts to account for the major experimental results obtained by Merzenich, Kaas, and their colleagues. This model is biologically based; i.e., it incorporates the basic elements of cortical anatomy and physiology, and it features a realistic synaptic modification rule that formally resembles the mechanism of operation of the NMDA (N-methyl-D-aspartate) receptor.[13,14] In addition to the question of the origin of topography, considerable attention is paid to receptive field properties, and these properties are directly measured in the model. Unlike previous models, the emphasis of the treatment is on simulation, rather than analytical techniques.

The network consists of 1512 cells, both excitatory and inhibitory, which make extensive local interconnections. A sheet of 1024 receptors represents the hand of the monkey and each cell receives extrinsic inputs from receptors on the front (glabrous surface) and back (dorsal surface) of the hand. The projections are arranged in a mirror-symmetric fashion such that the front and back of a given digit both project to the same region of the network. This double projection, together with the fact that each receptor contacts units distributed over approximately 10% of the network allows a given point on the hand to potentially be represented over a wide area of the network. Thus, many alternative physiological maps (those obtained by microelectrode recording of receptive fields) are possible given the anatomical substructure. The extrinsic projections are assumed to be topographically ordered at the start of the simulation (i.e., the developmental origins of topography are not studied). The problem of concern is to determine the rules governing the organization and reorganization of the observed physiological map.

While the extrinsic projections are anatomically ordered, their initial synaptic weightings are random, as are those of the intrinsic connections. The simulations commence with a protocol for stimulating the receptor sheet. Small (3×3 receptor) regions of the hand are stimulated in random sequences for several hundred to several thousand iterations (each location on the hand is stimulated approximately 10 times). Regardless of the size, shape, frequency, or pattern of stimulation (within rather large limits) cells self-organize into small neuronal groups. These are local collections of cells which have greatly increased the strengths of their connections to each other and have weakened their connections to cells not in the group. Once groups form they are stable, although they undergo a continual process of modification in which cells on the borders of adjacent groups can be traded back and forth.

The critical property of groups is their nonlinear cooperativity—the strong connections between the cells of a group lead to similar firing patterns among them, which in turn, leads to the development of highly overlapped receptive fields. In a sense, the receptive field becomes a collective property of the local group of cells, rather than the intrinsic property of individual units. (In this sense, it resembles the concept of the segregate as proposed by Whitsel and colleagues.[35]) Groups constantly modify the location and size of their receptive fields as a result of intergroup competition. The source of this competition is the amount and character of activation received by the group. The main aim of the model, then, is to examine how receptive field structure and map organization depend upon the characteristics of input stimulation to a network organized into neuronal groups.

The simulations show that if the receptor sheet is stimulated in different patterns, the resulting maps differ in detail, however, one always obtains a topographic representation of the hand which qualitatively resembles that found in cortical areas 3b and 1 of the monkey. The map is always topographic with discrete areas of representation of the front and back of the hand—the borders between these regions correspond to borders between neuronal groups.

Perturbations in the input received by the network result in shifts in map borders that closely correspond to those observed *in vivo*. For example, when the probability of stimulation of a particular hand region is increased, the size of the representation of that region correspondingly increases (by an order of magnitude). This expansion is at the expense of those regions that were formerly represented in adjacent areas of the network. Conversely, when stimulation of a hand region is decreased by cutting the median nerve (which subserves innervation of the median half of the front of the hand), the representation of that hand region disappears. In its place there emerges an intact, topographic representation of the corresponding back of the hand (subserved by the ulnar nerve), as would be expected from the underlying anatomy.

Neuronal groups play a critical role in these shifts in map borders, and one characteristic of the shifts argues strongly for the necessity of groups. It has been observed experimentally, that as map borders shift they always remain sharp. In other words, one does not see a "defocusing" of representation followed by a re-sharpening; rather, the borders shift sometimes by prodigious amounts, but at all times there is a relatively crisp border between, for example, the front and back of a digit. This "shift with sharpness" is difficult to explain with most models of map organization because they must pass through a state in which connection strengths are partially strengthened and weakened (with concomitant changes in the receptive fields). The sharpness of the map borders in the model is due to the fact that they occur at the borders between neuronal groups. Since group borders are one cell thick, the transitions between receptive fields will always be crisp. Shifts in map borders occur as groups change the size and location of their receptive field locations, but at all times, the transitions will be sharp. It is the cooperative nature of the intra-group dynamics coupled with the competitive processes operating between groups that then accounts for the observed map properties.

The properties of groups also provides an explanation for both the magnification rule (size of representation is directly proportional to input stimulation) and the inverse rule (size of representation is inversely proportional to receptive field size). The argument depends upon two assumptions: (1) that the anatomical size of a group changes slowly relative to changes in the size of its receptive field, and (2) that the relative overlap of receptive fields of adjacent groups remains fairly constant, regardless of actual receptive field size, due to the nature of group-group competition. Given these assumptions, which have been borne out in simulations, the size of a representation in cortex will depend upon the number of groups with receptive fields located in the area of interest. Since the overlap between receptive fields of adjacent groups is constant (e.g., 30%), the smaller the receptive fields the more groups required to span the space, and the larger the corresponding representation.

In a number of simulations, in fact, the correct workings of both the magnification and inverse rules were obtained. However, for the published simulations, receptive field sizes were driven to their minimal limits; i.e., they were only 1 receptor wide. This was done to allow the finest scale map possible with limited resources. The consequence of this action was that receptive fields could not decrease further in size, making it impossible fully to test the above rules. Grajsky and Merzenich[15] subsequently succeeded in demonstrating how the magnification rule can be simulated and we will discuss their findings below.

The model represents an attempt to simulate the emergence of a number of rules of map organization in a realistic, biologically based simulation. At its core is the idea that interactions among neuronal groups control the ongoing organization of the observed physiological map from the manifold of possible anatomical maps. The model also makes a number of experimental predictions, some of which have already been confirmed.[6]

THE GRAJSKI-MERZENICH MODEL

Grajski and Merzenich[15] have developed a simulation of somatotopic map organization which addresses the problem of the inverse rule. Their network uses a similar architecture to the Pearson et al. model, with excitatory and inhibitory cells, diverging afferents from the skin, and extensive local cortico-cortical connections. Altogether, 900 cells and 57,600 synapses are simulated, and the skin is represented by a 15×15 array of units divided into three "fingers." An important innovation is the inclusion of a subcortical layer of relay cells which both project to the cortical network and receive connections back from the cortical network. Such multilayer networks with reentrant connections offer the possibility of richer network dynamics. In fact, Grajski and Merzenich found that they could obtain similar results with a reduced two-layer network.

The model makes use of two types of normalization. First, the effect of each synaptic input on a cell is normalized by the total number of inputs of that class (i.e., from the same type of cell in the same network). Second, and more importantly, the total synaptic strength of each cell is normalized to a constant value. Synaptic normalization is a common assumption in models and may have some biological plausibility in that a cell must have a finite amount of resources that may be allocated among its synapses. However, synaptic normalization guarantees that strengthening of one set of connections to a cell obligately weakens all other sets, thus the assumption predisposes toward certain results.

Grajski and Merzenich start with a network which has been topographically refined through several cycles of stimulation (interestingly, they require 10–12 passes over the skin to reach a steady state which is identical to that required by the Pearson et al. model). The resulting map shows the required properties that receptive fields overlap considerably (up to 70%), receptive field overlap falls off monotonically with distance, and cortical magnification is proportional to the frequency of stimulation. A simulation was then performed in which the probability of stimulation to a small region of the skin was increased (by a factor of 5). As in the Pearson et al. model, this results in an expanded cortical representation. In addition, Grajski and Merzenich found a smaller but noticeable expansion in the subcortical representation. Most importantly, the cortical receptive fields located on the stimulated region were seen, on average, to decrease in size. Interestingly, subcortical receptive fields remained unchanged in size.

The converse experiment was also simulated in which a region of cortex is "lesioned;" i.e., the cells representing a given skin region are removed. The authors found that, under these conditions, with the Hebbian synaptic rule used, the intact representation is firmly entrenched. Thus, all cortico-cortical and cortical afferent connections are randomized and cortical excitation is artificially enhanced. Then, a new representation is seen to emerge in which receptive field sizes are markedly larger in the cortex and smaller in the subcortex.

Thus, the simulation appears generally to obey the inverse rule—although there are distinct differences in the behavior of the cortical layer and the subcortical layer. Moreover, there is no quantitative or qualitative relationship shown here; rather, just the sign of the changes are in the appropriate directions. Finally, the issue of synaptic normalization can be considered by the following argument. Under the action of a Hebbian rule, increasing the stimulation to a region of skin will cause strengthening of the afferent connections from that region. Cells which previously received only subthreshold inputs from the stimulated region will, after synaptic strengthening, include this region in their receptive fields. This accounts for the expansion of the representation (note that Kohonen achieved such a result using normalization). However, cells which previously had their receptive fields in the stimulated region will also undergo synaptic strengthening. If these cells have their synaptic strengths renormalized after the stimulation then all connections from nonstimulated regions will be automatically weakened. This will cause certain previously effective connections to become subthreshold, and will result in a decrease in receptive field area. Thus, synaptic renormalization, by itself, largely accounts for the inverse magnification rule.

Perhaps the major difference between the Grajski-Merzenich model and that of Pearson and colleagues, is the use of neuronal groups. Whether or not neuronal groups exist is a question which can only be answered experimentally. Topographic organization can emerge without groups, requiring only a structured (correlated) input, a reasonable anatomy, and a Hebbian-type synaptic rule. Groups are sufficient, and may be necessary for generating map shifts that retain sharpness. Finally, while the inverse rule has been shown to operate independently of groups under conditions of synaptic normalization, it remains to be demonstrated whether groups are necessary for the inverse rule if one does not assume synaptic normalization.

The Grajski-Merzenich model makes an important contribution in considering what processes are sufficient to account for the inverse rule. In addition, they make a number of nonintuitive experimental predictions about subcortical changes.

THE COMPETITIVE ACTIVATION MODEL OF REGGIA

Reggia and colleagues[3,30,33] have taken a different approach to accounting for many of these same findings on somatotopic plasticity. Their model is built on a novel mechanism for cell interactions explicitly based on competition.

The standard assumption in most cortical network models is that inhibition operates over a slightly broader spatial scale than excitation (e.g., the Pearson et al. model). Reggia proposes a new mechanism of "competitive activation" in which there are no explicit inhibitory connections. Rather, excitatory cells compete for activation in a "leaders-take-most" or "rich-get-richer" manner.

The output of cell i to cell j is given by

$$\text{out}_{ji}(t) = c_p \left\{ \frac{w_{ji}(t)[a_j(t) + q]}{\sum_k w_{ki}(t)[a_k(t) + q]} \right\} a_i(t)$$

where a_j is the activity of cortical element j, w_{ji} is the connection weight from cell i to cell j, and both q and c are constants. This rule allocates a specific fraction of a cell's output to each of its postsynaptic targets—more active targets draw more input. Thus, in place of cortico-cortical inhibition, this rule allows more strongly activated cells to increase their firing rate with respect to less activated cells. Cell activity is bounded between 0 and a maximum level, M, by an equation

$$\frac{d}{dt}a_j(t) = c_s a_j(t) + [M - a_j(t)]\, \text{in}_j(t)$$

where $\text{in}_j(t) = \sum \text{out}_{ji}(t)$ is the input to cell j.

Reggia and colleagues[30] cite a variety of anatomical studies which question the standard assumptions about direct or indirect peristimulus inhibition in cortex. And they demonstrate that this competitive activation rule works in much the same way as traditional network schemes of inhibition. However, unlike other inhibitory schemes, their rule acts as a type of AND gate—an active cell directs its output only toward other active cells. In Reggia et al.,[30] a possible neural implementation for the rule is proposed involving cortico-thalamic feedback loops. A thalamic cell can be envisaged as partitioning its output, $\text{out}_{ji}(t)$, among those cortical cells to which it projects. Feedback connections from cortex communicate the activation, $a_j(t)$, of each postsynaptic cortical cell. Thus, together with a normalization process (the denominator of the equation) the feedback loop can implement the competitive process. The role most commonly attributed to feedback is to enhance feature primitives related to attention and hierarchical processing. However, in Reggia's model, feedback serves to control the lateral interactions normally ascribed to intra-cortical inhibition.

In a series of papers, Reggia and colleagues demonstrate that the competitive activation mechanism can account for most of the experimental findings on soma-totopic maps. They use a network architecture with hexagonal arrays representing the thalamus and cortex. There is only one type of cell, all connections are exci-tatory, and toroidal boundary conditions are used to avoid the edge effects seen in the Pearson et al. model.

Simulations show that a refined map can be generated from initially coarse topography. Repetitive local stimulation leads to an increased magnification of the stimulated region. However, the average receptive field size in this stimulated region does not decrease, as required by the inverse magnification rule. Only cells whose receptive fields are located near the borders of the stimulated region show a decrease in size. Presumably, this is because thalamic inputs connect to a limited (60 cell) cortical neighborhood, and there is no competitive advantage to cells, all of whose neighbors see a uniform increase in stimulation. Simulations also show a shift in representational boundaries after "deafferentation." Since there is only one source of input (e.g., a glabrous surface), effects related to dorsal/glabrous competition could not be addressed.

Reggia also studied the effects of focal cortical lesions. After deactivating a region of the cortical network, a two-stage process of reorganization occurs. First, there is a shift, due to the competitive activation rule, of receptive fields originally located near the borders of the lesioned area. With continued stimulation, there is a second, slower synaptic reorganization which leads to improved topography and a reduced magnification of the affected region.

Grajski and Merzenich had also simulated effects of focal cortical lesions. But their network was more intransigent to reorganization, and required a renormalization of synaptic weights after lesioning, in order to show a shift in receptive field locations.

Reggia and colleagues have applied their model to a variety of other observations, including considerations of recovery from stroke. The competitive activation rule is an efficient means of expressing many of the observed results, and it reproduces many of the original findings. Most interestingly, it requires the presence of feedback from higher or lower areas.

RECENT EXPERIMENTAL FINDINGS

The successes of the models considered in accounting for the data of Merzenich and colleagues depends upon the assumptions made. However, a number of more recent experimental observations having to do with receptive field changes over very short and very long time periods suggest that additional processes to those considered may be at work. These experimental results were obtained after the publication of all of the models, with the exception of the competitive activation model. However, Reggia and colleagues explicitly state[33] that their model does not attempt to account for these findings.

In 1991, Pons and colleagues reported on the cortical effects of deafferentation in a set of monkeys who had survived over 10 years after total deafferentation of an arm. The effects Merzenich and others had previously observed had been limited to cortical shifts on the order of hundreds of microns (up to 1 mm in the most extreme cases). Pons reported map shifts on the order of 10–14 mm. In these animals, the cortical areas formerly devoted to representing the hand and arm were now occupied by a representation of the face. In the normal somatotopic map of the body, the representation of the hand and arm is interposed between that of the trunk and that of the face. The hand and face share a common border. In the deafferented monkeys, the representation of the lower face, chin and cheeks was stretched, as if on a rubber sheet, over the centimeter of cortex formerly devoted to the arm. The representation of the chest, which also borders the arm representation, was minimally if at all affected.

These observations imply that additional mechanisms to those considered in the above models must be at work. The branching thalamic arbors underlying the shifts seen in Merzenich's data extend over 1–2 mm^2 in cortex. There is no way in which these arbors can account for shifts over 15 mm, and perhaps the most likely explanation is that anatomical sprouting occurred over the decade of recovery. Sprouting has been documented in striate cortex following recovery from laser-induced retinal scotomas.[10] None of the models consider the possibility of new anatomical connections, and it remains to be determined whether such long-term changes follow similar rules.

Related map changes have been demonstrated in humans who had recently undergone arm amputation and were experiencing "phantom" limb syndrome.[27] In several of the patients examined by Ramachandran, there was a topographic representation of the amputated hand located on the lower portion of the face. When the chin and lower face was stimulated, the patients reported the sensation of touch on their missing hand (in addition to perceiving touch to their face). Thus, in their cortices, the representation of the face must have expanded into the region formerly devoted to the hand. Ramachandran argues that when the face is stimulated, it activates two sets of cells; the original, intact representation of the face, and also cells in the cortical region formerly devoted to the hand. Higher centers, viewing this latter activation, may "interpret" it as arising from the hand.

Patients could distinguish hot and cold stimuli, and water dripped down the face was perceived as if it was dripping down the arm. When patient's were instructed to mentally rotate their missing hands, the representation on the face shifted accordingly. Only some patients exhibited a map on the face, however, all patients had representations of the missing arm at one or more proximal sites on the remaining stump.

Further documentation for these findings comes from combined MEG-MRI studies (reviewed in Ramachandran[27]). In the hemisphere contralateral to the normal remaining arm, it is possible to observe a significant spacing between the representations of face-arm-hand. However, in the affected hemisphere, the representation of the face is seen to shift toward the site of representation of the missing arm and hand.

In perhaps the most striking finding, Ramachandran had amputee patients view a mirror image of their remaining hand from such an angle that it visually appeared to be the missing hand. Patients were instructed to move both "hands" symmetrically through a range of motions. Remarkably, patients reported the disappearance of painful phantom sensations, and eventually, the disappearance of the phantom limb perception itself.

These studies show that, in humans, large-scale map shifts can occur within days after deafferentation. Long-distance sprouting may not be necessary to account for the results, as the hand and face representations are fairly close, and it is not clear that the entire region formerly devoted to the arm is completely occupied. In fact, the occurrence of one or more representations on the remaining arm stump suggests that both arm and face are occupying the deafferented cortical territory. It is puzzling to contrast these results with an earlier finding of Merzenich. When a single digit is amputated, the deafferented cortical region is occupied by representations of the adjacent digits; but when two adjacent digits are removed, there remains an unresponsive cortical region. Perhaps over longer recovery times, the deafferented region would eventually fill-in. Different representations (face, chest, finger) may also have different abilities to expand.

A second set of surprising experimental observations concerns receptive field changes following digit amputation in the Australian flying fox[4] and related studies following creation of retinal scotomata in cat and monkey.[16] In the flying fox, there was a large, immediate expansion of the receptive field following digit amputation or application of local anesthesia. The changes occurred immediately—and in the absence of any correlated input to the skin surface. These changes were reversible when the anesthetic wore off. Analogous receptive field expansions are seen in visual cortex following creation of a focal scotoma.[16]

The magnitude of these receptive field changes are large compared to the anatomical scales of the thalamo-cortical connections in the early models. More importantly, the fact that the change occurs before any significant stimulation of the skin takes place, suggests a mechanism different from that mediating the chronic map changes due to altered patterns of stimulation. The immediate nature of the expansion leads these investigators to suggest that the change may be due to an alteration in the balance of excitation and inhibition, rather than due to synaptic plasticity, *per se*. In their article, Calford and Tweedale[5] point out that the Pearson et al. model makes the assumption that only excitatory cells receive a direct thalamic input. This was done for computational simplicity. But the result is that deafferentation affects excitatory cells directly but inhibitory cells only indirectly through the loss of inputs from local excitatory cells. Calford and Tweedale propose that receptive field expansion depends upon an effective release from inhibition. Thalamic drive to both excitatory and inhibitory cells is removed, but the effect on inhibition predominates. The lack of direct inhibitory connections in the computational model disallows the possibility of such a mechanism.

Some insight into possibly similar mechanisms comes from experiments by Pettet and Gilbert[24] involving the creation of an "artificial" scotoma. The artificial scotoma is formed by a dynamic textured stimulus that has a small, homogeneous grey region whose luminance equals the average luminance in the surround. The stimulus is placed such that the homogeneous region covers (and extends somewhat beyond) the classical receptive field of the recorded cell. Ramachandran and Gregory[28] showed psychophysically that when such a stimulus is viewed for several minutes, the texture appears to "fill-in" the homogeneous region (color also fills in with a slightly faster time course). Pettet and Gilbert reported that in cat cortex, there was a 5-fold expansion of receptive fields located within the scotoma region. This expansion was immediately reversible when the scotoma region was stimulated.

More recently, Freeman and colleagues[8] used reverse-correlation techniques to precisely determine receptive field changes during conditioning with similar artificial scotoma. They reported that many cells show no change in receptive field properties, and those that do change appear to have undergone a change in response gain. The overall responsiveness of a cell might be expected to change through contrast gain-control mechanisms as a result of decreasing the local contrast versus that in the surround. If response gain is increased, then inputs which were formerly subthreshold, can become suprathreshold, and thus weak distant inputs can now be included in the receptive field. deWeerd and colleagues[9] present data consistent with this interpretation. They recorded from cells during the conditioning phase of the artificial scotoma, and found that cells in the scotoma region gradually increase their firing rates.

The explanation offered for these results[8,9] is that the conditioning stimulus (the texture surrounding the homogeneous "hole") leads to an adaptation of the inhibition onto cells in the scotoma region. Reduced inhibition translates into increased gain response. Restimulation in the scotoma region itself activates local inhibition which reduces the gain. The long-distance inhibition could be mediated by horizontal cortico-cortical connections, or by basket cell networks, or by a combination of both. This is consistent with the recent computational model of Xing and Gerstein[37] who tested several related mechanisms and found that adaptation of inhibition best accounts for the types of receptive field expansions observed.

Regardless of the exact mechanism, decreased stimulation to the receptive field center, in the context of surround stimulation, may lead to an increase in cell gain—and cell gain can modulate receptive field size. Local excitatory recurrent connections may play an important role in this process. For example, Douglas and Martin[11] propose that thalamic input unleashes a cascade of recurrent cortical excitation that amplifies and sharpens the thalamic signal. Cortical inhibitory cells are activated earlier than excitatory cells due to a faster, highly myelinated set of inputs.[32] The extent of cortical activation depends upon the degree of amplification of the excitation before inhibition "wipes the slate clean." Deafferentation might alter the local balance between excitation and inhibition, resulting in a net increase in the positive feedback within the local recurrent circuits.

Other mechanisms are clearly possible, including effects mediated through inhibitory networks. Regardless, these results illustrate how the assumptions of the original models constrain their possible responses to altered stimulation conditions. Furthermore, it illustrates how the failure to incorporate a particular detail of cell physiology—in this case, overall cell responsiveness or gain—can have important implications for the ultimate success of the model.

CONCLUSIONS

Experimental observations by Merzenich and others over the last two decades have pointed toward certain basic rules of somatopic map organization. It appears that some type of competitive process influences the size and location of cortical representations, and that the amount of peripheral input determines the outcome of this competition. The models considered demonstrate that analogous competitive effects can occur in networks, under a variety of accompanying assumptions and conditions. In fact, the analogy has been made to competitive effects in ecology, where similar phenomena to the inverse magnification rule apply to niche dimensions.[12]

Since the argument is by analogy, much depends upon the degree to which the models accurately reflect what is going on in the nervous system. One clear limitation of all of these models, is that they are not pitched at the cellular level. It would be most interesting to see a model implemented at the level of Hodgkin-Huxley-type dynamics of cell firing, with simulated NMDA receptors and some degree of biochemical process-modeling; however, even an integrate-and-fire level model would represent a reasonable start. Such a platform would allow the effects of various transmitters, modulators, and drugs to be studied, and would force specific choices on which cellular mechanisms are critical.

The recent evidence reviewed suggests that synaptic plasticity may be only one component of the response to altered input. It is well known that cells adapt to continued stimulation, there are rapidly and slowly adapting cells throughout primary somatosensory cortex. In light of the multiplicity of known cellular processes, it appears somewhat optimistic to assume that the Hebb rule can account for everything. It is possible that immediate changes in cell responsiveness can be translated into longer-term changes in synaptic plasticity. And it is further possible that these synaptic facilitations play a role in the even longer-term axonal sprouting processes, possibly through the involvement of molecules such as CAM II kinase. CAM kinase is the major synaptic protein, it constitutes 1% of forebrain protein, is the major component of presynaptic and postsynaptic densities, and has been found, through pharmacological and genetic knock-out experiments, to be necessary for LTP.[18] Thus, there may be a continuum of changes, over different time scales, that allow the system to measure the permanence of the alteration in input characteristics. An improved model could be developed by adding direct thalamic→inhibitory cell connections, as suggested by Calford and Tweedale,[5] or by incorporating some type of local gain-control mechanism. But one wonders what future findings may depend upon other, yet to be incorporated processes. If modeling is to make a contribution, to be predictive, then it may be imperative to carry out thorough evaluation of the effects of known physiological processes.

Perhaps the most important conclusion comes from considering the history. Evidence for map plasticity was cited in Sherrington's original studies of motor cortex,[21] but overlooked for decades until high resolution recordings by Wall, Merzenich, and others documented variability across individuals and over time. At first controversial, dynamic organization of maps has come to be central to current thinking about the cortical computation. Now it appears that reorganization may occur through several different mechanisms over multiple time scales based on cell activity, synaptic modifications, and anatomical rewiring. The utility of any model rests in its ability to account for the greatest range of observations with the smallest number of assumptions—but also in its ability to generate new, nonintuitive predictions which can be tested experimentally and used to improve (or falsify) the model. It must be frankly admitted that, to date, the number and quality of experimental predictions generated by these computational models has not been overwhelming. But there is a need, perhaps now more pressing than ever, for the development and refinement of models that can integrate anatomical and physiological information into an explanation of how functional behavior emerges from the operation of cellular-level processes.

ACKNOWLEDGMENTS

This study was made possible by grants from Office of Naval Research N00014-93-1-0861, The Whitaker Foundation, and the McDonnell-Pew Program in Cognitive Neuroscience.

REFERENCES

1. Allard, T., and M. M. Merzenich. "Some Basic Organizational Features of the Somatosensory Nervous System." In *Connectionist Modeling and Brain Function: The Developing Inerface*, edited by S. J. Hanson and C. R. Olsen. Cambridge: MIT Press, 1990.
2. Allard, T., S. A. Clark, W. M. Jenkins, and M. M. Merzenich. "Reorganization of Somatosensory Area 3b Representations in Adult Owl Monkeys After Digital Syndactyly." *J. Neurophsiol.* **66** (1991): 1048.
3. Armentrout, S. L., J. A. Reggia, and M. Weinrich. "A Neural Model of Control Map Reorganization Following a Focal Lesion." *Art. Int. Med.* **6** (1994): 383–400.
4. Calford, M. B., and R. Tweedale. "Immediate and Chronic Changes in Responses of Somatosensory Cortex in Adult Flying-Fox after Digit Amputation." *Nature* **332** (1988): 446.
5. Calford, M. B., and R. Tweedale. "Acute Changes in Cutaneous Receptive Fields in Primary Somatosensory Cortex After Digit Denervation in Adult Flying Fox." *J. Neurophysiol.* **65** (1991): 178–187.
6. Clark, S. A., T. Allard, W. M. Jenkins, and M. M. Merzenich. "Receptive Fields in the Body-Surface Map in Adult Cortex Defined by Temporally Correlated Inputs." *Nature* **332** (1988): 444.
7. Cowan, W. M., and R. K. Hunt. "The Development of the Retinotectal Projection: An Overview." In *Molecular Basis of Neural Development*, edited by G. M. Edelman, W. E. Gall, and W. M. Cowan, 389–428. New York: Wiley, 1985.
8. DeAngelis, G. C., A. Anzai, I. Ohzawa, and R. D. Freeman. "Receptive Field Structure in the Visual Cortex: Does Selective Stimulation Induce Plasticity?" *Proc. Natl. Acad. Sci. USA* **92** (1995): 9682.
9. deWeerd, P., R. Gattass, R. Desimone, and L. G. Ungerlieder. "Responses of Cells in Monkey Visual Cortex During Perceptual Filling-In of an Artificial Scotoma." *Nature* **377** (1995): 731.
10. Darian-Smith, C., and C. D. Gilbert. "Axonal Sprouting Accompanies Functional Reorganization in Adult Cat Striate Cortex." *Nature* **368** (1994): 737.
11. Douglas, R. J. and K. A. C. Martin. "A Functional Microcircuit for Cat Visual Cortex." *J. Physiol.* **40** (1991): 735.
12. Edelman, G. M., and L. H. Finkel. "Neuronal Group Selection in the Cerebral Cortex." In *Dynamic Aspects of Neocortical Function*, edited by G. M. Edelman, W. E. Gall, and W. M. Cowan, 653–695. New York: Wiley, 1984.
13. Finkel, L. H. "A Model of Receptive Field Plasticity and Topographic Map Reorganization in the Somatosensory Cortex." In *Connectionist Modeling and Brain Function: The Developing Interface*, edited by S. J. Hanson and C. R. Olsen, 164–192. Cambridge: MIT Press, 1990.

14. Finkel, L. H., and G. M. Edelman. "Interaction of Synaptic Modification Rules Within Populations of Neurons." *Proc. Natl. Acad. Sci. USA* **82** (1985): 1291–1295.

15. Grajski, K. A., and M. M. Merzenich. "Hebb-Type Dynamics is Sufficient to Account for the Inverse Magnification Rule in Cortical Somatotopy." *Neural Comp.* **2** (1990): 71–84.

16. Gilbert, C. D., and T. N. Wiesel. "Receptive Field Dynamics in Adult Primary Visual Cortex." *Nature* **356** (1992): 150.

17. Jenkins, W. M., and M. M. Merzenich. "Reorganization of Neocortical Representations After Brain Injury: A Neurophysiological Model of the Bases of Recovery from Stroke." *Prog. Brain Res.* **71** (1987): 249.

18. Jones, E. G., G. W. Huntley, and D. L. Benson. "Alpha Calcium/Calmodulin-Dependent Protein Kinase II Selectively Expressed in a Subpopulation of Excitatory Neurons in Monkey Sensory-Motor Cortex: Comparison with GAD-67 Expression." *J. Neurosci.* **14** (1994): 611–629.

19. Kaas, J. H., M. M. Merzenich, and H. P. Killackey. "The Reorganization of Somatosensory Cortex Following Peripheral Nerve Damage in Adult and Developing Mammals." *Ann. Rev. Neurosci.* **6** (1983): 325–356.

20. Kohonen, T. "Self-Organized Formation of Topologically Correct Feature Maps." *Biol. Cybern.* **43** (1982): 59–69.

21. Leyton, A. S. F., and C. S. Sherrington. "Observations on the Excitable Cortex of the Chimpanzee, Orangutang, and Gorilla." *Quart. J. Exp. Physiol.* **11** (1917): 137–222.

22. Merzenich, M. M., G. Recanzone, W. M. Jenkins, T. T. Allard, and R. J. Nudo. "Cortical Representational Plasticity." In *Neurobiology of Neocortex*, edited by P. Rakic and W. Singer, 41–67. New York: Wiley, 1988.

23. Merzenich, M. M., and M. M. Jenkins. "Reorganization of Cortical Representations of the Hand Following Alterations of Skin Inputs Instead by Nature Injury, Skin Island Transfers, and Experience." *J. of Hand Therapy* **6** (1993): 89–104.

24. Pettet, M. W., and C. D. Gilbert. "Dynamic Changes in Receptive-Field Size in Cat Primary Visual Cortex." *Proc. Natl. Acad. Sci. USA* **89** (1992): 8366.

25. Pearson. J. C., L. H. Finkel, and G. M. Edelman. "Plasticity in the Organization of Adult Cerebral Cortical Maps: A Computer Simulation Based on Neuronal Group Selection." *J. Neurosci.* **7** (1987): 4209–4223.

26. Pons, T. P., P. E. Garraghty, A. K. Ommaya, J. H. Kaas, E. Taub, and M. Mishkin. "Massive Cortical Reorganization After Sensory Deafferentation in Adult Macaques." *Science* (1991): 1857.

27. Ramachandran, V. S. "Behavioral and Magnetoencephalographic Correlates of Plasticity in the Adult Human Brain." *Proc. Natl. Acad. Sci USA* **90** (1993): 10413.

28. Ramachandran, V. S., and R. L. Gregory. "Perceptual Filling-In of Artificially Induced Scotomas in Human Vision." *Nature* **350** (1991): 699.

29. Recanzone, G. H., M. M Merzenich, W. M. Jenkins, K. A. Grajski, and H. R. Dinse. "Topographic Reorganization of the Hand Representation in Cortical Area 3b of Owl Monkeys Trained in a Frequency-Discrimination Task." *J. Neurophysiol.* **67** (1992): 1031.

30. Reggia, J., C. D'Autrechy, G. Sutton, and M. Weinrich. "A Competitive Distribution Theory of Neocortical Dynamics." *Neural Comp.* **4** (1992): 287–317.

31. Ritter, H., and K. Schulten. "On the Stationary State of Kohonen's Self-Organizing Sensory Mapping." *Biol. Cyber.* **54** (1986): 99–106.

32. Somers, D. C., S. B. Nelson, and M. Sur. "An Emergent Model of Orientation Selectivity in Cat Visual Cortical Simple Cells." *J. Neurosci.* **15** (1995): 5448–5465.

33. Sutton, G. G., J. A. Reggia, S. L. Armentrout, and C. L. D'Autrechy. "Cortical Map Reorganization as a Competitive Process." *Neural Comp.* **6** (1994): 1–13.

34. Takeuchi, A., and S. Amari. "Formation of Topographic Maps and Columnar Microstructures in Nerve Fields." *Biol. Cyber.* **35** (1979): 63–72.

35. Whitsel, B. L., O. Favorov, M. Tommerdahl, M. Diamond, S. Juliano, and D. Kelly. "Dynamic Processes Govern the Somatosensory Cortical Response to Natural Stimulation." In *Organization of Sensory Processing*, edited by J. S. Lund, 84–116. New York: Oxford University Press, 1988.

36. Willshaw, D. J., and C. von der Malsberg. "How Patterned Neural Connections Can Be Set Up by Self-Organization." *Proc. Roy. Soc. (Lond.)* **B194** (1976): 431–445.

37. Xing, J., and G. L. Gerstein. "Simulations of Dynamic Receptive Fields in Primary Visual Cortex." *Vision Res.* **34** (1994): 1901.

Walter Fontana
Theoretical Division and Center for Nonlinear Studies, Los Alamos National Laboratory, Los Alamos, NM 87545 USA and Santa Fe Institute, 1120 Canyon Road, Santa Fe, NM 87501 USA

Functional Self-Organization in Complex Systems

This chapter originally appeared in *1990 Lectures in Complex Systems*, edited by L. Nadel and D. Stein, 407–426. Santa Fe Institute Studies in the Sciences of Complexity, Lect. Vol. III. Reading, MA: Addison-Wesley, 1991. Reprinted by permission.

A novel approach to functional self-organization is presented. It consists of a universe generated by a formal language that defines objects (=programs), their meaning (=functions), and their interactions (=composition). Results obtained so far are briefly discussed.

1. INTRODUCTION

Nonlinear dynamical systems give rise to many phenomena characterized by a highly complex organization of phase space, e.g., turbulence, chaos, and pattern formation. The structure of the interactions among the objects described by the variables in these systems is usually fixed at the outset. Changes in the phase portrait occur as coefficients vary, but neither the basic qualitative relationships among the variables, nor their number is subject to change.

In this contribution, I will be concerned with the class of systems that is, in some sense, complementary. This class contains systems that are "inherently constructive." By "constructive" I mean that the elementary interaction among objects includes the possibility of building new objects. By "inherently constructive" I want to emphasize that the generation of new objects from available ones is an intrinsic, specific, non-random property of the objects under consideration. It is *not* primarily caused by noise.

A prime example of such a system is chemistry. Molecules undergo reactions that lead to the production of specific new molecules. The formation of the product object is *instructed* by the interacting reactants. This is to be distinguished from a situation in which new objects are generated by chance events, as is the case with copying errors during the replication process of a DNA string (mutations).

Examples for complex systems belonging to the constructive category are chemical systems, biological systems like organisms or ecosystems, and economies. Clearly, nonlinearities and noise occur everywhere. In this note, I will be primarily concerned with the implications following from noiseless constructive properties. An example of a dynamical system built on top of these properties will be given, but. the formulation of a general theory (if it exists) that combines nonlinear dynamical *and* constructive aspects of complex systems is a major problem for the future.

The "manipulation" of objects through other objects, as it occurs with molecules, might in principle be reduced to the behavior of the fundamental physical forces relevant for the particular objects. Quantum mechanics is an example with respect to chemistry. At the same time, however, the very phenomenon of syntactical manipulation introduces a new level of description: it generates the notion of "functionality." Objects can be put into *functional* relations with each other. Such relations express which object produces which object under which conditions.

The main assumption of the present work is: *The action of an object upon other objects can be viewed as the application of a computable function to arguments from the functions' domain of definition (which can be other functions). Functional relations can then be considered completely independently from their particular physical realization.*

There is no doubt that this is a very strong assumption, but such an abstraction is useful if we want to focus on a classification of functional relations in analogy to a classification of attractors in dynamical systems theory. It is also useful in defining toy models that capture the feedback loop between objects and the functions acting on them defined by these very objects. It is such a loop that identifies complex constructive systems.

"Function" is a concept that—in some mathematical sense—is irreducible. Since 1936 mathematics has provided, through the works of Church,[4,5] Kleene,[16] Gödel,[12] and Turing,[23] a formalization of the intuitive notion of "effective procedure" in terms of a complete theory of particular functions on the natural numbers: the partial recursive functions.

The following is a very brief attempt to explore the possibility of establishing a useful descriptive level of at least some aspects of constructive complex systems by viewing their objects as being machines = computers = algorithms = functions. For a more detailed exposition see Fontana.[10,11]

---

2. WHAT IS A FUNCTION?

A function, f, can be viewed (roughly) in two ways:

- A function as an *applicative rule* refers to the *process*—coded by a definition—of going from argument to value.
- A function as a *graph*, refers to a set of ordered pairs such that if $(x, y) \in f$ and if $(x, z) \in f$, then $y = z$. Such a function is essentially a look-up table.

The first view stresses the computational aspect, and is at the basis of Church's λ-calculus (see Barendregt[2]). The theory λ is a formalization of the notion of computability in precisely the same sense as the Turing machine and the theory of general recursiveness. However, λ is a very different and much more abstract approach.

Stated informally, λ inductively defines expressions consisting of variables. A variable is an expression, and every *combination* of expressions, wrapped in parentheses, is again an expression. Furthermore, a variable can be *substituted* by *any* other expression that follows the expression to which the variable is bound. An elegant notational (syntactic) structure provides a means for defining the scope of the variables in an expression.

In an informal notation this means that, if $f[x]$ and $g[x]$ are expressions, then $(f[x]g[x])$ is also an expression, and $(f[x]g[x])$ is equivalent to $(f[x := g[x]])$, where the latter denotes the expression that arises if every occurrence of x in $f[x]$ is replaced by the expression $g[x]$.

Intuitively, what is captured here is just the notion of "evaluating" a function by "applying" it to the argument. That is, consider $f[x]$ as denoting a function, then the *value* of that function when applied to the argument expression a is obtained by literally substituting a for x in $f[x]$ *and* performing all further substitutions that might become possible as a consequence of that action. If all substitutions have been executed, then an expression denoting the value of a under f has been obtained. In this way functions that can express the natural numbers (numerals) or all computable operations on them, for example, addition and multiplication, can be defined.

Three features of functions in λ are important for the following:

- Functions are defined recursively in terms of other functions: imagine functions as being represented by trees with variables at the terminal nodes. This makes explicit that functions are "modular" objects, whose building blocks are again functions. This combinatorial representation, in which functions can be freely recombined to yield new functions, is crucial.

■ Objects in λ can serve both as arguments or as functions to be applied to these arguments.
■ There is no reference to any "machine" architecture.

Although the whole story is much more subtle, the preceding paragraphs should convey the idea of "function" that will be used in the framework described in the following sections. The point is that under suitable mathematical conditions λ-objects and composition give rise to a reflexive algebraic structure.

Turing's completely different, but equivalent, approach to computability worked with the machine concept. It was his "hardware" approach that succeeded in convincing people that a formalization of "effective procedure" had actually been achieved. The existence of a universal Turing machine implies the logical interchangeability of hardware and software. Church's world of λ is entirely "software" oriented. Indeed, in spite of its paradigmatic simplicity, it already contains many features of high-level programming languages, such as LISP. For a detailed account see Trakhtenbrot.[22]

3. THE MODEL

To set up a model that also provides a workbench for experimentation a representation of functions along the lines of the λ-calculus is needed. I have implemented a representation that is a somewhat modified and extremely stripped-down version of a toy-model of pure LISP as defined by Gregory Chaitin.[3] In pure LISP a couple of functions are predefined (six in the present case). They represent primitive operations on trees (expressions), for example, joining trees or deleting subtrees. This speeds up and simplifies matters as compared to the λ-calculus in which one starts from "absolute zero" using only application and substitution. Moreover, I consider for the sake of simplicity only functions in one variable. The way functions act on each other thereby producing new ones is essentially identical to the formalism sketched for λ in the previous section.

I completely dispense with a detailed presentation of the language (which I refer to as "AlChemy": a contraction of Algorithmic Chemistry). Figure 1 and its caption give a simple example for an evaluation that should depict what it is all about. The interested reader is referred to Fontana.[11]

The model is then built as follows.

1. *Universe.* A universe is defined through the λ-like language. The language specifies rules for building syntactically legal ("well-formed") objects and rules for interpreting these structures as functions. In this sense the language represents the "physics." Let the set of all objects be denoted by $\mathcal{F}$.
2. *Interaction.* Interaction among two objects, $f(x)$ and $g(x)$, is naturally induced by the language through function application, $f(g(x))$. The evaluation

of $f(g(x))$ results in a (possibly) new object $h(x)$. Interaction is clearly asymmetric. This could easily be repaired by symmetrizing. However, many objects like biological species or cell types (neurons, for example) interact in an symmetric fashion. I chose to keep asymmetry.

Note that "interaction" is just the name of a binary function $\phi(s,t)$ that sends any ordered pair of objects f and g into an object $h = \phi(f,g)$ representing the value of $f(g)$ (see Figure 2 for an example). More generally, $\phi(s,t) : \mathcal{F} \times \mathcal{F} \mapsto \mathcal{F}$ could be *any* computable function, not necessarily application, although application is the most natural choice. The point is that whatever the "interaction" function is chosen to be, it is itself evaluated according to the semantics of the language. Stated in terms of chemistry, it is the same chemistry that determines the properties of individual molecules and at the same time determines how two molecules interact.

3. *Collision rule*. While "interaction" is intrinsic to the universe as defined above, the collision rule is not. The collision rule specifies essentially three arbitrary aspects:

   a. What happens with f and g once they have interacted. These objects could be "used up," or they could be kept (information is not destroyed by its usage).

   b. What happens with the interaction product h. Some interactions produce objects that are bound to be inactive no matter with whom they collide. The so-called NIL-function is such an object: it consists of an empty expression. Several other constructs have the same effect, like function expressions that happen to lack any occurrence of the variable. In general such products are ignored, and the collision among f and g is then termed "elastic"; otherwise, it is termed "reactive."

   c. Computational limits. Function evaluation need not halt. The computation of a value could lead to infinite recursions. To avoid this, recursion limits, as well as memory and real-time limitations have to be imposed. A collision has to terminate within some pre-specified limits; otherwise, the "value" consists in whatever has been computed until the limits have been hit.

The collision rule is very useful for introducing boundary conditions. For example, every collision resulting in the copy of one of the collision partners might be ignored. The definition of the language is not changed at all, but identity functions would have now been prevented from appearing in the universe.

In the following I will imply that the interaction among two objects has been "filtered" by the collision rule. That is, the collision of f and g is represented by $\Phi(f,g)$ that returns $h = \phi(f,g)$ if the collision rule accepts h (see item (b) above); otherwise, the pair (f,g) is not in the domain of Φ.

4. *System*. To investigate what happens once an ensemble of interaction function "particles" is generated, a "system" has to be defined. The remaining sections will briefly consider two systems:

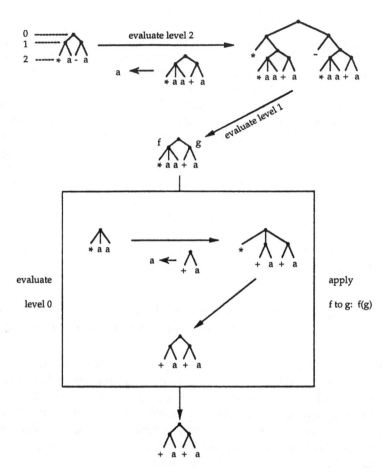

FIGURE 1 Evaluation example. The value of the expression $((+a)(-a))$ is computed when the variable a takes on the value $((*aa)(+a))$. The interpretation process follows the tree structure until it reaches an atom (leaf). In this case it happens at depth 2. The atoms are evaluated: the operators "+" and "−" remain unchanged, while the value of a is given by $((*aa)(+a))$. The interpreter backs up to compute the values of the nodes at the next higher level using the values of their children. The value of the left node at depth 1 is obtained by applying the unary "+"-operator to its sibling (which has been evaluated in the previous step). The "+" operation returns the first subtree of the argument, $(*aa)$ in this case. Similarly, the value at the right depth 1 node is obtained by applying the unary "−"-operator to its argument $((*aa)(+a))$. The "−" operation deletes the first subtree of its argument returning the remainder, $(+a)$. Now the interpreter has to assign a value to the top node. The left child's value is an expression representing again a function. This function, $(*aa)$, is labelled as f in the figure. Its argument is the right neighbor sibling, $(+a)$, labelled as g. Evaluating the top node means applying f to g. This is done by evaluating f while assigning to its variable a the value g. The procedure then recurs along a similar path as above, shown in the box. The result of $f(g)$ is the expression $((+a)(+a))$. This is the (continued)

FIGURE 1 (continued) value of the root (level 0) of the original expression tree, and therefore the value of the whole expression, given the initial assignment. The example can be interpreted as "function $((+a)(-a))$ applied to function $((*aa)(+a))$."

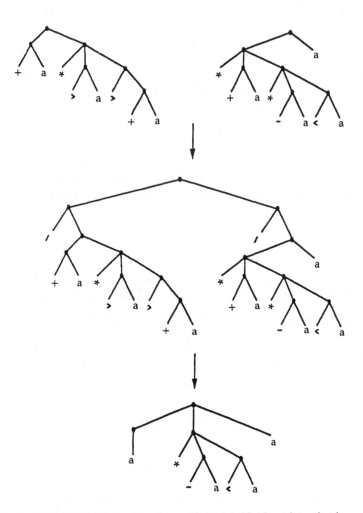

FIGURE 2 Interaction between functions. Two algorithmic strings (top) represented as trees interact by forming a new algorithmic string (middle) that corresponds to a function composition. The new root with its two branches and ′-operators is the algorithmic notation for composing the functions. The action of the unary ′-operator ("quote"-operator) consists in preventing the evaluation of its argument. The interaction expression is evaluated according to the semantics of the language and produces an expression (bottom) that represents a new function.

a. *An iterated map acting on sets of functions.* Let $\mathcal{P}$ be the power set, $2^{\mathcal{F}}$, of the set of all functions $\mathcal{F}$. Note that $\mathcal{F}$ is countably infinite, but $\mathcal{P}$ is uncountable. Let $\mathcal{A}_i$ denote subsets of $\mathcal{F}$, and let $\Phi[\mathcal{A}]$ denote the set of functions obtained by all $[\mathcal{A}]^2$ pair-interactions (i.e., pair-collisions) $\Phi(i, k)$ in $\mathcal{A}$, $\Phi[\mathcal{A}] = \{j : j = \Phi(i, k), (i, k) \in \mathcal{A} \times \mathcal{A}\}$. The map M is defined as

$$M : \mathcal{P} \mapsto \mathcal{P}, \mathcal{A}_{i+1} = \Phi[\mathcal{A}_i]. \tag{1}$$

Function application induces a dynamics in the space of functions. This dynamics is captured by the above map M. An equivalent representation in terms of an interaction graph will be given in the next section.

b. *A Turing gas.* The Turing gas is a stochastic process that induces an additional dynamics on the nodes of an interaction graph. Stated informally, individual objects now acquire "concentrations" much like molecules in a test-tube mixture. However, the graph on which this process lives changes as reactive collisions occur. Section 5 will give a brief survey on experiments with the Turing gas.

4. AN ITERATED MAP AND INTERACTION GRAPHS

The interactions between functions in a set $\mathcal{A}$ can be represented as a directed graph G. A graph G is defined by a set $V(G)$ of vertices, a set $E(G)$ of edges, and a relation of incidence, which associates each edge with two vertices (i, j). A directed graph, or digraph, has a direction associated with each edge. A labelled graph has in addition a label k assigned to each edge (i, j). The labelled edge is denoted by (i, j, k).

The action of function $k \in \mathcal{A}$ on function $i \in \mathcal{A}$ resulting in function $j \in \mathcal{A}$ is represented by a directed labelled edge (i, j, k):

$$(i, j, k): \quad i \xrightarrow{k} j, \quad i, j, k \in \mathcal{A}. \tag{2}$$

Note that the labels k are in $\mathcal{A}$. The relationships among functions in a set are then described by a graph G with vertex set $V(G) = \mathcal{A}$ and edge set $E(G) = \{(i, j, k) := j = k(i)\}$.

A useful alternative representation of an interaction is in terms of a "double edge,"

$$(i, j, k): i \xrightarrow{(i,k)} j \xrightarrow{(i,k)} k, \quad i, j, k \in \mathcal{A}, \tag{3}$$

where the function k acting on i and producing j has now been connected to j by an additional directed edge. The edges are still labelled but no longer with an element of the vertex set. The labels (i, k) are required to uniquely reconstruct the edge set from a drawing of the graph. The graph corresponding to a given edge set is uniquely specified. Suppose, however, that a function j is produced by two different interactions. The corresponding vertex j in the graph then has four inward edges. Uniquely reconstructing the edge set, or modifying the graph, for example by deleting a vertex, requires information about which pair of edges results from which interaction. Some properties of the interaction graph can be obtained while ignoring the information provided by the edge labels. The representation in terms of double-edges (i, j, k) has the advantage of being meaningful for any interaction function Φ mapping a pair of functions (i, k) to j, and not only for the particular Φ representing application. The double edge suggests that both i and k are needed to produce j. In addition, the asymmetry of the interaction is relegated to the label: (i, k) implies an interaction $\Phi(i, k)$ as opposed to $\Phi(k, i)$. This representation is naturally extendable to n-ary interactions $\Phi(i_1, i_2, \ldots, i_n)$. In the binary case considered here, every node in G must therefore have zero or an even number of incoming edges.

The following gives a precise definition of an interaction graph G. As in Eq. (1) let $\Phi[\mathcal{A}]$ denote the set of functions obtained by all possible pair collisions $\Phi(i, k)$ in $\mathcal{A}$, $\Phi[\mathcal{A}] = \{j : j = \Phi(i, k), (i, k) \in \mathcal{A} \times \mathcal{A}\}$. The interaction graph G of set $\mathcal{A}$ is defined by the vertex set

$$V(G) = \mathcal{A} \cup \Phi[\mathcal{A}] \tag{4}$$

and the edge set

$$E(G) = \{(i, j, k) : i, k \in \mathcal{A}, j = \Phi[(i, k)]\}. \tag{5}$$

The graph G is a function of $\mathcal{A}$ and Φ, $G[\mathcal{A}, \Phi]$. The action of the map

$$M : \mathcal{A}_{i+1} = \Phi[\mathcal{A}_i] \tag{6}$$

on a vertex set $\mathcal{A}_i$ leads to a graph representation of M. Let

$$G^{(i)}[\mathcal{A}, \Phi] := G[\Phi^i[\mathcal{A}], \Phi] \tag{7}$$

denote the ith iteration of the graph G starting with vertex set $\mathcal{A}$; $G^{(0)} = G$.

A graph G and its vertex set $V(G)$ are closed with respect to interaction, when

$$\Phi[V(G)] \subseteq V(G); \tag{8}$$

otherwise, G and $V(G)$ are termed innovative.

Consider again the map M, Eq. (6). What are the fixed points of $\Phi[\cdot]$? $\mathcal{A} = \Phi[\mathcal{A}]$ is equivalent to (1) $\mathcal{A}$ is closed with respect to interaction and (2) the set $\mathcal{A}$ reproduces itself under interaction. That is,

$$\forall j \in \mathcal{A}, \quad \exists i, k \in \mathcal{A} \text{ such that } j = \Phi(i, k). \tag{9}$$

Condition (9) states that all vertices of the interaction graph G have at least one inward edge (in fact, an even number). Such a self-maintaining set will also be termed "autocatalytic," following M. Eigen[7] and S. A. Kauffman[14,15] who recognized the relevance of such sets with respect to the self-organization of biological macromolecules.

Consider a set $\mathcal{F}_i$ for which Eq. (9) is still valid, but which is not closed with respect to interaction. $\mathcal{F}_{i+1}$ obviously contains $\mathcal{F}_i$, because of Eq. (9), and, in addition, it contains the set of new interaction products $\Phi[\mathcal{F}_i] - \mathcal{F}_i$. These are obviously generated by interactions with $\mathcal{F}_i \subset \Phi[\mathcal{F}_i]$. Therefore Eq. (9) also holds for the set $\Phi[\mathcal{F}_i]$, implying that the set $\mathcal{F}_{i+1}$ is autocatalytic. Therefore, if $\mathcal{A}$ is autocatalytic, it follows that

$$G[\mathcal{A}, \Phi] \subseteq G^{(1)}[\mathcal{A}, \Phi] \subseteq G^{(2)}[\mathcal{A}, \Phi] \subseteq \ldots \subseteq G^{(i)}[\mathcal{A}, \Phi] \subseteq \ldots . \tag{10}$$

In the case of strict inclusion, let such a set be termed "autocatalytically self-extending." Such a set is a case of innovation, in which

$$\Phi[V(G)] \supseteq V(G) \tag{11}$$

holds, with equality applying only at closure of the set.

An interesting concept arises in the context of finite, closed graphs. Consider, for example, the autocatalytic graph G in Figure 3(b), and assume that G is closed. The autocatalytic subset of vertices $V_1 = \{A, B, D\}$ induces an interaction graph $G_1[V_1, \Phi]$. Clearly, $G[V, \Phi] = G_1^{(2)}[V_1, \Phi]$, which means that the autocatalytic set V_1 regenerates the set V in two iterations. This is not the case for the autocatalytic graph shown in Figure 3(a). More precisely, let G be a finite-interaction graph, and let $G_\alpha \subseteq G$ be termed a "seeding set" of G, if

$$\exists i, \text{ such that } G \subseteq G_\alpha^{(i)}, \tag{12}$$

where equality must hold if G is closed. Seeding sets turn out to be interesting for several reasons. For instance, in the next section a stochastic dynamics (Turing gas) will be induced over an interaction graph. If a system is described by a graph that contains a small seeding set, the system becomes less vulnerable to the accidental removal of functions. In particular cases a seeding set can even turn the set it seeds into a limit set of the process. Such a case arises when every individual function f_i in $\mathcal{A}$ is a seeding set of $\mathcal{A}$:

$$\begin{aligned} f_{i+1} &= \Phi(f_i, f_i), \qquad i = 1, 2, \ldots, n-1 \\ f_1 &= \Phi(f_n, f_n) . \end{aligned} \tag{13}$$

Furthermore, suppose that G is finite, closed, and autocatalytic. It follows from the above that all seeding sets G_α must be autocatalytically self-extending, for example, as in Figure 3(b). If G is finite, closed, but not autocatalytic, there can be no seeding set. Being closed and not autocatalytic implies $V(G^{(1)}) \subset V(G)$.

The vertices of G that have no inward edges are lost irreversibly at each iteration. Therefore, for some i either $G^{(i)} = \emptyset$, or $G^{(i)}$ becomes an autocatalytic subset of G.

In the case of innovative, not autocatalytic sets, i.e., sets for which

$$\Phi[\mathcal{A}] \not\subseteq \mathcal{A} \wedge \Phi[\mathcal{A}] \not\supseteq \mathcal{A} \tag{14}$$

holds, no precise statement can be made at present.

A digraph is called connected if, for every pair of vertices i and j, there exists at least one directed path from i to j and at least one from j to i. An interaction graph G that is connected not only implies an autocatalytic vertex set, but in addition depicts a situation in which there are no "parasitic" subsets. A parasitic subset is a collection of vertices that has only incoming edges, like the single vertices C and E in Figure 3(b), or the set $\{C, E\}$ in Figure 3(a). As the name suggests, a parasitic subset is not cooperative, in the sense that it does not contribute to generate any functions outside of itself.

All the properties discussed in this section are independent of the information provided by the edge labels (in the double-edge representation). Note, furthermore, that the above discussion is independent of any particular model of "function." It never refers to the implementation in the LISP-like AlChemy. The representation of function in terms of that particular language is used in the simulations reported briefly in the next section.

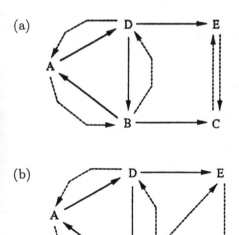

FIGURE 3 Interaction graph and seeding set. Two self-maintaining (autocatalytic) graphs. As outlined in Section 4, a particular function, say D, is produced through interaction between A and B. Therefore, D has two incoming edges from A and B. The edge labels are omitted. The dotted line indicates that B is applied to the argument A (solid line). The graphs are self-maintaining because every vertex has incoming edges. The lower graph, (b), can be regenerated from the vertex subset $\{A, B, D\}$, in contrast to the upper graph, (a). Both contain parasitic subsets: $\{E, C\}$ in (a), and $\{C\}, \{E\}$ in (b).

5. A TURING GAS

The interaction graph, and, equivalently, the iterated map, describe a dynamical system induced by the language on the power set of functions. This graph dynamics is now supplemented by a mass action kinetics leading to a density distribution on the set of functions in a graph. The kinetics is induced through a stochastic process termed "Turing gas."

A Turing gas consists of a fixed number of function particles that are randomly chosen for pairwise collisions. In the present scheme a reactive collision keeps the interaction partners in addition to the reaction product. When a collision is reactive, the total number of particles increases by one. To keep the number constant, one particle is chosen at random and erased from the system. This mimics a stochastic unspecific dilution flux. The whole system can be compared to a well-stirred chemical flow-reactor.

Three versions of the Turing gas have been studied. In one version the time evolution of the gas is observed after its initialization with N (typically $N = 1000$) randomly generated functions. In the second version the collision rule is changed to forbid reactions resulting in a copy of one of the collision partners. In the third version the gas is allowed to settle into a quasi-stationary state, where it is perturbed by injecting new random functions.

The following summarizes very briefly some of the results.

- *Plain Turing Gas.* Ensembles of initially random functions self-organize into ensembles of specific functions sustaining cooperative interaction pathways. The role of a function depends on what other functions are present in the system. A function, for example, that copies itself and some, but not all, others, acts as a pure self-replicator in the absence of those particular functions that it could copy. If some of them were present, the copy function would suddenly act "altruistically." The dynamics of the system is shaped by self-replicators (functions that copy themselves, but not the others present in the system), parasites (see section 4), general copy functions (identity functions), and partial copiers (functions that copy some, but not all functions they interact with). The "innovation rate," i.e., the frequency of collisions that result in functions not present in the system, decreases with time indicating a steady closure with respect to interactions (mainly due to the appearance of identity functions). If the stochastic process is left to itself after injecting the initial functions, fluctuations will eventually drive it into an absorbing barrier characterized by either a single replicator type, or by a possibly heterogeneous mixture of nonreactive functions ("dead system"), or by a self-maintaining set where each individual function species is a seeding set (Section 4). The system typically exhibits extremely long transients characterized by mutually stabilizing interaction patterns. Figure 4 shows an interaction graph (in a slightly different representation than described in Section 4; see caption) of a very stable self-maintaining set that evolved during the first 3×10^5 collisions starting from 1000 random functions. All functions present in the system differ from that initial set. The numbers refer to the function expressions (not shown) as they rank in lexicographic order. Function 17 is an identity function, although—for the sake of a less congested picture—only the self-copying interaction is displayed. Patterns like those in Figure 4 often include a multitude of interacting self-maintaining sets. In Figure 4, for example, deleting group I on the upper left still leaves a self-maintaining system (group II). Several other parts can be deleted while not destroying the cooperative structure. Sometimes these subsets are disconnected from each other with respect to interconversion pathways (solid arrows), but connected with respect to functional couplings (dotted lines). Figure 4 shows two groups of functions (indicated by I and II) that are not connected by transformation pathways (solid arrows). That is, no function of I is acted upon by any other function in the system such that it is converted into a function of group II—and *vice versa*. Group I, however, depends on group II for survival. The introduction of a "physical boundary" between I and II, cutting off all functional couplings among them, would destroy group I, but not group II.

- *Turing Gas Without Copy Reactions.* Copy reactions, i.e., interactions of the type $f(g) = g$ or f, strongly influence the patterns that evolve in the Turing gas as described above. Forbidding copy reactions (by changing the collision rule, section 3) results in a rather different type of cooperative organization as compared to the case in which copy reactions and therefore self-replicators were allowed. The system switches to functions based on a "polymeric" architecture that entertain a web of mutual synthesis and degradation reactions. "Polymer" functions are recursively defined in terms of a particular functional "monomer" (Figure 5). The individual functions are usually organized into disjoint subsets of polymer families based on distinct monomers. As in the case of copy reactions, these subsets interact along specific functional pathways leading to a cooperativity at the set level. Figure 5(a) and 5(b) show an example of two interacting polymer families. Neither family could survive without the other. Due to the polymeric structure of the functions the Turing gas remains highly innovative. A much higher degree of diversity and stability is achieved than in systems that are dominated by individual self-replicators. The high stability of the system is due to very small seeding sets. For example, everything in the system shown in Figure 5(a) and 5(b) follows from the presence of monomer 1 of type A (Figure 5(a)) and monomer 1 of type B (Figure 5(b)). Almost the whole system can be erased, but as long as there is one monomer A1 and one monomer B1 left, the system will be regenerated.

- *Turing Gas With Perturbations.* The experiments described so far kept the system "closed" in the sense that at any instant the system's population can be described by series of applications expressed in terms of the initially present functions. An open system is modeled by introducing new random functions that perturb a well-established ecology. In the case without copy reactions, the system underwent transitions among several new quasi-stationary states

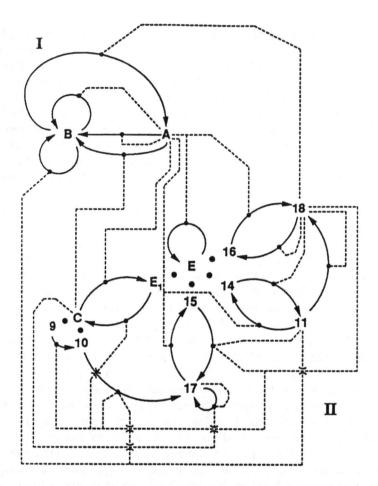

FIGURE 4 Interaction graph of a metastable Turing gas transient. The interaction graph of the functions present in the system after 3×10^5 collisions is shown. The system started with 1000 random functions, and conserves the total number of particles (1000). The numbers denote the individual functions according to their lexicographic ordering (not shown). Capital letters denote sets, where $A = \{1, 2, 3, 4\}, B = \{5, 6, 7, 8\}, C = \{9, 10\}, E = \{12, 13, 14, 15, 16\}$, and $E_1 = \{12, 13\} \in E$. Solid arrows indicate transformations and dotted lines functional couplings. A dotted line originates in a function, say k, and connects (filled circle) to a solid arrow, whose head is j and whose tail is i. This is to be interpreted as $j = k(i)$. Large filled circles indicate membership in a particle set. Function 17 is an identity function. Note: all dotted lines and solid arrows that result from 17 copying everything else in addition to itself have been omitted. The function set is closed with respect to interaction.

(a)

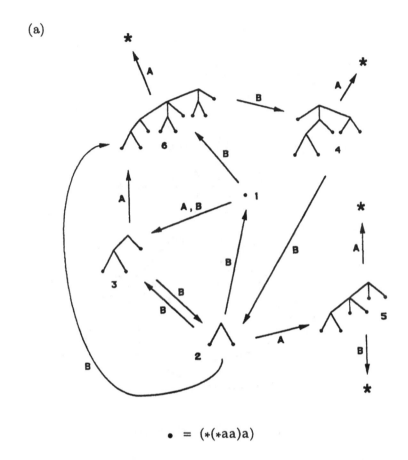

$$\bullet \ = \ (*(*aa)a)$$

FIGURE 5   (a) Interaction graph of a metastable Turing gas state without copy reactions. The figures show the interaction pathways among two polymer families established after 5×10^5 collisions starting from random initial conditions. The tree structure of the functions is displayed. The leaves are "monomers" representing the functional group indicated at the bottom of each graph. Solid arrows indicate transformations operated by function(s) belonging to the family denoted by the arrow label(s). Stars in the transformation pathways represent functions that were not present in the system at the time of the snapshot ("innovative reactions"). Due to the polymeric architecture of the functions, the system remains highly innovative. There are always (at least) two functions in the system that a polymerizing function (like the monomer 1 in 5(a)) can combine in order to produce a third one not in the system. (continued)

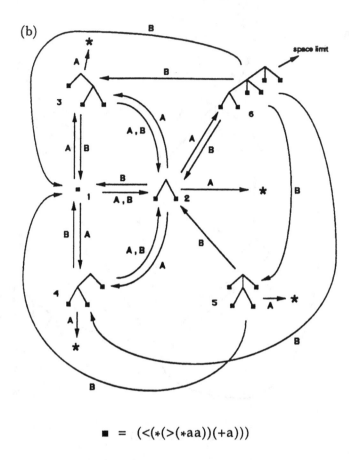

$$\blacksquare \;=\; (<(*(>(*aa))(+a)))$$

FIGURE 5 (continued.) The "space limit" tag in (b) indicates that the corresponding reaction product would hit the length limitation imposed on each individual function expression (300 characters in the present case).

(metastable transients), each characterized by an access to higher diversity. Systems with copy reactions were more vulnerable to perturbations and lost in the long run much of their structure.

A detailed analysis is found in Fontana.[11]

6. CONCLUSIONS

The main conclusions are:

1. A formal computational language captures basic qualitative features of complex adaptive systems. It does this because of:

   a. a powerful, abstract, and consistent description of a system at the "functional" level, due to an unambiguous mathematical notion of function;

   b. a finite description of an infinite (countable) set of functions, therefore providing a potential for functional open-endedness; and

   c. a natural way of enabling the construction of new functions through a consistent definition of interaction between functions.

2. Populations of individuals that are both an object at the syntactic level and a function at the semantic level, give rise to the spontaneous emergence of complex, stable, and adaptive interactions among their members.

7. QUESTIONS AND FUTURE WORK

The main questions and directions for the future can be summarized as follows.

1. Is there an equivalent of a "dynamical systems theory" for functional interactions? Can the dynamical behavior of the iterated map, Eq.(1), be characterized? Can examples be found that exhibit attractors other than fixed points? Can a classification of all finite self-maintaining sets of unary functions be made?

2. What is beyond replicator (Lotka-Volterra) equations? The standard replicator equation[8,13] on the simplex $S_n = \{\vec{x} = (x_1, x_2, \ldots, x_n) \in \mathbf{R}^n : \sum x_i = 1, x_1 > 0\}$,

$$\dot{x}_i = x_i \left(\sum_j a_{ij} x_j - \sum_{k,l} a_{kl} x_k x_l \right), \quad i = 1, \ldots, n, \tag{15}$$

considers objects i that are individual replicators. The Turing gas represents a stochastic version of the deterministic (infinite-population) equation[19]

$$\dot{x}_i = \sum_{j,k} a_{ijk} x_j x_k - x_i \sum_{r,s,t} a_{rst} x_s x_t, \quad i = 1, \ldots, n, \tag{16}$$

where in the present case the entry $a_{ijk} = 1$ iff function k acting on function j produces function i, and $a_{ijk} = 0$ otherwise, and $x_i, 0 \leq x_i \leq 1$, represents the frequency of function i in the system.

What can be said about the behavior of Eq. (16)? What can be deduced from it for the finite population Turing gas? Note that the use of a formal language, like λ, allows a finite description of an infinite matrix $a_{ijk}, i, j, k = 1, 2, \ldots$.

3. An obvious extension of the present work is the multivariable case. The restriction to unary functions implied that only binary collisions could be considered. n-ary functions lead to $(n+1)$-body interactions. Suppose the interaction is still given by function composition. A two-variable function $f(x, y)$ then interacts with functions g and h (in this order) by producing $i = f(g, h)$. f acts with respect to any pair g and h precisely like a binary interaction law expression Φ. However, f can now be modified through interactions with other components of the same system. This might have significant consequences for the architecture of organizational patterns that are likely to evolve. The extension to n variables is currently in preparation.

4. Future work includes a systematic investigation of the system's response to noise (section 5, item 3), either in the form of a supply of random functions, or in the form of a "noisy evaluation" of functions. What are the properties of the (noisy) Turing gas?

5. If some of the questions above are settled, then an extension to more sophisticated, typed languages (distinction among various object/data "types") that enable a compact codification of more complicated—even numerical—processes could be envisioned; always keeping in mind that processes should be able to construct new processes by way of interaction. The combination with a spatial extension or location of these processes then leads to Chris Langton's[17] vision of a "process gas." I leave it to the reader to further speculate where such intriguing tools might lead.

8. RELATED WORK

The coupling of a dynamics governing the topology of an interaction graph with a dynamics governing a frequency distribution on its vertices is a common situation in biological systems. The immune system,[6] development (ontogenesis),[20] and prebiotic molecular evolution[1] are but a few areas in which some modeling has been done.

The approach sketched here is related in particular to the pioneering work of S. A. Kauffman,[14,15] D. Farmer et al.,[9] R. Bagley et al.,[1] S. Rasmussen et al.,[20] and J. McCaskill.[18]

Kauffman, Farmer et al., and Bagley et al. consider a system of polymers, intended to be polynucleotides or polypeptides, each of which specifically instructs (and catalyzes) the condensation of two polymers into one or the splitting of one into two. This sets up the constructive part of the system and the related dynamics of interaction graphs. The production rate of individual polymers is then described by differential equations in terms of enzyme kinetics[1,9] providing a (nonlinear) dynamical system living on the interaction graphs. This approach represents one of the most advanced attempts to model a specific stage in prebiotic evolution.

Rasmussen and McCaskill made a first step toward abstraction, and the approach described here was prompted by their investigations. Rasmussen's system consists of generalized assembler code instructions that interact in parallel inside a controlled computer memory giving rise to cooperative phenomena. This intriguing system lacks, however, a clear cut and stable notion of function, except at the individual instruction level. McCaskill uses binary strings to encode transition-table machines of the Turing type that read and modify bit strings.

A model like the present Turing gas cannot provide much *detailed* information about a particular *real* complex system whose dynamics will highly depend on the physical realization of the objects as well as on the scheme by which the functions or interactions are encoded into these objects. Nevertheless, many phenomena that emerge, for example, in the polymer soup of Kauffman, Farmer et al., and Bagley et al., appear again within the approach described here. The hope then is that an abstraction cast purely in terms of functions might enable a quite general mathematical classification of cooperative organization. The question is if such an abstract level of description still can capture *principles* of complex *physical* systems. How much—in the case of complex systems—can we abstract from the "hardware" until a theory loses any explanatory power? I think the fair answer at present is: we don't know.

ACKNOWLEDGMENTS

This work is ultimately the result of many "reactive collisions" with John McCaskill, David Cai, Steen Rasmussen, Wojciech Zurek, Doyne Farmer, Norman Packard, Chris Langton, Jeff Davitz, Richard Bagley, David Lane, and Stuart Kauffman. Thanks to all of them!

REFERENCES

1. Bagley, R. J., J. D. Farmer, S. A. Kauffman, N. H. Packard, A. S. Perelson, and I. M. Stadnyk. "Modeling Adaptive Biological Systems." *BioSystems* **23** (1989): 113–138.
2. Barendregt, H. P. *The Lambda Calculus*. Studies in Logic and the Foundations of Mathematics, vol. 103. Amsterdam: North-Holland, 1984.
3. Chaitin, G. J. *Algorithmic Information Theory*. Cambridge: Cambridge University Press, 1987.
4. Church, A. "An Unsolvable Problem of Elementary Number Theory." *Am. J. Math* **58** (1936): 345–363.
5. Church, A. *The Calculi of Lambda-Conversion*. Princeton, NJ: Princeton University Press, 1941.
6. deBoer, R., and A. Perelson. "Size and Connectivity as Emergent Properties of a Developing Immune Network." *J. Theor. Biol.* (1990).
7. Eigen, M. "Self-Organization of Matter and the Evolution of Biological Macromolecules." *Naturwissenschaften* **58** (1971): 465–523.
8. Eigen, M, and P. Schuster. *The Hypercycle*. Berlin: Springer-Verlag, 1979.
9. Farmer, J. D., S. A. Kauffman, and N. H. Packard. "Autocatalytic Replication of Polymers." *Physica D* **22** (1986): 50–67.
10. Fontana, W. "Algorithmic Chemistry." Technical Report LA-UR 90-1959, Los Alamos National Laboratory, 1990. To appear in *Artificial Life II*, edited by C. G. Langton et al. Santa Fe Institute Studies in the Sciences of Complexity, Proc. Vol. X. Redwood City, CA: Addison-Wesley, 1991.
11. Fontana, W. "Turing Gas: A New Approach to Functional Self-Organization." Technical Report LA-UR 90-3431, Los Alamos National Laboratory, 1990. Submitted to *Physica D*.
12. Gödel, K. "Turing's Analysis of Computability and Major Applications of it." In *The Universal Turing Machine: A Half-Century Survey*, edited by R. Herken, 17–54. Oxford: Oxford University Press, 1988. Presented in his 1934 lectures at the Institute for Advanced Study. Quoted from: S. Kleene.
13. Hofbauer, J., and K. Sigmund. *The Theory of Evolution and Dynamical Systems*. Cambridge: Cambridge University Press, 1988.
14. Kauffman, S. A. *J. Cybernetics* **1** (1971): 71–96.
15. Kauffman, S. A. "Autocatalytic Sets of Proteins." *J. Theor. Biol.* **119** (1986): 1–24.
16. Kleene, S. "Lambda-Definability and Recursiveness." *Duke Math. J.* **2** (1936): 340–353.
17. Langton, C. G. Personal communication, 1990.
18. McCaskill, J. S. Unpublished manuscript.
19. Mjolsness, E., D. H. Sharp, and J. Reinitz. "A Connectionist Model of Development." Technical Report YALEU/DCS/RR-796, Yale University, 1990.

20. Rasmussen, S., C. Knudsen, R. Feldberg, and M. Hindsholm. "The Coreworld: Emergence and Evolution of Cooperative Structures in a Computational Chemistry." *Physica D* **42** (1990): 111–134.
21. Stadler, P., W. Fontana, and J. H. Miller. "Random Catalytic Reaction Networks." *Physica D* **63** (1993): 378–392.
22. Trakhtenbrot, B. A. "Comparing the Church and Turing Approaches: Two Prophetical Messages." In *The Universal Turing Machine: A Half-Century Survey*, edited by R. Herken, 603–630. Oxford: Oxford University Press, 1988.
23. Turing, A. M. "On Computable Numbers with an Application to the Entscheidungs Problem." *P. Lond. Math. Soc. (2)* **42** (1936–1937): 230–265.

Raymond E. Goldstein
Department of Physics, Jadwin Hall, Princeton University, Princeton, NJ 08544;
e-mail gold@davinci.princeton.edu

Nonlinear Dynamics of Pattern Formation in Physics and Biology

This chapter originally appeared in *1992 Lectures in Complex Systems*, edited by L. Nadel and D. Stein, 401–424. Santa Fe Institute Studies in the Sciences of Complexity, Lect. Vol. V. Reading, MA: Addison-Wesley, 1993. Reprinted by permission.

This chapter summarizes recent work on the nonlinear dynamics of pattern formation in a broad range of dissipative and Hamiltonian systems. We focus on four central issues: the laws of motion of fields and surfaces; the consequences of geometrical or topological constraints; the existence of variational principles; and the mechanisms of pattern selection. These are illustrated in the context of three classes of problems: (i) the motion of shapes governed by bending elasticity, (ii) interface dynamics in two-dimensional systems with nonlocal interactions, and (iii) integrable curve dynamics. We begin with a discussion of the mathematical structure of curve motion in the plane, and a dynamical formalism for shape evolution with global geometric constraints. Some two-dimensional systems are considered, with emphasis on the complex configuration space which exists when short-range and long-range interactions compete in the presence of constraints. Experimental and theoretical work on patterns found in systems with long-range electromagnetic interactions (Langmuir monolayers, magnetic fluids, Type I superconductors) and their connection with reaction-diffusion patterns are reviewed. Finally, the mathematics of some integrable soliton systems is shown to be equivalent to a hierarchy of dynamics of closed curves in the plane, shedding light on the geometry of Euler's equation.

INTRODUCTION

In fields as diverse as developmental biology and fluid dynamics there is a recurring notion of a link between *form* and *motion*. The present lecture is a summary of recent theoretical and experimental work that has endeavored to shed light on this general question of the interplay of geometry and dynamics in a class of systems. It is a common feature in nature that patterns arise from a competition between thermodynamic driving forces and global topological and geometrical constraints. While much is known concerning the static aspects of problems of this type (e.g., the elastic basis for the shapes of red blood cells), little is known about the *dynamics* by which arbitrary shapes and patterns relax to the global or local minimum of free energy under such constraints.

In the first section, we elaborate on a central theme for this investigation: the notion that the evolution of patterns is often described by d-dimensional surfaces (or interfaces) that move in response to forces derivable from an energy functional. Often, there is a competition between short-range and long-range interactions between points on the surface. The necessary differential geometry for discussing the simplest problems, those involving curve motion in the plane, is then described, followed by a variational principle for the dissipative evolution of boundaries of two-dimensional domains subject to global geometric constraints. The formalism is then applied to the motion of incompressible domains which relax to an accessible energetic minimum driven by line tension, elasticity, or nonlocal interactions as found in polymers, amphiphilic monolayers, and magnetic fluids. We seek to answer such questions as: Is an observed time-independent shape a unique energetic ground state or does the energy functional contain multiple metastable local minima? Are these minima roughly equivalent in energy? How can they be organized and classified? What kinetic considerations force a relaxing system into a metastable local minimum instead of the true ground state? Such questions are of course not confined to these particular examples of pattern formation, but also arise in systems such as spin glasses[4] and in protein folding.[14]

One class of systems we describe in detail is that in which there are long-range interactions of electromagnetic origin. These include Langmuir monolayers of dipolar molecules at the air-water interface, Type I superconductors in the intermediate state, and magnetic fluids in Hele-Shaw flow. All are known to exhibit "labyrinthine" pattern formation, in which an interface between coexisting phases or immiscible fluids develops a convoluted and branched space-filling pattern. Recent work suggests a unifying picture of the interface motion in these systems, and intriguing connections with experiments and theoretical studies of chemical reaction-diffusion systems.

In the penultimate section we give a brief description of a recent problem of *surface* motion—the dynamics of the "pearling" instability in lipid vesicles. Finally, we close with a very brief discussion of how aspects of the differential geometry of curve dynamics are shown to provide a new interpretation of the mathematics of integrable Hamiltonian systems. In particular, certain hierarchies of integrable systems are shown to be equivalent to a hierarchy of chiral shape dynamics of closed curves in the plane. These purely local dynamics conserve an infinite number of global geometric properties of the curves, such as perimeter and enclosed area. They in turn are related to the motion of iso-vorticity surfaces in ideal incompressible flow in two dimensions.

ENERGETICS AND CONSTRAINTS OF SURFACES

Here we describe some typical examples of patterns defined by surfaces, their energetics and constraints. Perhaps the simplest is a drop of incompressible fluid surrounded by a second immiscible fluid. The energy of interest is just that associated with surface tension γ, written as

$$\mathcal{E}_\gamma = \gamma \oint dS \,, \tag{1}$$

where dS is the differential of surface area. Such a drop moves with conserved volume, but nonconserved surface area. A lipid membrane, like that of a biological cell or artificial vesicle, is described by an elastic energy of the form[27]

$$\mathcal{E}_k = \oint dS \frac{1}{2} k_c \left(H - H_0 \right)^2 + \oint dS \bar{k}_c K, \tag{2}$$

where $H = 1/R_1 + 1/R_2$ is the mean curvature and $K = 1/R_1 R_2$ is the Gaussian curvature, with R_1 and R_2 the local principle radii of curvature of the surface. By hypothesis, this surface moves with fixed total area, and may be impermeable to the flow of fluid (and hence have conserved enclosed volume) or leaky (with a volume determined by the osmotic pressure difference across the membrane).

More complicated and nonlocal potential energies arise when surfaces are allowed to interact with themselves. For instance, a membrane may have a long-range van der Waals attraction which leads to adhesion, described by some pairwise interaction ϕ between points on the surface,

$$\mathcal{E} = \mathcal{E}_k + \frac{1}{2} \oint dS \oint dS' \phi \left(|\mathbf{r}(S) - \mathbf{r}(S')| \right) . \tag{3}$$

Such two-body interactions may be more complex, involving not just the positions of points on the surface but also their orientation. Indeed, we shall see below both for two-dimensional dipolar domains and ideal fluids that these interactions may involve the *tangent vectors* $\hat{\mathbf{t}}$ to a curve in the plane,

$$\mathcal{E} = \mathcal{E}_\gamma + \oint ds \oint ds' \hat{\mathbf{t}}(s) \cdot \hat{\mathbf{t}}(s') \Phi\left(|\mathbf{r}(s) - \mathbf{r}(s')|\right). \tag{4}$$

If we turn to a molecule like DNA which has an internal helical structure, an additional internal degree of freedom must enter the energy,

$$\mathcal{E} = \oint ds \left[\frac{1}{2}k_c\kappa^2 + \frac{1}{2}g\left(\omega - \omega_0\right)^2\right], \tag{5}$$

where κ is the local curvature, ω describes the local rate of *twist* of the helix and ω_0 the natural twist rate. A closed (e.g., circular) DNA molecule may adopt a complex three-dimensional *supercoiled* shape as a consequence of internal elastic strains associated with an excess or deficiency of windings of one edge of the helix about the other, a conserved *topological* quantity known as the "linking number deficit." The length is, of course, fixed, and barring the action of certain enzymes, so is the *knottedness* of the molecule.

Finally, turning to hydrodynamics, and in particular ideal inviscid flows, we find again the appearance of surface motion. Classical examples are the dynamics of vortex lines[3] and patches.[53] The important conservation laws in such systems include, of course, energy and momentum, but also quantities related to the vorticity through the Kelvin circulation theorem,

$$\oint_C d\mathbf{l} \cdot \mathbf{v} = \text{const.}, \tag{6}$$

where C is a contour in the fluid flow and $\mathbf{v}$ is the local velocity.

With these examples in mind, the central question we would like to address is: Given some energy functional $\mathcal{E}$, what is the motion it determines for the shape? More specifically, we may ask how a shape relaxes to some local (or global) minimum of that energy if it is prepared initially in some nonequilibrium state. To begin, we require some basic results from differential geometry and Lagrangian mechanics.

DYNAMICS OF CURVES IN THE PLANE
GEOMETRICAL PRELIMINARIES

For the remainder of this discussion, we will restrict our attention to motion of the simplest kind of surfaces, closed curves in the plane. Such geometrical objects, along with space curves, possess arclength as a natural parametrization, unlike their generic higher-dimensional counterparts. Nevertheless, it is often pedagogically useful to imagine an arbitrary variable $\alpha \in [0,1]$ labelling points $\mathbf{r}(\alpha)$ along the curve. Using a subscript to denote differentiation, the metric factor $\sqrt{g} = |\mathbf{r}_\alpha|$ is the Jacobian of the transformation between α and distance along the curve. Thus, for the *arclength parametrization* $s(\alpha)$ of the curve, there is the differential relation $ds = \sqrt{g}\,d\alpha$. The primary global geometrical quantities of interest, the length L and area A, are then given by

$$L = \int_0^1 d\alpha\,\sqrt{g}, \qquad A = \frac{1}{2}\int_0^1 d\alpha\,\mathbf{r}\times\mathbf{r}_\alpha. \tag{7}$$

Here, $\mathbf{a}\times\mathbf{b} \equiv \epsilon_{ij}a_i b_j$.

At each point along the curve there is a local coordinate system defined by the unit tangent vector $\hat{\mathbf{t}}$ and normal $\hat{\mathbf{n}}$. The former is defined as $\hat{\mathbf{t}}(\alpha) \equiv g^{-1/2}\mathbf{r}_\alpha$, the latter is rotated by $-\pi/2$ with respect to it. Traversing the shape in a counterclockwise direction, the curvature κ is defined through the arclength derivatives of these vectors by the Frenet-Serret equations,

$$\partial_s \begin{pmatrix} \hat{\mathbf{t}} \\ \hat{\mathbf{n}} \end{pmatrix} = \begin{pmatrix} 0 & -\kappa \\ \kappa & 0 \end{pmatrix} \begin{pmatrix} \hat{\mathbf{t}} \\ \hat{\mathbf{n}} \end{pmatrix}. \tag{8}$$

The curvature in turn is related to the angle $\theta(s)$ between the tangent vector and some arbitrary fixed axis by $\kappa = \theta_s$.

KINEMATIC CONSTRAINTS

A dynamics for the shape may be specified in terms of the components of the velocity in the local Frenet-Serret frame as

$$\mathbf{r}_t = U\hat{\mathbf{n}} + W\hat{\mathbf{t}}, \tag{9}$$

where normal and tangential velocities U and W are arbitrarily complicated local or nonlocal functions of $\mathbf{r}(\alpha)$. For closed curves, these functions must be periodic functions of s. Constraints such as length or area conservation may now be recast as constraints on the velocities U and W. Let us remark, however, that whatever the particular forms of U and W, the time evolution of κ and θ follow from Eq. (9) as[5]

$$\theta_t = -U_s + \kappa W, \tag{10}$$

and

$$\kappa_t = -\left(\partial_{ss} + \kappa^2\right)U + \kappa_s W. \tag{11}$$

The intrinsically nonlinear nature of these evolution equations reflects the interdependence of the internal coordinate s and the vector $\mathbf{r}(s)$.

If we consider length first among the global conservation laws, we observe that it may be conserved either globally or locally, the latter implying the former, but not vice versa. Local conservation means that the distance along the curve between two points labelled by α and α' remains constant in time. Since this distance is determined by the metric, local length conservation means $\partial_t |\mathbf{r}_\alpha| = 0$, or, equivalently, $\hat{\mathbf{t}} \cdot \partial_s \mathbf{r}_t = 0$. Using Eq. (8) we obtain the local arclength conservation constraint in differential form, $W_s = -\kappa U$, or integral form,

$$W(s) = -\int^s ds' \kappa U \equiv -\partial^{-1}\kappa U. \qquad \text{(local)} \qquad (12)$$

If we demand only the global conservation condition $L_t = 0$, then only the integral over s of $\partial_t \sqrt{g}$ need vanish. This can be shown to yield the global constraint

$$\oint ds\, \kappa U = 0. \qquad \text{(global)} \qquad (13)$$

Note that the local constraint relates both components U and W of the velocity, whereas the global constraint leaves the tangential component free. Given a normal velocity U, the local constraint determines W only up to an additive time-dependent function independent of s. A non-zero value of this function simply reparametrizes the curve, without changing its shape.

We appeal to reparametrization invariance to choose a convenient tangential velocity W. For systems with conserved *total* arclength the natural choice would be one which conserves local arclength, i.e., the metric $\sqrt{g}$. When the total arclength is not constant, a useful choice is still that which maintains uniform spacing of points on the curve, the *relative arclength* gauge. The condition $\partial_t(s/L) = 0$ determines W as

$$W(s) = \frac{s}{L} \oint ds'\, \kappa U - \int_0^s ds'\, \kappa U. \qquad (14)$$

The surface of an incompressible two-dimensional domain moves with fixed enclosed area, a constraint which is again nonlocal in U,

$$\oint ds\, U = 0, \qquad (15)$$

the form of which is clear on an intuitive level. In general, a *local* normal velocity $U(s)$, that is, one which depends only on the local geometry of the curve at the point s, will not satisfy these global integral constraints, so the motion will be intrinsically nonlocal (and mathematically complex!). Exceptions to this occur if, for instance, U and κU are total derivatives with respect to s. In the dissipative dynamics formulation, just as in equilibrium statistical mechanics, the mathematical freedom to satisfy these constraints arises from Lagrange multipliers present in an augmented free energy, as we now describe.

DISSIPATIVE DYNAMICS OF CURVES
A VARIATIONAL FORMALISM

In many of the systems of physical and biological interest, the surface motion is dominated by viscous drag—inertial effects are unimportant. Thus, we shall focus on strongly overdamped dynamics, and hence seek *first-order* equations of motion which relax a shape to a minimum of an energy functional $\mathcal{E}$. It is convenient to derive the motion from an action principle using Lagrange's formalism for dissipative processes.[16] In constructing a Lagrangian $\mathcal{L}$, the generalized coordinates q_α are the positions of the points $\mathbf{r}(\alpha)$ on the curve and the potential energy is just the energy functional $\mathcal{E}$. In general, the equations of motion are

$$\frac{d}{dt}\frac{\partial \mathcal{L}}{\partial \dot{q}_\alpha} - \frac{\partial \mathcal{L}}{\partial q_\alpha} = -\frac{\partial \mathcal{R}}{\partial \dot{q}_\alpha}, \tag{16}$$

where the Rayleigh dissipation function $\mathcal{R}$ is proportional to the rate of energy dissipation by the viscous forces. For the typical viscous forces linear in the velocity, $\mathcal{R}$ is quadratic in $\mathbf{r}_t$, and so its derivative is linear in $\mathbf{r}_t$.

In order to study the interplay of geometry, dynamics, and constraints, we make a model for $\mathcal{R}$ which assumes *local dissipation with isotropic drag*, and write

$$\mathcal{R} = \frac{1}{2}\eta \int_0^1 d\alpha \sqrt{g}\,|\mathbf{r}_t - \Theta(\alpha,t)\hat{\mathbf{t}}|^2, \tag{17}$$

where η is some friction coefficient. For motions that must be invariant under arbitrary time-dependent reparametrizations, we need the "gauge function" $\Theta(\alpha,t)$ to ensure that the reparametrizations do not contribute to the dissipation. Under the transformation $\alpha \to \alpha'(\alpha,t)$, the velocity transforms as $\mathbf{r}_t \to \mathbf{r}_t + \mathbf{r}_{\alpha'}\alpha'_t$, and the dissipation function is unchanged if we let $\Theta \to \Theta + \sqrt{g}\alpha'_t$.

In the viscous limit we neglect the kinetic energy terms in the Lagrangian, so $\mathcal{L} = -\mathcal{E}[\mathbf{r}]$, and by absorbing η into a rescaled time, we may rewrite Eq. (14) in terms of functional derivatives as

$$\mathbf{r}_t = -\frac{1}{\sqrt{g}}\frac{\delta \mathcal{E}}{\delta \mathbf{r}} + \Theta\hat{\mathbf{t}}. \tag{18}$$

The gauge function Θ is a tangential velocity, showing that it is indeed a reparametrization of the curve. Eq. (16) has the appearance of the time-dependent Ginzburg-Landau equation of dynamic critical phenomena,[28] and is also a version of the Rouse model of polymer dynamics.[10]

CONSTRAINED DYNAMICS

To include the possibility of imposing global conservation laws, we introduce time-dependent Lagrange multipliers Π and Λ conjugate to the area and length in an augmented energy functional

$$\mathcal{E} = \mathcal{E}_0 - \int_0^1 d\alpha \sqrt{g} \Lambda(\alpha) - \Pi A, \tag{19}$$

where $\mathcal{E}_0$ is the energy of the unconstrained system. Π and Λ are determined as follows: let U_0 and W_0 be the velocities derived from $\mathcal{E}_0$,

$$-\frac{1}{\sqrt{g}} \frac{\delta \mathcal{E}_0}{\delta \mathbf{r}} = U_0(\alpha, t) \hat{\mathbf{n}} + W_0(\alpha, t) \hat{\mathbf{t}}. \tag{20}$$

Differentiation of the augmented free energy in Eq. (19), with L and A given in Eq. (7), yields dynamics as in Eq. (9) with

$$U(s) = U_0 + \Lambda \kappa + \Pi \qquad \text{and} \qquad W(s) = W_0 - \Lambda_s + \Theta. \tag{21}$$

It is now necessary to consider two classes of motion, distinguished from each other by the way in which the unknown functions Θ and Λ are determined. In the *reparametrization-invariant* (RI) class, only the curve itself has physical meaning; the points α are simply labels. Given dynamics that conserves global arclength, we may always find a reparametrization Θ so that local arclength is conserved as well. A consistent value of $\Lambda(s)$ is then the constant determined at each instant of time by the global length constraint. This implies

$$\oint ds \kappa U_0 + \Lambda \oint ds \kappa^2 + 2\pi \Pi = 0, \tag{22}$$

where we have used $\oint ds\kappa = 2\pi$. In the *non-reparametrization-invariant* (NRI) class, we require $\Theta = 0$ and local arclength conservation is accomplished by choosing Λ to satisfy a differential equation

$$\left(\partial_{ss} - \kappa^2 \right) \Lambda(s, t) = \kappa U_0 + \partial_s W_0 + \kappa \Pi. \tag{23}$$

If area is also conserved, then Λ and Π also satisfy

$$\oint ds U_0 + \oint ds \kappa \Lambda + \Pi L = 0. \tag{24}$$

We can see from Eqs. (22) and (24) that even though the dissipation function was taken to be local, global constraints ultimately do lead to nonlocality through the Lagrange multipliers.

EXAMPLES

Here we illustrate the dissipative dynamics formalism developed above in the context of three systems; tense interfaces, elastic filaments, and dipolar domains. The first two examples are primarily pedagogical. In the discussion of the third class we will also summarize the salient results from recent experiments and highlight the general principles that have emerged from the study of a variety of systems.

TENSE INTERFACES. Consider again the incompressible fluid drop mentioned in the introduction. In the two-dimensional problem, the motion is driven by the *line tension* γ. The energy $\mathcal{E}_0 = \gamma L$ is clearly minimized by shapes having the smallest perimeter consistent with the prescribed area, i.e., circles. We find by functional differentiation that the normal velocity is $U_0(s) = -\gamma\kappa$, and W_0 vanishes. Since the perimeter is clearly not conserved, we need only determine the area Lagrange multiplier Π. It is $\Pi = (2\pi/L)\gamma$. The entire dynamics reduces to two coupled differential equations for the length and curvature,

$$L_\tau = \frac{(2\pi)^2}{L} - \oint ds\kappa^2, \qquad \kappa_\tau = \kappa_{ss} + \kappa^3 - \frac{2\pi}{L}\kappa^2 + \kappa_s W, \qquad (25)$$

with W given by Eq. (14), $U = U_0 + \Pi$, and $\tau = \gamma t$. Apart from the nonlocality associated with the tangential velocity, Eq. (25) is an area-conserving version of the well-known "curve-shortening equation."[15] If we neglect the tangential velocity and the area constraint, the dynamics is just

$$\mathbf{r}_\tau = \mathbf{r}_{ss}, \qquad (26)$$

a diffusion equation. This linearity is deceptive since, as mentioned earlier, $\mathbf{r}$ and s are not independent. The full dynamics in Eq. (25) relaxes the shape to one of uniform curvature $\kappa = 1/R_0$, with $L = 2\pi R_0$, where πR_0^2 is the area of the initial shape. This is illustrated in Figure 1, where we see the curvature evolution associated with an ellipse relaxing to a circle. The corresponding shape evolution is shown in Figure 2, along with the perimeter relaxation. In this simple example, it is plausible that the circle is the unique minimum in the energy functional, and that all initial conditions will relax to it.

CLOSED ELASTIC FILAMENTS. If we now endow the planar curves with an elastic energy we have a simple model of two-dimensional vesicles or ring polymers. Those with constrained enclosed area are "impermeable," whereas permeable vesicles have an area set by specifying an osmotic pressure Π. We may then ask the question: How does a closed polymer or vesicle in two dimensions "fold" itself into an energetic minimum without self intersections?[19]

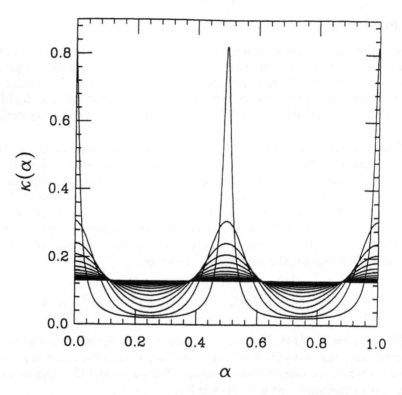

FIGURE 1 Curvature evolution for the relaxation of an ellipse to a circle, according to the dissipative dynamics formalism.

The conventional elastic energy is a quadratic form in the curvature,[35]

$$\mathcal{E}_k = \frac{1}{2}k_c \oint ds\, \kappa^2, \tag{27}$$

where k_c is the rigidity of the membrane. It is also possible to consider a more general form in which there exists a quenched-in local preferred curvature $\kappa_0(s)$, reflecting the specific monomer sequence along the polymer chain. The functional derivative of Eq. (27) combines with the Lagrange multipliers to provide the normal velocity

$$U = k_c\left(\kappa_{ss} + \frac{1}{2}\kappa^3\right) + \Lambda\kappa + \Pi, \tag{28}$$

and the bare tangential velocity W_0 vanishes. Interestingly, the dynamics in Eq. (28) appears as an expansion in the curvature and its derivatives, like the "geometrical"

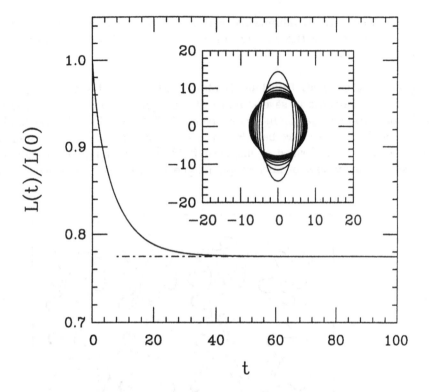

FIGURE 2 Shape evolution associated with Figure 1, and perimeter relaxation.

models of pattern formation used in the study of crystal growth.[5] Its presence here, however, is a consequence of a variational formulation not envisioned in the nonequilibrium crystal growth process.

In the presence of pairwise interactions as in Eq. (3), the normal velocity has an additional contribution

$$U_0 = \hat{\mathbf{n}}(s) \cdot \oint ds' \frac{(\mathbf{r}(s) - \mathbf{r}(s'))}{|\mathbf{r}(s) - \mathbf{r}(s')|} \, \phi'\left(|\mathbf{r}(s) - \mathbf{r}(s')|\right), \tag{29}$$

which is simply the (nonlocal) normal force at point s due to the rest of the chain. A repulsive core to the potential ϕ will prevent self-crossings and combines with the attractive tail to produce a minimum in ϕ.

Figure 3 illustrates how two initial shapes relax under these dynamics to a common local energetic minimum, in the presence of a spontaneous curvature with three narrow peaks.[19] The two differ by the relationship between the initial bends and the locations of the peaks in κ_0. Independent of the initial condition, we obtain "hairpin loops" localized at the peaks in $\kappa_0(s)$. These are the natural compromise between the attractive membrane interactions favoring local parallelism of the chain segments, and the curvature energy, which disfavors the bends necessitated by the constraint of closure.

MAGNETIC FLUIDS AND DIPOLAR DOMAINS. Motivated by the appearance of "labyrinthine" patterns in several quite distinct physical systems, thin magnetic films,[50] amphiphilic "Langmuir" monolayers,[1,36,46,49] and Type I superconductors in magnetic fields,[29] we have been led to investigate both theoretically and experimentally the patterns formed by magnetic fluids[47] ("ferrofluids"). These materials are colloidal suspensions of microscopic magnetic particles. When placed between

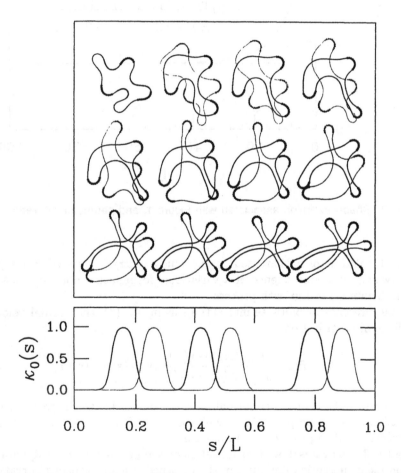

FIGURE 3 Folding of a closed elastic polymer in two dimensions, in the presence of pairwise interactions and a spontaneous curvature (lower panel). Two initial conditions, differing by the relation between initial folds and the peaks in $\kappa_0(s)$, find their way to the same ground state. From Goldstein et al.[21]

closely spaced parallel glass plates and magnetized by an external magnetic field normal to the plates, they are macroscopic examples of two-dimensional dipolar domains. As such, they constitute a convenient system in which to study the competition between short-range and long-range forces.[51]

We are interested in understanding whether the similarities found in the pattern formation of such distinct physical systems as mentioned above do actually reflect a common mechanism. To the extent that the observed patterns reflect the underlying energetics of the shapes, the similarities are indeed understandable. In each case, the labyrinth is formed of the boundary between two thermodynamic phases (up- and down-magnetized domains, expanded and condensed dipolar phases, normal and superconducting regions, and magnetic fluid against water), and has an associated surface tension which favors minimizing the contour length. Each system also possesses long-range bulk interactions of various origins. In the ferrofluid example studied here, the interaction is the dipole-dipole force between suspended magnetic particles aligned with the applied field, and is repulsive, tending to extend the fluid along the plates. Similar interactions exist in amphiphilic monolayers, the dipolar molecules of which are aligned perpendicular to the air-water interface. In solid-state magnetic systems the spontaneous magnetization produces the long-range interactions, while in superconducting thin films the in-plane Meissner currents interact via the Biot-Savart force.

Figure 4 illustrates the shape evolution of a ferrofluid domain after the magnetic field is brought rapidly to a fixed value.[9] The pattern evolution lasted approximately 60 s after the application of the field; the figure at the lower right is essentially time-independent and locally stable to small perturbations. In general we find that the branching process displays sensitive dependence on initial conditions in the sense that two initially circular shapes, indistinguishable to the eye, evolve under identical applied fields to trees differing in the shapes, lengths, and connectivity of their branches. Therefore a vast number of geometrically different minima may be reached by the system.

It has been recognized for some time, especially in the context of amphiphilic systems,[1] that the competition between these long-range forces and surface tension can result in a variety of *regular* patterns, such as lamellar stripe domains, hexagonal arrays, etc. The more widely encountered *irregular*, or disordered, patterns such as those in Figure 4 are, however, poorly understood.

In the simplest theory of this pattern formation, we obtain an energy functional like that in Eq. (4), where the nonlocal term represents the field energy associated with the oriented dipoles. The calculation of this is presented elsewhere.[32] Its form is understandable from the usual association between magnetization and current loops; it is just the self-energy of a ribbon of current flowing around the boundary of the domain. Viewed this way, the scalar product in the energy reflects the attraction (repulsion) between parallel (antiparallel) current-carrying wires. The pair interaction has a complex form because of the finite height h of the sample;

$$\Phi(\xi) = -\frac{\mu^2}{h}\left\{\sinh^{-1}(1/\xi) + \xi - \sqrt{1+\xi^2}\right\}. \tag{30}$$

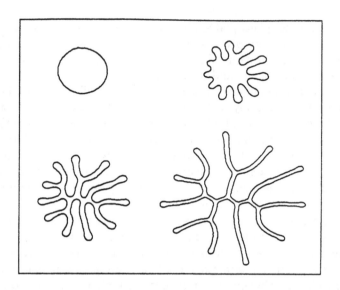

FIGURE 4   Snapshots of the fingering instability of an initially circular ferrofluid domain.
From Goldstein and Jackson.[17]

Here, $R = |\mathbf{R}| = |\mathbf{r}(s') - \mathbf{r}(s)|$ is the in-plane distance between points at positions s and s' on the boundary and μ is the dipole density per unit area. For $\xi \gg 1$, the function Φ is essentially Coulombic ($\Phi \simeq 1/2\xi$), whereas for $\xi \lesssim 1$ it is less singular, varying as $\ln(2/\xi)$. This crossover to logarithmic behavior occurs because of the finite thickness h of the slab, and prevents the integrals from diverging without additional cutoffs. From these results we deduce that the full energy is determined by one dimensionless parameter, the "magnetic Bond number"[47] $2\mu^2/\gamma$, and by the shape of the dipolar region.

The velocity arising from the functional derivative of the total energy is

$$U_0(s) = -\gamma\kappa + \frac{2\mu^2}{h^2} \oint ds' \, \hat{\mathbf{R}} \times \hat{\mathbf{t}}' \left[\sqrt{1 + (h/R)^2} - 1 \right], \tag{31}$$

where $\hat{\mathbf{R}}$ is the unit vector pointing from the point s toward s'. The nonlocal term is essentially a Biot-Savart force due to a wire (of finite height) carrying an effective current $I = E_0 hc/4\pi$ around the boundary. The tangential force $W_0(s)$ vanishes.

Analytic progress can be made in the linear stability analysis about a circular shape,[30,32,51] but the nonlinear regime requires numerical study; some results are shown in Figure 5. This simple dynamics satisfactorily reproduces the essential features of the experimental pattern formation, most notably the existence of many local minima, sensitive dependence on initial conditions, and the gross geometric features of the trees. A more detailed study[30] of the hydrodynamics of this

FIGURE 5 Numerical solution of contour dynamics for a dipolar domain.[30]

interfacial motion shows that the competing Young-Laplace and Biot-Savart forces in Eq. (31) enter as boundary conditions on the pressure at the domain boundary. In Langmuir monolayers, where the domain thickness is truly microscopic, the "ultra-thin" limit of the energy functional in Eq. (4) can be used to understand observed shape relaxation and thermal fluctuations of dipolar domains.[17] Finally, thin slabs of Type I superconductors in perpendicular magnetic fields display branched flux domains quite similar to that shown in Figure 4. They, too, may be understood by a model of the form (4), with a global flux conservation constraint.[18]

A REACTION-DIFFUSION SYSTEM. Several years ago, Lee, McCormick, Ouyang, and Swinney discovered a new mechanism by which space-filling patterns may appear in reaction-diffusion systems.[33,34] As shown in Figure 6, they observed the growth and repeated fingering of the boundary between regions of differing chemical composition (pH). This instability requires a finite-amplitude perturbation; the homogeneous black or white states are linearly stable. In this sense, the observed behavior is quite unlike Turing's scenario[52] of a linear instability taking a homogeneous state to a periodic space-filling pattern, confirmed in other recent experiments.[6,42] The patterns are seen to have sharp fronts between adjacent regions of differing composition, and these fronts are generally observed to be self-avoiding. This behavior was termed "pattern formation by interacting chemical fronts,"[33] and bears a remarkable resemblance to that found in the dipolar systems of the previous section.

A natural question that arises is: What law of motion governs the interfacial evolution from compact configurations to space-filling labyrinths? Ultimately, one would like to answer this on the basis of the underlying chemical kinetics of the experimental system. As a first step toward that, we have studied how a model reaction-diffusion system may be reduced to such a contour dynamics, and have shown that its behavior is very similar to that of the experiments.[20,44]

The FitzHugh-Nagumo model[13] describes two chemical species, an *activator* u and an *inhibitor* v. By suitable rescalings the PDEs for the pair are

$$u_t = D\nabla^2 u - F'(u) - \rho(v - u) , \qquad \epsilon v_t = \nabla^2 v - (v - u) . \tag{32}$$

The diffusion constant for u is assumed much smaller than that of v, so in our rescaled units this means $D \ll 1$. The fronts between the coexisting states of u will then be relatively sharp. Underlying the dynamics is the assumed bistability of u, with the two states ($u = 0$ and $u = 1$) being the minima of a function $F(u)$. We may think of these states as corresponding to the experimental ones with different pH values in Figure 6. The derivative of F is defined to be $F'(u) = u(u - r)(u - 1)$, so the control parameter r tunes the relative depth of the minima, with $\Delta F \equiv F(1) - F(0) = (r - 1/2)/6$. The coupling constant ρ controls the strength of inhibition of u by v, while the term in the v dynamics due to u represents the stimulation of inhibitor by activator. Finally, the appearance of v

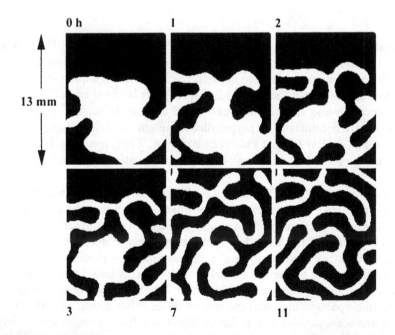

FIGURE 6   Pattern formation in the iodide-ferrocyanide-sulfite reaction in a gel reactor. Times shown are in hours following a perturbation. An indicator is used to show regions of low (white) and high (black) pH. Adapted from Lee et al.[33]

on the RHS of the inhibitor dynamics represents a self-limiting effect that prevents runaway of v in regions of high u. These kinds of cross-couplings are typical of what one finds in biochemical systems.

The model (32) is a linear combination of a gradient flow and a Hamiltonian system, a feature responsible for the remarkable diversity of patterns that it can produce. When the time scale $\epsilon \gg 1$ (slow inhibition) rotating spiral waves form.[37] The fast-inhibitor limit $\epsilon \ll 1$ was first considered by Koga and Kuramoto[31] for one-dimensional problems, and later by others in higher dimensions,[41] and allows for the existence of "localized states." These are compact configurations of $u = 1$ in a background of $u = 0$, formed by nonlinear excitations to homogeneous states that are linearly stable to the Turing bifurcation. Instabilities of localized states lead to the formation of labyrinths.

In the fast-inhibitor limit ($\epsilon \to 0$), the inhibitor equation is just the Poisson equation for a (screened) Coulomb "potential" v due to a distribution of "charge" u. Solving for v in terms of the free-space Green's function $\mathcal{G}$ and substituting back into the activator equation gives a single *nonlocal* equation of motion for u. This is a gradient flow, $u_t = -\delta\mathcal{E}/\delta u$, with energy functional[40]

$$\mathcal{E} = \int d\mathbf{x} \left\{ \frac{D}{2}|\nabla u|^2 + F(u) - \frac{\rho}{2}u^2 \right\} + \frac{\rho}{2} \int d\mathbf{x} \int d\mathbf{x}' u(\mathbf{x})\mathcal{G}(\mathbf{x} - \mathbf{x}')u(\mathbf{x}') . \quad (33)$$

Figure 7 shows the results of a direct numerical solution of the fast-inhibition dynamics. The initial condition, a compact domain of $u = 1$ surrounded by $u = 0$, expands and fingers, ultimately filling the (periodic) computational domain without interface self-crossings. The black regions, where $u \simeq 1$, act as sources for the inhibitor, whose concentration falls off exponentially into the nearby white regions. The overlap of those tails from adjacent black areas leads to a repulsive interaction between the fronts. The pattern apparently settles down to a time-independent state. Similar behavior has been seen in other models.[25,43]

We now wish to recast the energy functional $\mathcal{E}$ in Eq. (33) in terms of the closed chemical front surrounding a single domain of $u \simeq 1$ in a sea of $u \simeq 0$. The term involving $|\nabla u|^2$ is large only near $\mathcal{C}$, so if L is the length of $\mathcal{C}$ this contribution is γL, with γ a (positive) line tension. The polynomial terms $F(u) - \frac{1}{2}\rho u^2$ contribute proportionally to the area A of the domain. Finally, from the definition of the Green's function, $(\nabla^2 - 1)\mathcal{G} = -\delta$, the nonlocal contribution may be rewritten in terms of an area term and a pairwise interaction between tangent vectors $\hat{\mathbf{t}}(s)$ to the boundary in the form of the self-induction. The result is an energy functional

$$\Delta\mathcal{E}[\mathbf{r}] = \gamma L + \Delta F A - \frac{1}{2}\rho \oint ds \oint ds' \hat{\mathbf{t}}(s) \cdot \hat{\mathbf{t}}(s')\mathcal{G}(\mathbf{r} - \mathbf{r}') . \quad (34)$$

Remarkably, in a system with no dipoles or physical currents, we find again the self-induction common to all of the labyrinth-forming systems discussed earlier.

FIGURE 7   Labyrinth formation in the FitzHugh-Nagumo model, from numerical solution of the fast-inhibitor dynamics. Images are thresholded contour plots of the activator, with $u < 1/2$ shown white and $u > 1/2$ shown black. Evolution proceeds from upper left to lower right.

The boundary dynamics is then simple generalization of the u dynamics, a gradient flow of the form $\mathbf{r}_t = -\Gamma \delta \mathcal{E}/\delta \mathbf{r}$, where Γ is a kinetic coefficient derivable from the original PDEs. This normal velocity of the curve is as in Eq. (31),

$$\hat{\mathbf{n}} \cdot \mathbf{r}_t = -\Gamma \left\{ \gamma \kappa(s) + \Delta F - \rho \oint ds' \hat{\mathbf{R}} \times \hat{\mathbf{t}}' \mathcal{G}'(R) \right\}, \qquad (35)$$

where $\hat{\mathbf{R}}(s, s') = (\mathbf{r}(s) - \mathbf{r}(s'))/|\mathbf{r}(s) - \mathbf{r}(s')|$. This "contour dynamics" involves an important generalization of the familiar law of motion by mean curvature[48] in reaction-diffusion systems, namely a *nonlocal* coupling in the form of a screened Biot-Savart interaction. Figure 8 shows the two opposing tendencies of these forces. The first tends to relax a fingered structure to a circle, while the second leads to repulsion between interface segments on adjacent fingers. While different from the usual unscreened interaction between true current-carrying wires, it nevertheless has the same sign, so that *antiparallel sections of the interface repel.*

Figure 9 shows how a compact initial condition repeatedly fingers to produce a space-filling, but non-self-crossing pattern. These and other more quantitative studies confirm that the contour dynamics accurately captures the behavior of the FitzHugh-Nagumo model in the limit of fast-inhibition.

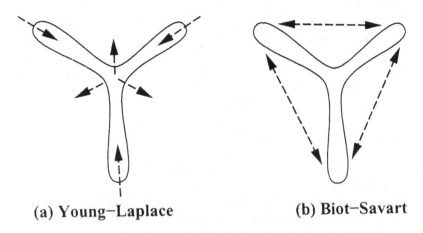

(a) Young–Laplace   (b) Biot–Savart

FIGURE 8   Schematic illustration of the shape-dependent forces acting on chemical fronts in the FitzHugh-Nagumo model. (a) The Young-Laplace force acts inward in regions of positive curvature and outward when $\kappa < 0$. (b) The nonlocal Biot-Savart force between nearby interfaces is repulsive.

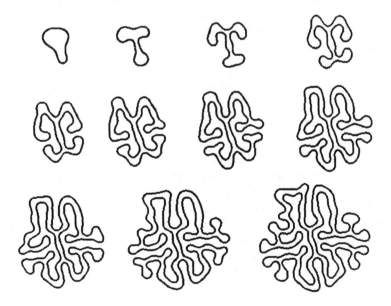

FIGURE 9   Contour dynamics evolution from a compact initial condition (upper left) to a labyrinthine pattern (lower right). From Goldstein et al.[20]

A PROBLEM IN SURFACE MOTION

All of the examples of pattern formation described thus far have involved curves in the plane. Here we give a brief discussion of a problem in membrane physics involving axisymmetric surface motion. It serves to illustrate the strongly nonlinear features of surface motion, as well as the feature of front propagation that plays an important role in many areas of pattern formation. The crucial experimental observation is due to Bar-Ziv and Moses,[2] who discovered the "pearling instability" of tubular lipid vesicles acted on by an optical trap (laser tweezers). It was observed that a vesicle that is initially cylindrical, with small thermal fluctuations, would become tense when illuminated by the laser, and develop a peristaltic perturbation that propagates outward from the laser spot at some 50 μm/sec. After some time, the entire vesicle achieves a stationary shape best described as a string of pearls: roughly spherical forms connected by thin tethers. Removal of the laser allows the vesicle ultimately to return to its original shape.

The basic physics of this shape transformation is believed to be like the Rayleigh instability of a column of fluid with tension,[45] that tension being provided by the laser trap; the intense electric fields in the trap attract dielectric particles (e.g., lipid molecules). This translates into a tension Σ that is estimated to be quite large relative to the scale k_c/R_0^2 set by the membrane elasticity and initial radius.[39] A straightforward hydrodynamic calculation then shows that the most unstable wavelength is set by the tube radius, consistent with experiment. Because of the large 2-d bulk modulus of the membrane, this tension is expected to propagate much faster than the actual shape deformation.[21] Thus, the propagation of the "pearling" may be understood as the relaxation dynamics of a surface endowed with a *uniform* tension and bending elasticity, and thus with an energy functional

$$\mathcal{E} = \int dS \left\{ \Sigma + \frac{1}{2} k_c H^2 \right\} , \qquad (36)$$

with $\Sigma R_0^2 / k_c \gg 1$.

Since the straight tense cylinder is linearly unstable, this shape propagation involves a stable state (pearls) invading an unstable one, a type for which there are no rigorous *general* predictions for the propagation velocity. The principle of "marginal stability"[8] is often predictive in these cases, and appears to give a semiquantitative account of the selected velocity and its dependence on laser power.[21] To test the accuracy of this method, which is based on the linear stability spectrum, we have investigated a very simple model for the shape relaxation. It is best described as "deformable pipe flow," and derives first from the continuity equation for the time evolution of the tube radius $r(z,t)$,

$$\frac{\partial}{\partial t} \pi r^2 = -\frac{\partial J}{\partial z} , \qquad (37)$$

where J is the axial current. We use the results from Poiseuille flow to express the flux in terms of the gradient of pressure,

$$J = -\frac{\pi}{4\eta}r^4\frac{\partial P}{\partial z} \ . \tag{38}$$

The pressure is just that determined by the energy functional (36). Rescaling both r and z with the unperturbed radius R_0 and introducing the rescaled time $\tau = (\eta R_0^3/4\kappa)t$, we obtain the partial differential equation[21]

$$r\frac{\partial r}{\partial \tau} = -\frac{\partial}{\partial z}\left[r^4\frac{\partial}{\partial z}\left(-\sigma H + \nabla^2 H + 2H^3 - 2HK\right)\right] \ , \tag{39}$$

where σ is the rescaled tension $\Sigma R_0^2/k_c$.

The dynamics in Eq. (39) is a gradient flow in which the energy functional $\mathcal{E}$ evolves as

$$\frac{d\mathcal{E}}{d\tau} = \int dz \frac{\delta \mathcal{E}}{\delta r}\frac{\partial r}{\partial \tau} = -\int dz\, r^4\partial_z\left(\frac{1}{r}\frac{\delta\mathcal{E}}{\delta r}\right)^2 \leq 0 \ . \tag{40}$$

Provided the interface does not pinch off (so $r > 0$), $\mathcal{E}$ decreases monotonically. When $\mathcal{E}$ is constant in time the functional derivative $\delta\mathcal{E}/\delta r = 0$ and the system is at an energetic extremum. One can easily show that this model exhibits a linear instability toward pearling, and thus it is a dynamics that connects the basic Rayleigh-like instability and the stationary final states, all the while obeying the relevant hydrodynamic conservation laws. It also has a form similar to that of models for topology transitions and singularities in viscous flows.[12,22]

Figure 10 shows the evolution of a cylindrical vesicle with $\sigma = 10$ perturbed initially with a localized distortion. The figure clearly shows that model supports propagating fronts of peristalsis moving outward from the initial perturbation. Many of the features of this front (e.g., speed, sharpness) are in semiquantitative agreement with experiment.[2,21]

CURVE DYNAMICS AND SOLITONS

The problems in pattern formation considered thus far have been strongly dissipative. Quite remarkably, however, many of the mathematical issues and techniques brought to bear on those problems also have application in integrable Hamiltonian dynamics, as we now describe.

Returning to the integral constraints required by perimeter and area conservation, Eqs. (13) and (15), one notices that such conservation is automatic if U and κU are total derivatives, with respect to arclength, of any periodic functions.

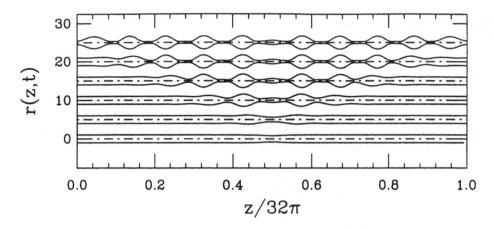

FIGURE 10 The pearling instability of a tense tubular vesicle, from Eq. (39).[21]

Moreover, with the tangential velocity determined by local arclength conservation, W is then determined *locally* from U. It follows then that the curvature evolution is determined entirely from the normal velocity as $\kappa_t = -\Omega U$, where $\Omega = \partial_{ss} + \kappa^2 + \kappa_s \partial^{-1} \kappa$. The simplest pair of velocity functions which conserve length and area is $(U^{(1)} = 0, W^{(1)} = -c)$, with c a constant. The motion is simply a reparametrization, as can be seen by the curvature evolution

$$\kappa_t = c\kappa_s. \tag{41}$$

A second choice, due to Constantin,[7] $(U^{(2)} = \kappa_s, W^{(2)} = -(1/2)\kappa^2)$, yields

$$\kappa_t = -\kappa_{sss} - \frac{3}{2}\kappa^2 \kappa_s. \tag{42}$$

These two curvature dynamics happen to be the first two members of a hierarchy of integrable systems,[23] an infinite set of $(1 + 1)$-dimensional partial differential equations, each with a common infinite number of conserved quantities.[11] Equation (41) is known as the "modified Korteweg-de Vries (mKdV) equation" related to the KdV equation which describes solitons in narrow channels of fluid.[11] Each of these dynamics is *chiral*, breaking the symmetry between s and $-s$ which we saw in the dissipative dynamics in earlier sections.

In addition to conserving length and area, by construction, the additional conserved quantities of Eq. (33) are of the form $H_k = \oint ds\, h_k$ with h_k obeying a continuity equation $\partial_t h_k + \partial_s j_k = 0$, for some currents j_k. For mKdV, which is already in the form of a continuity equation, the successive conserved quantities are

$$H_1 = \oint ds\kappa, \quad H_2 = -\frac{1}{2}\oint ds\kappa^2, \quad H_3 = \oint ds\left\{-\frac{3}{8}\kappa^4 + \frac{1}{2}\kappa_s^2 - \kappa\kappa_{ss}\right\}, \tag{43}$$

etc. For $k \geq 2$ these are just the tangential velocities of the hierarchy, while H_1 is just the "winding angle"—the angle through which the tangent vector rotates as the curve is traversed. Thus, starting from the conservation of perimeter and enclosed area, we have ended up with dynamics with an infinite number of conservation laws!

The mKdV hierarchy parallels the more familiar KdV hierarchy which is based on the KdV equation itself, $u_t + u_{sss} - 3uu_s = 0$. The two hierarchies are connected by the Miura transformation[38]

$$u = -\frac{1}{2}\kappa^2 - i\kappa_s, \tag{44}$$

such that if $\kappa(s,t)$ satisfies the nth order mKdV equation, then u satisfies the nth order KdV equation. Since the variable of the mKdV hierarchy is the curvature, it is natural to inquire about the geometrical significance of u.[24] Consider then the curve in the complex plane given by $z(s,t) = x(s,t) + iy(s,t)$, with $z_s(s,t) = e^{i\theta(s,t)}$ being the tangent vector. Using the associated representation of the curvature, $\kappa = -iz_{ss}/z_s$, we find

$$u = -\left[\left(\frac{z_{ss}}{z_s}\right)_s - \frac{1}{2}\left(\frac{z_{ss}}{z_s}\right)^2\right] \equiv -\{z,s\}. \tag{45}$$

We recognize the quantity $\{f,x\}$ as the Schwarzian derivative of a function f with respect to its argument x. This quantity has the property of being invariant under fractional linear transformations in the complex plane; that is, $\{z,s\} = \{w,s\}$ under transformations of the form $z \to w = (az+b)/(cz+d)$, which takes circles to circles. Thus, not only do the KdV curve dynamics have an infinity of conservation laws, they also have very strong invariance properties under mappings of the complex plane.

In addition to the conservation of enclosed area, which we naturally associate with an incompressible fluid, a second aspect of the KdV dynamics suggests that the curves whose motion is described by them are associated with ideal fluid flow, and hence with solutions of Euler's equation. Among the conserved quantities for each member of the hierarchy is the tangential velocity W. This is actually just the Kelvin circulation theorem in Eq. (6).

We have found[24] that the relationship between the KdV dynamics and the motion of ideal fluids with vorticity mirrors a well-known result in three-dimensional ideal fluid flow, the connection[26] between the Nonlinear Schrödinger (NLS) equation, $i\psi_t = -\psi_{xx} - (1/2)|\psi|^2\psi$, and the motion of a vortex filament. The NLS is the geometric evolution equation in a *local approximation* to the full nonlocal dynamics governed by the Biot-Savart law. Unlike the Euler equations themselves, the NLS is an integrable system with an infinite number of conserved quantities.

In two dimensions, the idealized distribution of vorticity analogous to the filament is a *vortex patch*, a bounded region of constant vorticity surrounded by irrotational fluid. The known exact equation of motion[53] for the boundary of such a domain is very nonlocal (again reflecting an underlying Biot-Savart law). Under a local approximation like that used in the NLS, the evolution equation for the curvature of the boundary is the mKdV equation.

Among the most interesting features of these results is that the integrable curve dynamics obey a variational principle of the form

$$\hat{\mathbf{n}} \cdot \mathbf{r}_t = \partial_s \left(\hat{\mathbf{n}} \cdot \frac{\delta \mathcal{H}}{\delta \mathbf{r}} \right) , \qquad (46)$$

remarkably similar to the dissipative dynamics result in Eq. (18). This suggests that it may be possible to develop a common language with which to describe both dissipative and Hamiltonian pattern formation.

CONCLUSIONS

Our emphasis here has been to illustrate issues and techniques which arise in the study of pattern formation in biological and physical systems. Whether for dissipative or Hamiltonian systems, some common aspects of these systems are emerging, including the stucture of variational principles, the competition between short- and long-range interactions, and the unifying point of view stemming from the differential geometry of surface motion. It is hoped that these investigations may provide a framework for deeper study of particular systems of physical or biological interest, and for addressing the striking complexity of the patterns seen in nature.

ACKNOWLEDGMENTS

I am grateful to my colleagues Andrejs Cebers, Akiva Dickstein, Alan Dorsey, Shyamsunder Erramilli, Stephen Langer, David Muraki, Philip Nelson, Thomas Powers, and Udo Seifert for the collaborations described in this review. All of this research has benefitted enormously from valuable discussions with Michael Shelley. This work has been supported by NSF Grants CHE-9106240 and Presidential Faculty Fellowship DMR-9350227, and by the Alfred P. Sloan Foundation.

REFERENCES

1. Andelman, D., F. Brochard, and J.-F. Joanny. "Phase Transitions in Langmuir Monolayers of Polar Molecules." *J. Chem. Phys.* **86** (1987): 3673–3681.
2. Bar-Ziv, R., and E. Moses. "Instability and 'Pearling' States Produced in Tubular Membranes by Competition of Curvature and Tension." *Phys. Rev. Lett.* **73** (1994): 1392–1395.
3. Batchelor, G. K. *An Introduction to Fluid Dynamics.* Cambridge: Cambridge University Press, 1967.
4. Binder, K., and A. Young. "Spin Glasses. Experimental Facts, Theoretical Concepts, and Open Questions." *Rev. Mod. Phys.* **58** (1986): 801–976.
5. Brower, R.C., D.A. Kessler, J. Koplik, and H. Levine. "Geometric Models of Interface Evolution." *Phys. Rev. A* **29** (1984): 1335-1342.
6. Castets, V., E. Dulos, J. Boissonade, and P. De Kepper. "Experimental Evidence of a Sustained Standing Turing-Type Nonequilibrium Chemical Pattern." *Phys. Rev. Lett.* **64** (1990): 2953–2956.
7. Constantin, P. Private communication.
8. Dee, G., and J. Langer. "Propagating Pattern Selection." *Phys. Rev. Lett.* **50** (1983): 383–386.
9. Dickstein, A. J., S. Erramilli, R. E. Goldstein, D. P. Jackson, and S. A. Langer. "Labyrinthine Pattern Formation in Magnetic Fluids." *Science* **261** (1993): 1012–1015.
10. Doi, M., and S. F. Edwards. *The Theory of Polymer Dynamics.* New York: Oxford University Press, 1986.
11. Drazin, P. G., and R. S. Johnson. *Solitons: An Introduction.* New York: Cambridge University Press, 1989.
12. Eggers, J. "Universal Pinching of 3D Axisymmetric Free-Surface Flow." *Phys. Rev. Lett.* **71** (1993): 3458–3461.
13. FitzHugh, R. "Impulses and Physiological States in Theoretical Models of Nerve Membrane." *Biophys. J.* **1** (1961): 445–466.
14. Frauenfelder, H., S. G. Sligar, and P. G. Wolynes. "The Energy Landscape and Motions of Proteins." *Science* **254** (1991): 1598–1603, and references therein.
15. Gage, M. E. "Curve Shortening Makes Convex Curves Circular." *Invent. Math.* **76** (1984): 357–364.
16. Goldstein, H. *Classical Mechanics.* Reading, PA: Addison-Wesley, 1980.
17. Goldstein, R. E., and D. P. Jackson. "Domain Shape Relaxation and the Spectrum of Thermal Fluctuations in Langmuir Monolayers." *J. Phys. Chem.* **98** (1994): 9626–9636.
18. Goldstein, R. E., D. P. Jackson, and A. T. Dorsey. "Current-Loop Model for the Intermediate State of Type-I Superconductors." *Phys. Rev. Lett.* **76** (1996): 3818–3821.

19. Goldstein, R. E., and S. A. Langer. "Nonlinear Dynamics of Stiff Polymers." *Phys. Rev. Lett.* **75** (1995): 1094–1097.
20. Goldstein, R. E., D. J. Muraki, and D. M. Petrich. "Interface Proliferation and the Growth of Labyrinths in a Reaction-Diffusion System." *Phys. Rev. E* **53** (1996): 3933–3957.
21. Goldstein, R. E., P. Nelson, T. Powers, and U. Seifert. "Front Propagation in the Pearling Instability of Tubular Vesicles." *J. Phys. II France* **6** (1996): 767–796.
22. Goldstein, R. E., A. I. Pesci, and M. J. Shelley. "Topology Transitions and Singularities in Viscous Flows." *Phys. Rev. Lett.* **70** (1993): 3043–3046.
23. Goldstein, R. E., and D. M. Petrich. "The Korteweg-de Vries Hierarchy as Dynamics of Closed Curves in the Plane." *Phys. Rev. Lett.* **67** (1991): 3203–3206.
24. Goldstein, R. E., and D. M. Petrich. "Solitons, Euler's Equation, and Vortex Patch Dynamics." *Phys. Rev. Lett.* **69** (1992): 555–558.
25. Hagberg, A., and E. Meron. "From Labyrinthine Patterns to Spiral Turbulence." *Phys. Rev. Lett.* **72** (1994): 2494–2497.
26. Hasimoto, H. "A Soliton on a Vortex Filament." *J. Fluid. Mech.* **51** (1972): 477-485.
27. Helfrich, W. "Elastic Properties of Lipid Bilayers—Theory and Possible Experiments." *Z. Naturforsch.* **28c** (1973): 693–703.
28. Hohenberg, P. C., and B. I. Halperin. "Theory of Dynamic Critical Phenomena." *Rev. Mod. Phys.* **49** (1977): 435–479.
29. Huebener, R. P. *Magnetic Flux Structures in Superconductors.* New York: Springer-Verlag, 1979.
30. Jackson, D. P., R. E. Goldstein, and A. O. Cebers. "Hydrodynamics of Fingering Instabilities in Dipolar Fluids." *Phys. Rev. E* **50** (1994): 298–307.
31. Koga, S., and Y. Kuramoto. "Localized Patterns in Reaction-Diffusion Systems." *Prog. Theor. Phys.* **63** (1980): 106–121.
32. Langer, S. A., R. E. Goldstein, and D. P. Jackson. "Dynamics of Labyrinthine Pattern Formation in Magnetic Fluids." *Phys. Rev. A* **46** (1992): 4894–4904.
33. Lee, K. J., W. D. McCormick, Q. Ouyang, and H. L. Swinney. "Pattern Formation by Interacting Chemical Fronts." *Science* **261** (1993): 192–194.
34. Lee, K. J., and H. L. Swinney. "Lamellar Structures and Self-Replicating Spots in a Reaction-Diffusion System." *Phys. Rev. E* **51** (1995): 1899–1915.
35. Love, A. E. H. *A Treatise on the Mathematical Theory of Elasticity*, 4th ed. London: Cambridge University Press, 1965.
36. McConnell, H. M., and V. T. Moy. "Shapes of Finite Two-Dimensional Lipid Domains." *J. Phys. Chem.* **92** (1988): 4520–4525.
37. Meron, E. "Pattern Formation in Excitable Media." *Phys. Reports* **218** (1992): 1-66.
38. Miura, R. S. "Korteweg-de Vries Equation and Generalizations. I. A Remarkable Explicit Nonlinear Transformation." *J. Math. Phys.* **9** (1968): 1202–1204.

39. Nelson, P., T. Powers, and U. Seifert. "Dynamic Theory of Pearling Instability in Cylindrical Vesicles." *Phys. Rev. Lett.* **74** (1995): 3384–3387.
40. Ohta, T., A. Ito, and A. Tetsuka. "Self-Organization in an Excitable Reaction-Diffusion System: Synchronization of Oscillatory Domains in One Dimension." *Phys. Rev. A* **42** (1990): 3225–3232.
41. Ohta, T., M. Mimura, and R. Kobayashi. "Higher Dimensional Localized Patterns in Excitable Media." *Physica D* **34** (1989): 115–144.
42. Ouyang, Q. and H.L. Swinney. "Transition from a Uniform State to Hexagonal and Striped Turing Patterns." *Nature* **352** (1991): 610-612.
43. Pearson, J. E. "Complex Patterns in a Simple System." *Science* **261** (1993): 189–192.
44. Petrich, D. M., and R. E. Goldstein. "Nonlocal Contour Dynamics Model for Chemical Front Motion." *Phys. Rev. Lett.* **72** (1994): 1120–1123.
45. Rayleigh, Lord. "On the Instability of Jets." *Proc. Lond. Math. Soc.* **10** (1879): 4–13.
46. Rice, P. A., and H. M. McConnell. "Critical Shape Transitions of Monolayer Lipid Domains." *Proc. Natl. Acad. Sci.* **86** (1989): 6445–6448.
47. Rosensweig, R. E. *Ferrohydrodynamics.* Cambridge: Cambridge University Press, 1985.
48. Rubinstein, J., P. Sternberg, and J. Keller. "Fast Reaction, Slow Diffusion, and Curve-Shortening." *SIAM J. Appl. Math.* **49** (1989): 116–133.
49. Seul, M., and M. J. Sammon. "Competing Interactions and Domain-Shape Instabilities in a Monomolecular Film at an Air-Water Interface." *Phys. Rev. Lett.* **64** (1990): 1903–1906.
50. Seul, M., L. R. Monar, L. O'Gorman, and R. Wolfe. "Morphology and Local Structure in Labyrinthine Stripe Domain Phase." *Science* **254** (1991): 1616–1618.
51. Tsebers, A. O., and M. M. Maiorov. "Magnetostatic Instabilities in Plane Layers of Magnetizable Fluids." *Magnetohydrodynamics* **16** (1980): 21–28.
52. Turing, A. M. "The Chemical Basis of Morphogenesis." *Philos. Trans. R. Soc. London Ser. B* **237** (1952): 37–72.
53. Zabusky, N. J., M. H. Hughes, and K. V. Roberts. "Contour Dynamics for the Euler Equations in Two Dimensions." *J. Comp. Phys.* **30** (1979): 96–106.

Charles M. Gray, Ph.D.
The Center for Neuroscience, Department of Neurobiology, Physiology and Behavior, University of California, Davis, CA 95616; e-mail: cmgray@ucdavis.edu

Synchronous Oscillations in Neuronal Systems: Mechanisms and Functions

This paper originally appeared in *Journal of Computational Neuroscience* 1 (1994): 11–38. Reprinted with permission of Kluwer Academic Publishers, The Netherlands.

INTRODUCTION

How are the functions performed by one part of the nervous system integrated with those of another? This fundamental issue pervades virtually every aspect of brain function from sensory and cognitive processing to motor control. Yet from a physiological perspective we know very little about the neural mechanisms underlying the integration of distributed processes in the nervous system. Even the simplest of sensorimotor acts engages vast numbers of cells in many different parts of the brain. Such actions require coordination between a host of neural systems, each of which must carry out parallel functions involving large populations of interconnected neurons.

It seems reasonable to assume that a mechanism or class of mechanisms has evolved to temporally coordinate the activity within and between subsystems of the central nervous system. For several reasons, neuronal rhythms have long been thought to play an important role in such coordination. Since the discovery of the Electroencephalogram (EEG) over 60 years ago it has been known that a number of structures in the mammalian brain engage in rhythmic activities. These patterned neuronal oscillations take many forms. They occur over a broad range of frequencies, and are present in a multitude of different systems in the brain, during a variety of different behavioral states. They are often the most salient aspect of observable electrical activity in the brain and typically encompass widespread regions of cerebral tissue.

With the advent of new techniques in multielectrode recording and neural imaging, it is now within the realm of possibility to record from 100 single neurons simultaneously,[170] to optically measure the activity in a cortical area,[20,164] or to noninvasively image the pattern of electric current flow in an alert human being performing a task.[127] These new techniques have revealed that spatially and temporally organized activity among distributed populations of cells often takes the form of synchronous rhythms. When combined with cellular neurophysiological and anatomical studies these findings provide new insights into the behavior and mechanisms controlling the coordination of activity in neuronal populations.

In this article I will review recent advances in four areas of mammalian systems neurophysiology where synchronous rhythmic activity has been observed and investigated. I have made no attempt to exhaustively review the literature on the occurrence of neuronal oscillations, rather I have chosen to focus on particular areas where it appears that coordinated activity may play a functional role. Out of neccessity I have left out a number of interesting systems. Those discussed here include the olfactory bulb, the visual cortex, the hippocampus, and the somatomotor cortex. In each case I have attempted to combine a discussion of the macroscopic behavior of the system and its relation to behavior with the cellular mechanisms thought to control the rhythmic activity and its synchronization.

THE OLFACTORY SYSTEM: THE INDUCED WAVE AND THE SPATIAL CODING HYPOTHESIS

THE INDUCED WAVE

In an early series of studies investigating the coding properties of neuronal activity in the mammalian olfactory bulb Adrian[2,3,4,5] described what has come to be known as the "induced wave," a rhythmic fluctuation of voltage in the EEG that is evoked by natural stimulation of the olfactory receptor sheet (Figure 1A). Adrian showed that these oscillatory bursts of activity range in frequency from 40–80 Hz and are associated with increased neuronal firing in mitral cells, whose axons form the lateral olfactory tract (LOT) and project to the piriform cortex.[39] Other studies revealed that the induced wave is a general phenomenon. It has been observed in the olfactory bulbs of amphibia and fish,[103,156] a variety of mammalian species,[26,27,61] and humans.[90] Its occurrence in the bulb requires input from the primary olfactory nerve, itself showing no evidence of oscillatory activity during natural stimulation (Figure 2).[61,126]

The induced wave is also prevalent in central olfactory structures in mammals such as the anterior olfactory nucleus, the piriform and entorhinal cortices, where it has a broader and somewhat lower range of frequencies (Figure 1B).[22,27,28,29,57,58,61] In these nuclei the induced wave occurs in response to bulbar input. If this input is blocked by severing the LOT, the main output pathway of the bulb, the induced waves are no longer evoked in the piriform cortex,[14,28,29] but persist in the bulb[76] (Figure 2).

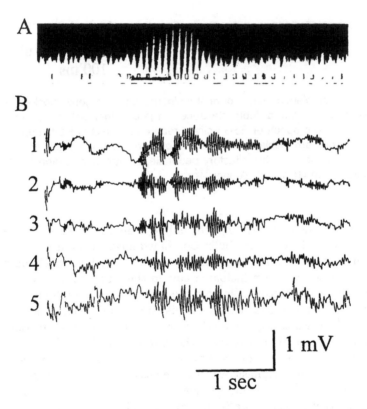

FIGURE 1 A. "Induced Waves" due to olfactory stimulation (amyl acetate). The record is from a rabbit, anesthetized deeply with urethane. Electrode in surface layer of olfactory bulb. Time mark (black line) gives 0.1/sec. The frequency of the waves is 60/sec. Modified from Adrian.[3] B. Bursts of EEG activity from olfactory bulb (1) and prepyriform cortex (2–5) in waking cat. Modified from Freeman.[58]

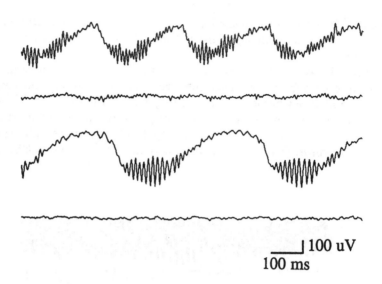

‚──┘ 100 uV
100 ms

FIGURE 2 The combined effects of nostril closure and cryogenic blockade on the bulbar EEG recorded from a depth electrode in the olfactory bulb of an alert rabbit. Each trace is a 1 sec epoch of EEG activity recorded at 1–300 Hz bandpass. Four separate conditions are displayed from top to bottom: control, nostrils lightly pinched shut, cryogenic blockade of the olfactory peduncle, cryogenic blockade combined with nostril closure. From Gray and Skinner.[174]

These findings demonstrate that the bulbar induced wave is generated by mechanisms intrinsic to the olfactory bulb. Combined anatomical and physiological evidence gradually led to a working model of a candidate cellular mechanism. Anatomical studies revealed that a principal circuit in the bulb consists of a recurrent inhibitory feedback onto mitral cells from the granule cells, an inhibitory interneuron population.[39,144] An essential feature of this circuit is the dendrodendritic reciprocal synapse.[134] Mitral and granule cells were found to make bidirectional synaptic contact on their dendritic processes in the external plexiform layer of the bulb.[144] The organization of this circuit was found to be consistent with the time course and laminar profile of a triphasic waveform evoked in the bulb by antidromic stimulation of the LOT.[135] These findings suggested that sustained activity in the bulb would lead to oscillations whose frequency depends on the time delay inherent in the negative feedback circuit.

Indirect evidence supporting this prediction was obtained by combined recording of mitral cell activity and the local field potential from the bulb. Mitral cells were found to fire at low rates, usually less than 10 spikes/sec, and typically showed no evidence of rhythmicity when examined using autocorrelation and interspike interval analysis. However, when compared to the local field potential, using conditional spike probability analysis,[61,76] a pronounced rhythmicity in firing probability was apparent. Mitral cells were found to fire most often shortly before the surface negative peak of the oscillation in the induced wave.[61,76] This lead in firing probability was approximately 1/4 of the period length of the induced wave, consistent with the prediction that bulbar rhythmicity was generated by a recurrent inhibitory interaction between the mitral and granule cell populations.[61,135]

Subsequent studies confirmed the basic tenets of this model leading to the following scheme thought to underlie the rhythmicity in the induced wave.[60] Afferent activity in the olfactory nerve axons excites the apical dendritic tufts of the mitral cells. This depolarization propagates along the mitral cell dendritic membrane resulting in both spike discharge at the soma and the synaptic excitation of the granule cells at the dendrodendritic synapses. Activity in the granule cells leads to a recurrent inhibition of the mitral cell population, the decreased activity of which results in a disexcitation of the granule cell population. This latter effect in turn results in a disinhibition of the mitral cell population. Sustained excitation arising from the olfactory nerve or from within the bulbar network causes the mitral cells to be reexcited resulting in a repetition of the sequence.[60,61,69]

SPATIAL DISTRIBUTION OF ACTIVITY

On the basis of his studies Adrian[4] proposed what has come to be known as the spatial coding hypothesis. He reasoned that for each discriminable odor there should exist a unique spatial pattern of activity representing a given odor quality that persists transiently throughout the olfactory system. Early attempts to address this hypothesis were hampered by technical limitations. In subsequent years, however, methods were developed to measure the spatial distribution of activity over the bulb and cortex in behaving animals.[61,62,63] This approach was founded on two assumptions: First, the representation of an odor in the olfactory bulb and cortex should exist during the inspiratory phase of respiration when the induced wave is present. Second, the close correlation between the phase of the induced wave and the firing patterns of cells in both the bulb and cortex[61] should enable one to indirectly measure the spatial pattern of activity of large populations of cells by recording the induced wave at many locations simultaneously.

On the basis of a spatial frequency analysis of the bulbar induced wave, Freeman developed a recording array for measuring the induced wave over the lateral surface of the bulb or the cortex at 64 locations simultaneously.[62,63] Each recording site was separated by 0.5 mm. This provided a sample of activity from roughly 20% of the bulbar surface and 30–50% of the piriform cortical surface at an optimal spatial resolution. Using this technique it was discovered that the oscillations in the induced wave are synchronous over broad regions of the bulb and cortex (Figure 3).[27,62] Phase gradients were found to vary from burst to burst. The frequencies in the signals changed over time within and between bursts but were found to covary across space. Fluctuations in the signal at one site were accompanied by similar changes across the structure. The most notable inhomogeneities were in the amplitude of the signals. The induced wave was often found to have one or more foci, or hot spots, where the amplitude of the signals over a local region were significantly higher than in surrounding regions. These spatial patterns varied slightly from burst to burst but often took the form of a characteristic signature for each animal.[62]

The mechanisms underlying the widespread synchrony in these two structures are not fully understood. However, long-range synaptic interactions are prevalent in both the bulb and cortex. In the bulb a sparse system of excitatory axon collaterals has been indentified in both the mitral and tufted cell populations.[93,141,168] But it is conceivable that the extensively branching dendritic arbors of the mitral cells could make a major contribution to the synchronization. Propagation of activity along the dendritic tree would act to excite large groups of inhibitory granule cells that would in turn deliver inhibitory feedback onto groups of mitral cells thereby controlling the timing of their ouput. The most likely candidate for long-range coupling in the piriform cortex is the excitatory axon collaterals of the layer 2 and 3 pyramidal cells.[85,86,87] The terminals of these axons make synaptic contact in layer 1 on the apical dendrites of other pyramidal cells, immediately below the excitatory inputs coming from the mitral cell axons forming the LOT. This structural arrangement not only provides a mechanism for establishing synchrony[169] but may also enable the association of piriform cortical output with the incoming pattern established in the bulb.[87]

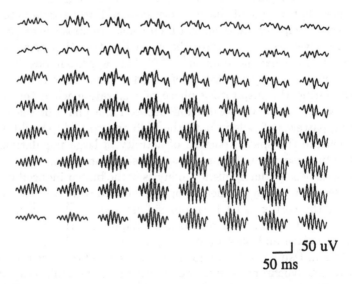

50 uV

50 ms

FIGURE 3   Induced wave recorded at 64 locations over the lateral surface of the olfactory bulb in an alert rabbit breathing purfied air. The inter-electrode spacing is 0.5 mm.

INFLUENCE OF ODORS ON SPATIAL PATTERNS

Initial studies of the influence of odor stimuli on the spatial patterns of bulbar and piriform cortical activity initially failed to reveal a relation. The spatial patterns of phase, frequency, and root-mean-square (rms) amplitude of the induced wave were measured during the passive presentation of several different odors. None of the measures revealed a significant change in the patterns related to the odor stimuli.[62] Spatial phase changed unpredictably in the bulb showing no consistent relation to the odors. In the piriform cortex the principal anteroposterior phase gradient of the induced wave was found to be the result of conduction delays in the propagation of activity in the LOT. Outside the trajectory of the LOT the phase patterns bore no relation to the odors presented. The frequency content of the signals was equally uninformative. The induced wave consisted of multiple frequencies in both the bulb and cortex and no relation could be found between frequency content and odor presentation. Similarly for amplitude the pattern of rms power of the induced wave showed no consistent relation to the odor.[62]

To improve the resolution of the measurements the experiments were repeated with two significant changes in the paradigm. The animals were engaged in a classical discriminative conditioning protocol in order to present the odors within a behaviorally significant context, and a new paradigm of statistical tests were devised to measure amplitude pattern differences with much greater sensitivity. These modifications revealed significant differences between the control amplitude pattern of the induced wave and the pattern evoked during the response to an odor.[65,166] This difference was transient and most pronounced during the first three inspirations following odor presentation. If odors were presented without reinforcement, the animals rapidly habituated their behavioral responses[77] and no reliable spatial pattern differences could be detected.[62,65]

A second, surprising result was revealed by these experiments. The mean spatial rms amplitude pattern of the induced wave was observed to change during the course of behavioral training (Figure 4). For example, after acquisition of discrimination behavior to a pair of odors, the foci in the control patterns were seen to shift in position in a contiguous fashion.[166] Such spatial shifts in the control pattern continued for as long as the animal learned new behavioral discriminations. If an inital discrimination set was extinguished and then later retrained after an intervening sequence of different discriminations, the control pattern did not revert back to its original form. This result revealed clear, long-term, learning-dependent changes in the organization of the bulbar activity. Moreover, it suggested that the control pattern reflected a sum over all learning and not some simple spatial representation of the currently trained odors.[65,66,166]

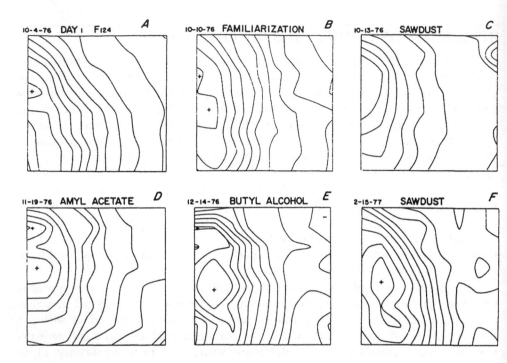

FIGURE 4 Contour plots of the mean EEG root-mean-square amplitude during the induced wave from a trained rabbit over a period of four months. Under aversive or appetitive conditioning the patterns changed with each new set of stimulus response contingencies. Examples are shown of five such changes. In the last stage of conditioning the presentation of the odorant "sawdust" did not result in the return of the EEG pattern of the first stage. From Freeman and Schneider.[65]

In spite of these findings it was not possible to confirm or reject Adrian's hypothesis on the basis of the available data. A difference in the pattern of activity produced by a reinforced and an unreinforced odor could not be distinguished. To resolve the uncertainty a number of data processing techniques designed to remove sources of noise from the signals, both behavioral and neural were applied to the data.[66,68] Analysis was restricted to those signals recorded on sessions after the acquisition of discrimination behavior was complete, and only to those trials in which a correct behavioral response was given by the animals. Within each trial only the first three inspiratory bursts (80 msec in duration) were evaluated. Here it was reasoned that the relevant odor-specific patterns should be present during the burst at the beginning of a sniffing response.[166]

These studies revealed differences in the spatial amplitude pattern of the bulbar induced wave evoked by reinforced and unreinforced odors.[66] The conditions under which this process was detected were different from what Adrian had hypothesized.

Rather than a labeled-line type of process where each odor evokes a unique spatial pattern of activity, these studies suggested that the spatial coding of odor quality in the bulb is a learning-dependent process that relies on experience in a behaviorally meaningful context. If odors are presented without reinforcement a process of habituation ensues and the spatial specificity of the patterns of the induced wave are reduced making them undetectable with current methodology.

So what is the role of neuronal oscillations in these complicated olfactory processes? The results of Freeman and colleagues suggest that information concerning odor quality is likely to be encoded in the spatial amplitude pattern of activity evoked during inspiration. Thus the spatial pattern of firing rates among the mitral cells in the bulb is the relevant quantity. The fact that these patterns occur during states of widespread synchrony suggests an important role for the oscillatory time structure of the activity. However, it is highly unlikely that the oscillations per se play any direct role in the actual coding of information, such as a temporal code. They are simply too variable and unspecific. It is more likely that the oscillations provide a local mechanism to enable the establishment of synchrony among a much larger population of synaptically coupled cells. In this context, oscillations of firing probability convey no information but provide a carrier wave for the coupling of large distributed populations of cells into functionally coherent patterns.

THE VISUAL CORTEX AND FEATURE INTEGRATION

Within the context of sensory integration the mammalian visual system has solved a particularly difficult task, the recognition of complex patterns in visual images. During a period that may be as brief as 200 milliseconds the visual system is capable of grouping combinations of visual features into a collection of independent objects. The number of possible interrelationships that features and objects can have within visual scenes is of nearly infinite variety. Yet the visual system has adapted effective mechanisms to cope with the combinatorial explosion of information that is present within day-to-day visual images.

The neuronal mechanisms underlying these recognition processes are poorly understood. It is thought that an early step in the process requires the establishment of relations among disparate visual features. Sets of features making up an object in an image must be grouped according to some criteria and segregated from those belonging to other objects in order to prevent improper feature conjunctions. How this might be achieved is not readily apparent. Mammalian visual cortex is organized into a collection of interconnected areas,[53,129,137,143] each known to process different or overlapping featural attributes of visual stimuli.[42,46,105,149] Given this organization it is apparent that most visual images are likely to evoke activity in a large population of cells distributed throughout the different cortical areas in the two cerebral hemispheres. The establishment of relations among features thus requires a mechanism for the integration of this distributed activity.

To achieve this integration at least two problems must be overcome. First, active cells distributed across the cortical network representing different parts of the same visual object must be identified as belonging together. This is commonly referred to as the "binding problem." An example would be linking the different contours of an object, that because of retinotopic organization would be represented by the activity of different populations of cells. Second, a mechanism is required to avoid interference between coexisting distributed activation patterns. This issue is often referred to as the "superposition problem," and an example would be the segregation of figure from background in a visual image.

Although a number of mechanisms have been proposed to solve these problems, one recent model suggests that temporal correlations in activity occurring on a millisecond time scale may provide a plausible solution.[110,111,112,117,147] The principal argument suggests that the binding and superposition problems cannot be solved if the system has to rely solely on the firing rates of the constituent neurons. If this were the case the relations among the patterns of activity evoked by each feature in an image would be lost. It would become difficult to identify which neurons in the overall population of active cells code for which feature and object. To solve this problem it was postulated that the firing patterns of cells responding to different parts of the same object, but not different objects, would be temporally synchronous.[110,111,117,147] Such a mechanism could achieve binding and avoid superposition in one step and allow for the coexistence of multiple activity patterns in the same set of cortical areas without interference.[151]

The temporal correlation model makes a number of specific predictions regarding the spatial extent and stimulus dependence of synchronous activity. In order to represent local features and establish relations between similar, but spatially separate features, synchronized firing should extend over a range of spatial scales among the neurons located in the same cortical area. Similarly, to provide for the integration of different visual features belonging to the same object (e.g., color, form, and motion), synchronized activity should exist between neurons in different cortical areas. The same argument applies to the linkage of information present in the two visual hemifields, suggesting that synchrony should exist between neurons in the same or different cortical areas of the two cerebral hemispheres. Finally, because the relations among visual features making up objects in visual images exist in nearly infinite variety, the patterns of synchrony among cells within and between cortical areas should be dynamically flexible. A given cell should be capable of transiently and selectively synchronizing its activity with a large number of other cells in a way that depends on the properties of the visual stimulus.

The temporal correlation model has received renewed interest with the recent discovery of synchronous oscillatory activity in the visual cortex. These data indicate that neuronal synchronization is more widespread and robust throughout the visual cortex than had previously been thought. Moreover, the experiments provide some support for each of the predictions discussed above.

SYNCHRONOUS OSCILLATIONS IN VISUAL CORTEX

It has long been known that the spike trains of many visual cortical neurons exhibit repetitive and semi-periodic patterns in response to visual stimuli.[88] The potential significance of this behavior became apparent following the discovery that such firing patterns are often synchronous over a range of spatial scales in the visual cortex of the cat.[47,75,78,79] In one set of experiments Gray and Singer[75,78] recorded multiple unit activity and local field potentials from single electrodes placed in the striate cortex of anesthetized and paralyzed cats. Using autocorrelation and spectral analysis they discovered that a significant fraction of the recorded signals displayed an irregular oscillation ranging in frequency from 30 to 60 Hz. These rhythmic patterns were readily apparent on single trials without the aid of signal averaging (Figure 5). They occurred almost exclusively during the presentation of visual stimuli, decreased in magnitude in response to nonoptimal stimuli, and showed no evidence of being time-locked to the onset of stimulation. Using spike-triggered averaging of the local field potential they found that the probability of neuronal firing was greatest during the negative phase of the field potential oscillation (Figure 5), suggesting that this signal is generated by the synchronous synaptic activation of a population of cells near the recording electrode.

Analysis of spike trains at the single unit level revealed that these periodic firing patterns most often consisted of sequences of repetitive burst discharges (Figure 6).[80] The interburst intervals ranged from 15–30 ms while the intraburst firing rate ranged from 300–700 Hz. In the same study the frequency of the oscillatory discharge was found to increase with increasing stimulus velocity. Recently each of these properties, stimulus dependent and locally synchronous oscillations of 30–60 Hz, repetitive burst discharges in single cells, and a dependence of oscillation frequency on stimulus velocity, have been confirmed in single electrode recordings in area 17 of alert cats (Figure 6).[83]

In area 17 of the monkey the occurrence and properties of synchronous oscillatory discharge is the subject of some debate. In one report Young et al.[173] found little or no evidence for oscillations in multiunit and field potential recordings in area V1 of macaque monkeys. These negative findings occurred in spite of the fact that the experimental and data analysis methods were similar to those used in the earlier cat studies.[49,78] In contrast, Livingstone[104] found convincing evidence in area V1 of squirrel monkeys for synchronous oscillatory responses in both single unit and field potential recordings. Similarly, a recent study by Eckhorn et al.[48] demonstrated oscillatory responses in multiunit and field potential recordings in area V1 of the awake macaque monkey. Both groups reported that cortical oscillations in the monkey range in frequency from 40–90 Hz. Eckhorn et al.[48] reported that oscillatory responses occur in the monkey with greater probability and magnitude than in the cat, further contrasting the absence of these signals reported by Young et al.[173]

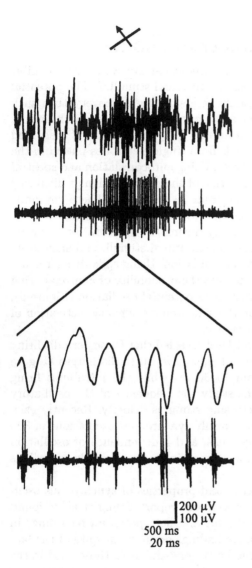

FIGURE 5   Multiunit activity and local field potential responses recorded from area 17 in an adult cat to the presentation of an optimally oriented light bar moving across the receptive field of the recorded cells. Oscilloscope records of a single trial showing the response to the preferred direction of movement. In the upper two traces, at a slow time scale, the onset of the response is associated with an increase in high-frequency activity in the local field potential. The lower two traces display the activity at an expanded time scale. Note the presence of rhythmic oscillations in the local field potential and the multiunit activity that are correlated in phase. Modified from Gray and Singer.[78]

Together, these discoveries prompted further investigations to determine the spatial distribution, temporal properties and stimulus dependence of the synchronization. In the cat, simultaneous recordings of multiunit activity demonstrated synchronized oscillatory responses over distances ranging from a few hundred microns up to 7 mm.[49,79,140] Nearby cells having spatially overlapping receptive fields (0.5–2.0 mm) were found to synchronize with the same probability irrespective of their orientation preferences (Figure 7).[49] But if the cells had nonoverlapping receptive fields and were separated by greater than 2 mm, synchronization occurred

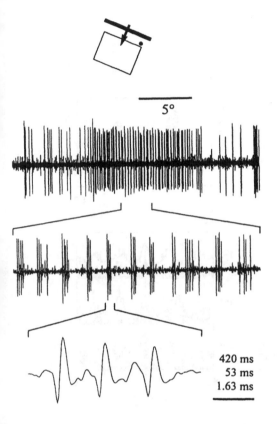

FIGURE 6 Repetitive burst-firing cell recorded extracellularly from area 17 of an awake behaving cat. The upper plot shows a schematic view of the fixation spot, the receptive field of the recorded neuron, and the visual stimulus. Below are shown three traces of the unit activity at three different time scales recorded during a single presentation of the stimulus. Note the presence of repetitive burst discharges at a frequency near 40 Hz. The interspike interval within a burst ranges from 400–700 Hz. From Gray and Viana Di Prisco.[83]

less frequently and was primarily found between cells with similar orientation preference.[49,79,162] In the squirrel monkey, oscillatory activity was found to be synchronous over distances up to 5 mm.[104] As with the single electrode recordings these rhythmic patterns occurred exclusively during periods of visual stimulation but showed no evidence of being time-locked to the stimulus. Similar measurements of the local field potential revealed this signal to be less specific. The magnitude of correlations decreased with spatial separation but showed no dependence on the orientation preference of the cells at each recording site.[49]

In order to determine the temporal properties of the synchronous activity Gray et al.[81] applied a moving window correlation measure to pairs of local field potential signals, recorded from area 17, exhibiting a significant average correlation. Cross-correlograms were calculated on 100-ms epochs of activity, at intervals of 30 ms, on each trial recorded in response to an optimal visual stimulus. A measure of the correlation magnitude, time lag and joint frequency of the field potentials was derived. The results revealed a high degree of dynamic variability not demonstrated by the previous analyses. The amplitude, frequency and phase of the synchronous

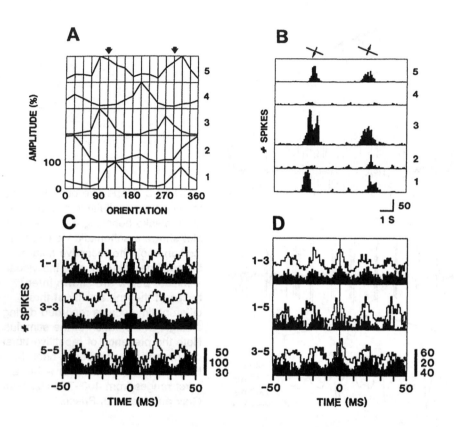

FIGURE 7　Intercolumnar synchronization of oscillatory neuronal responses in area 17 of an adult cat. A. Normalized orientation tuning curves of the neuronal responses recorded from 5 electrodes spaced 400 um apart. Response amplitudes are expressed as a percentage of the maximum response on each electrode. The arrows indicate the stimulus orientation at which the responses were recorded in B, C, and D. B. Post-stimulus-time histograms recorded simultaneously from the same five electrodes at an orientation of 112 degrees. C. Auto-correlograms of the responses recorded at sites 1 (1-1), 3 (3-3) and 5 (5-5). D. Cross-correlograms computed for the three possible combinations (1-3, 1-5, 3-5) between responses recorded on electrodes 1, 3 and 5. Correlograms computed for the first direction of stimulus movement are displayed with unfilled bars with the exception of comparison 1-5 in D. From Gray et al., 1989.

oscillations fluctuate over time. The onset of the synchrony is variable and bears no fixed relation to the stimulus. Multiple epochs of synchrony can occur on individual trials and the duration of these events also fluctuates from one stimulus presentation to the next. Most importantly, the results demonstrated that response synchronization can be established within 50–100 ms,[81] a time scale consistent with behavioral performance on visual discrimination tasks.

Experiments addressing the issue of interareal synchronization in the visual system were conducted by Eckhorn et al.[47] They demonstrated that oscillatory field potentials recorded in area 18 of the cat were often synchronous with unit activity recorded simultaneously in area 17. Subsequently, Engel and colleagues demonstrated synchronized oscillatory responses among cells recorded in areas 17 and PMLS (postero medial lateral suprasylvian sulcus),[50] and among cells in area 17 of the two cerebral hemispheres.[51] The properties of these interareal and inter-hemispheric interactions were similar to those observed within area 17. The correlated rhythmic activity was stimulus dependent, variable in frequency and not time-locked to the stimulus presentation. Synchrony occurred most often and with greatest magnitude among cells having overlapping receptive field locations and tended to favor cells having similar orientation preferences. Moreover, the synchronization occurred on average with little or no phase lag in spite of the long conduction delays between these cortical regions. Evidence that these interactions are mediated by intracortical connections was demonstrated by the absence of interhemispheric synchrony in animals with a severed corpus callosum.[51]

STIMULUS DEPENDENCE OF RESPONSE SYNCHRONIZATION

If the synchronization of neuronal activity contributes to the binding of distributed features in the visual field, then the incidence and magnitude of synchrony should not only depend on the receptive field properties of the recorded cells, but should change dynamically with variations of the visual stimulus.[110,111] A number of examples of stimulus-dependent changes in correlated firing among neurons in visual cortex have been observed. In an early study Ts'o et al.[162] demonstrated that the correlated firing between two nearby cells in area 17 of the cat increased when they were coactivated by a single light bar as compared to two independent bars presented together. A similar result was obtained by Gray et al.[79] when observing oscillatory neuronal responses in cat area 17. They recorded multiunit activity from two locations separated by 7 mm. The cells showed no evidence of correlated firing when activated by two light bars moving in opposite directions. Weak correlation was present when the light bars moved in the same direction and speed. And the correlation was strongly enhanced when the cells were activated by a single continuous contour.

In a nearly identical experiment Engel et al.[50] demonstrated that the synchronization of activity between cells in areas 17 and PMLS of the cat also depends on the properties of the visual stimulus. They recorded from cells having nonoverlapping receptive fields with similar orientation preference that were aligned colinearly. They found little or no correlation when the cells were activated by oppositely moving contours and a robust synchronization of activity when the cells were stimulated by a single long bar moving over both fields (Figure 8).[50]

A more detailed analysis of the influence of visual stimuli on response synchronization was conducted by Engel et al.[52] In this study multiunit activity was recorded from up to 4 electrodes having a spacing of approximately 0.5 mm. The proximity of the electrodes yielded recordings in which all the cells had overlapping receptive fields and a range of orientation preferences. They compared the incidence of correlated activity among cells coactivated by either one or two moving bars. In a number of cases cells having different orientation preferences were found to fire

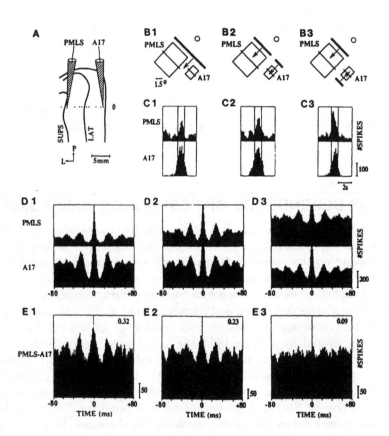

FIGURE 8   Interareal synchronization is sensitive to global stimulus features. A. Position of the recording electrodes. A17, area 17; LAT, lateral sulcus; SUPS, suprasylvian sulcus; P, posterior; L, lateral. B(1-3). Plots of the receptive fields of the PMLS and area 17 recordings. The diagrams depict the three stimulus conditions tested. The circle indicates the visual field center. C(1-3). Peristimulus-time histograms for the three stimulus conditions. The vertical lines indicate 1-sec windows for which auto-correlograms and cross-correlograms were computed. D(1-3). Comparison of the auto-correlograms computed for the three stimulus conditions. E(1-3). Cross-correlograms computed for the three stimulus conditions. From Engel et al.,1991a.

asynchronously when activated by two independent bars of differing orientation, whereas the same cells fired synchronously when coactivated by a single bar of intermediate orientation.[52]

That this process is general and not confined to the visual cortex of anesthetized cats was recently demonstrated by Kreiter et al.[96] Recordings of multiunit activity were made from two electrodes in area MT of a macaque monkey. The electrode separation was less than 0.5 mm yielding cells with nearly completely overlapping receptive fields but often differing direction preferences. Under these conditions little or no correlated activity was observed when the cells were activated by two independently moving bars, but synchronized firing was observed when responses were evoked by a single bar moving over both fields.[96] Repeated measures of the effect from the same cells under identical conditions revealed the effect to be stable.

These examples demonstrate that under appropriate conditions the synchronization of activity of two or more groups of neurons in cat and monkey visual cortex can be influenced by the properties of visual stimuli. The results suggest that response sychronization occurs preferentially when the cells are activated by stimuli having coherent properties. The data thus provide preliminary support for the hypothesis that synchrony can act as a mechanism for signaling relations between spatially distributed features.

MECHANISMS UNDERLYING THE GENERATION OF OSCILLATORY RESPONSES IN VISUAL CORTEX

There are several mechanisms likely to contribute to the generation of synchronized oscillations in the visual cortex. First, it is conceivable that the temporal structure of cortical responses could be simply due to an oscillatory input from the lateral geniculate nucleus (LGN), perhaps in a manner related to the dependence of piriform cortical activity on input from the olfactory bulb. Indeed numerous studies have demonstrated the existence of robust oscillatory activity in the frequency range of 30–60 Hz in both the retina and the LGN.[10,13,15,45,70,73,100,120] Such input to the cortex could provide a bias or even a driving influence for oscillatory responses. However, the data on retinogeniculate oscillations don't appear to account for the observations of intra- and interareal synchronization.[47,49,50,51] To provide the common drive for this synchrony there would either have to be a broad divergence of connectivity from LGN to cortex or a synchronization of activity within the LGN. Thalamic afferents to the cortex from the LGN rarely span more than 3 mm[54] and are exclusively ipsilateral and therefore cannot account for the interareal and interhemispheric synchrony.[50,51] Synchronous oscillatory activity has been observed within the LGN[10,100] but its stimulus dependence appears to differ from that observed in cortex and it is most prominent among cells having overlapping receptive fields. (see also Sillito et al.[145]).

Alternatively, cortical oscillatory firing could arise soley from intracortical network interactions. Evidence for this has come from recent intracellular recordings from cat striate cortex demonstrating robust oscillations of membrane potential in response to visual stimuli.[31,91] These fluctuations are stimulus dependent, largely absent during periods of spontaneous activity and increase in amplitude in response to visual input when the cells are hyperpolarized, suggesting that the oscillations arise from intracortical excitatory synaptic input. How the rhythmic EPSPs are generated remains open. One possible substrate could be the negative feedback arising from local recurrent inhibition. In this scheme oscillations would arise through the interaction of excitatory and inhibitory neurons in much the same manner as that described above for the olfactory bulb.[61] The resulting synchronous firing of excitatory neurons could then explain the oscillatory EPSPs recorded intracellularly.[91] Although direct evidence for such a mechanism is limited[55,82,108] such a prediction appears plausible since local circuit inhibitory interneurons exist in abundance in the neocortex,[109,113] and artificial neuronal networks containing recurrent inhibitory connections readily exhibit oscillatory activity.[33,59,61,94,171,169]

Cortical oscillatory activity may also arise as a consequence of activity in a subpopulation of cells that are intrinsically oscillatory,[107] much as it does in thalamocortical relay neurons[114] and cells of the thalamic reticular nucleus.[12,153] Support for this conjecture has been obtained from both *in vitro* and *in vivo* intracellular recordings. In the former case a subpopulation of inhibitory interneurons in layer 4 of rat frontal cortex were reported to exhibit subthreshold, voltage-dependent 10–50 Hz oscillations in membrane potential in response to depolarizating current injection.[108] These cells, if present visual cortex, would be expected to produce rhythmic IPSPs in their postsynaptic targets, a result for which at present there is little experimental evidence.[55,91] Recent intracellular recordings *in vivo* from cat striate cortex have revealed a subpopulation of cells that fire in regular repetitive bursts at 20–70 Hz in response to visual stimulation and intracellular depolarizing current injection.[115] It is not known if these cells are excitatory or inhibitory but in either case it is conceivable that their activation could serve to drive a local network of cells into a pattern of synchronous oscillation.

FUNCTIONAL SIGNIFICANCE OF VISUAL CORTICAL SYNCHRONOUS OSCILLATIONS

At present the evidence reviewed briefly above provides support for several predictions of the temporal correlation hypothesis. Synchronized oscillatory activity occurs over a range of spatial scales, it is present in abundance in the striate cortex of alert cats and monkeys, and its occurrence and magnitude are influenced by the properties of visual stimuli. It is not clear, however, what the functional role of the oscillations themselves are. They appear to be too variable in frequency and amplitude to serve as a carrier of information.[73,78,81] Rather it has been argued that they may provide an important mechanism for enabling long-range synchronization to occur.[95] Although synchronous oscillations have been observed in each of the visual cortical areas so far investigated in the cat, the occurrence and properties of this phenomenon in the extrastriate cortex of the monkey are the subject of debate.[11,96,158,173] Moreover, synchronous activity is not only limited to firing patterns having an oscillatory temporal structure. There exist many examples of this in the literature.[30,32,124,159,160,162] Thus, oscillations are associated with a prevalent and powerful form of synchronization but do not provide the only means by which synchrony can be established.

The more important issue concerns the functional significance of synchronization. A number of attempts have been made to address this issue but much more work is needed. The correlation hypothesis makes strong predictions that the patterns of synchronization should be stimulus specific, that multiple synchronized ensembles should be capable of coexisting in the same cortical network, and that the patterns of correlated activity in the cortex should be related to what an animal or person visually perceives. These predictions remain largely untested.

HIPPOCAMPAL RHYTHMS: CARRIER WAVES FOR SYNAPTIC PLASTICITY?

The hippocampus is an archicortical structure in the limbic system, known for the similarity of its shape to a seahorse. During a number of different behavioral states it is known to exhibit some of the most robust forms of synchronous rhythmic activity to be observed in the central nervous system. Foremost among these is the theta rhythm, a sinusoidal-like oscillation of neuronal activity at 4–10 Hz that occurs during particular behavioral states. In addition two other neuronal rhythms have been discovered, one having a frequency range of 40–100 Hz that co-occurs with the theta rhythm,[25,34,102] and another more recently discovered signal having a frequency around 200 Hz associated with alert immobility and the presence of sharp waves in the hippocampal EEG.[38,172]

THETA RHYTHM

The theta rhythm or rhythmic slow activity (RSA) was originally discovered in 1938 by Jung and Kornmuller and later investigated more extensively by Green and Arduini.[84] It occurs primarily in nonprimate mammals and is broadly distributed throughout the hippocampus during two behavioral states.[19] Type I or movement-related theta[165] is prominent during exploratory behaviors such as walking, shifts in posture, or the manipulation of objects with the forelimbs. It occurs throughout the hippocampus, is less well documented in cats and has been intensively studied in rats. Type II theta occurs during periods of behavioral immobility or in response to sensory stimulation.[19] It can also be elicited under anesthesia making it quite amenable to detailed analysis.

In both behaving and anesthetized animals theta activity is localized largely to the hippocampus,[17] the surrounding entorhinal cortex,[1,6,118] and the medial septal nucleus.[132] The generation of hippocampal RSA has long been thought to depend largely on input from the medial septal nucleus and entorhinal cortex, structures that are thought to act as pacemakers.[6,19,34,132] Both structures appear to be capable of intrinsically generating RSA either through network interactions[19,157] or by intrinsic cellular mechanisms.[8] Stellate cells in layer 2 of the entorhinal cortex, for example, are intrinsically oscillatory.[8] These neurons project via the perforant path to area CA1 and the dentate gyrus of the hippocampus where they act to drive the system into 4- to 10-Hz activity.[34] Although it is clear that both the medial septum and entorhinal cortex provide a rhythmic afferent drive to hippocampus, recent intracellular recordings indicate that hippocampal pyramidal neurons are also capable of generating intrinsic oscillations of membrane potential in the 4- to 10-Hz range.[101,125] Thus there exist a redundant set of mechanisms in these structures capable of generating theta activity.

During periods of theta activity, single cells in the hippocampus, entorhinal cortex, and medial septal nucleus exhibit a broad range of synchronized activity. Some cells show a rather precise pattern of repetitive burst firing synchronized to the phase of the RSA observed in the field potential. In the hippocampus these neurons are termed theta cells and are largely thought to be inhibitory interneurons.[19,56] In the entorhinal cortex such cells have been classed as rhythmic[7] whereas in the medial septum they are termed "B" cells.[131] In each of these structures there are significant numbers of cells that display little or no rhythmicity in autocorrelation histograms of their discharge patterns. However, a majority of these cells show a clear relation with the phase of the theta rhythm, as revealed by spike triggered averaging of the local field potential.[7,18,131] This result is similar to that observed for mitral cells and pyramidal neurons in the olfactory bulb and cortex, respectively (see above, Freeman[61]). Thus, although often not readily apparent in the activity of single cells, the hippocampal theta rhythm represents the coordinated activation of very large populations of synchronously active cells, as is seen in multielectrode recordings. In one early study Bland et al.[16] demonstrated that theta activity recorded in the CA1 region was highly synchronous over a region spanning 8 mm by 6 mm along the longitudinal and transverse axes, respectively. Correlation coefficients of theta field potentials were often near 1.0 and the widely separate signals showed little or no phase difference within CA1. Similar measurements have demonstrated that the hippocampal theta rhythm is also synchronous between the two hemispheres.[36] In another study Kuperstein et al.,[99] using multiple microelectrode recording techniques, demonstrated that groups of neurons in the CA3 region exhibit coherent synchronous patterns of firing during periods of theta activity.

HIPPOCAMPAL FAST ACTIVITY

Another form of rhythmic activity in the hippocampus about which much less is known is the so-called fast or gamma activity.[25,102] This oscillatory activity, best documented in field potential recordings in the rat, has a frequency ranging from 30–100 Hz, and occurs under a number of behavioral conditions, including under anesthesia, but is most prominent during behaviors such as exploratory walking that result in theta activity. Figure 9 shows an example of theta, gamma, and unit activity recorded simultaneously in the hilar region of a behaving rat. During periods of exploration gamma activity is greatly increased in amplitude (Figure 9A). It occurs with somewhat greater amplitude during the positive phase of the theta rhythm and is closely associated with the firing of single units. When the animal ceases walking and sits motionless the theta rhythm is virtually absent and the gamma activity is reduced by a factor of 5 (data not shown).

Although gamma activity has been recorded in CA1 and the dentate, it is most prominent in the hilus of the dentate gyrus,[34] where the signals exhibit a high degree of local and long-range coherence. Locally the signals are most often correlated with the activity of interneurons and granule cells.[34] The gamma signals show a sharp fall off in amplitude outside the hilar region but exhibit pronounced coherence along the longitudinal axis of the hippocampus. Figure 9B shows an example of the theta and gamma frequency coherence values computed from a multisite recording in the hippocampus of a behaving rat. Gamma coherence values as high as 0.7 occur at distances of up to 2.1 mm, but fall off sharply along the transverse axis (not shown). Finally, as in the visual and motor cortices (see below), gamma activity has been observed to be synchronous between the two hippocampi.[34]

Although little else is known regarding the functional significance of this form of activity in the hippocampus,[102] these fast rhythms are indicative of the propensity of the hippocampus to engage in large-scale cooperative patterns of synchronous activity.

HIPPOCAMPAL SHARP WAVES

During the transition from exploratory behavior to alert immobility, associated with acts such as eating and grooming, hippocampal activity in the rat shows another remarkable change. Highly synchronous theta activity vanishes and is replaced by a pattern of irregular activity intermingled with large amplitude (1–3 mV) sharp waves of voltage lasting 40–100 ms.[34,35,36] These events are synchronous in the two hippocampi (Figure 10A) and show a high degree of coherence across the transverse and longitudinal axes of the hippocampus.[36] Moreover, each sharp wave is associated with a synchronous burst discharge of action potentials in a population of pyramidal cells.[36]

Recently these signals were examined at a higher degree of spatial and temporal resolution using multielectrode recording techniques.[37,172] These studies reveal that the sharp wave consists of a synchronous oscillation of activity in a population

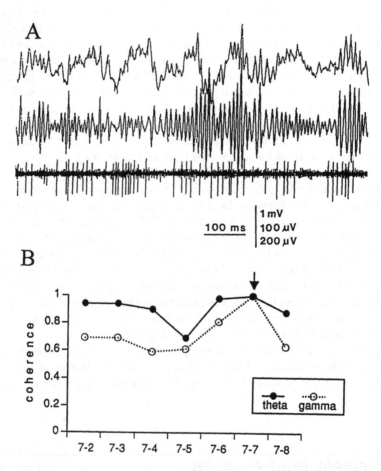

FIGURE 9 Gamma activity in the Hilar region of the hippocampus. Microelectrode recording during exploratory walking (A). Upper trace: wide band recording. Middle and lower traces: gamma activity (40–150 Hz) and spike train of an identified interneuron (500 Hz–10 kHz), respectively. Note the theta and gamma related modulation of the firing rate of the isolated neuron. B. Coherence of theta and gamma activity along the longitudinal axis of the hippocampus. Recordings were made from 8 electrodes along the long axis, each having a 300-um spacing. Coherence functions were computed for each electrode combination and plotted for the frequency components at 8 Hz (theta) and 80 Hz (gamma). There was high coherence and near zero phase shift (not shown) of both the theta and gamma activity along the long axis and a steep decrease of coherence in the transverse direction (not shown). From Bragin et al.[25]

of cells at a frequency near 200 Hz (Figure 10B). Individual cells fire at low rates but their coordinated activation yields an emergent pattern of oscillation. Single cells recorded in isolation in the pyramidal cell layer often fire only one or two spikes but these usually occur in phase with the 200-Hz field potential oscillation. In a few recordings the activity of identified interneurons was also measured. These cells fired consistently at 180 degrees out of phase from that of the pyramidal cell population, suggesting a role for the interneuron population in the generation of the

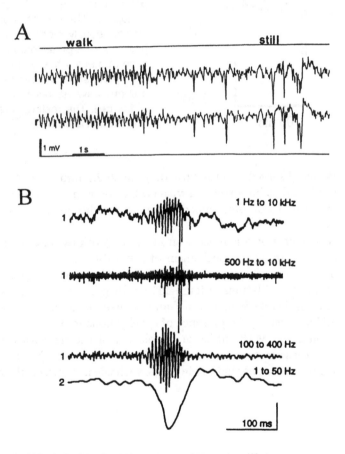

FIGURE 10 A. EEG recorded from the stratum radiatum of the left (l) and right (r) CA1 region of the hippocampus during walk-immobility (still) transition. Note regular theta waves during walking and large monophasic sharp waves during immobility. Note also the bilaterally synchronous nature of sharp waves. Modified From Buzsaki.[36] B. Fast field oscillation in the CA1 region of the dorsal hippocampus. Simultaneous recordings from the CA1 pyramidal cell layer (1) and stratum radiatum (2). Note the simultaneous occurrence of fast field oscillations, unit discharges, and sharp wave. Calibrations: 0.5 mV (trace 1), 0.25 mV (traces 2 and 3), and 1.0 mV (trace 4). From Buzsaki et al.[37]

extra

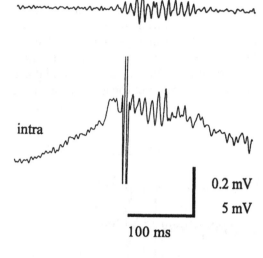

intra

0.2 mV

5 mV

100 ms

FIGURE 11   Intracellular correlate of the high-frequency oscillations in a CA1 pyramidal neuron. The upper trace shows the extracellular field potential recorded at 50- to 250-Hz bandpass. The lower trace shows the simultaneously recorded membrane potential of a pyramidal neuron, the action potential has been truncated. Both plots show the oscillations during a sharp wave associated depolarization of the cell. The resting potential was 65 mV. From Ylinen et al.[172]

oscillations. When observed at sites extending up to 2.1 mm apart the activities of single cells were found to be correlated with each other and with the oscillatory field potential. It remains to be determined if similar synchronous interactions extend over larger distances.

The striking degree of coherence seen in the network oscillations raises a number of questions regarding its mechanism of generation and coordination. Recent intracellular recordings of sharp wave activity *in vivo* have revealed pronounced membrane potential oscillations (Figure 11).[172] In pyramidal neurons these oscillations range in amplitude from 1–5 mV and reverse at membrane potentials near −70 to −80 mV suggesting the presence of synchronous inhibitory synaptic potentials. This is consistent with the close correlation of the network oscillation and the activity of interneurons. Moreover, in their recent report Ylinen et al.[172] describe identified interneurons that fire in phase with the network oscillation recorded extracellularly.

FUNCTIONAL SIGNIFICANCE OF SYNCHRONOUS HIPPOCAMPAL RHYTHMS

Although the functional significance of these synchronous rhythms is largely unknown such stereotyped and global patterns of activity are suggestive of an important function. Multiple, redundant mechanisms have evolved to enable the generation of various forms of rhythmic activity[34,161] and the anatomical organization of hippocampus clearly lends itself to the establishment of macroscopic patterned states. Hippocampal pyramidal cells have widely arborizing axonal collaterals that extend over many millimeters including the contralateral hippocampus.[153] These long-range connections in combination with the intrinsic membrane properties of cells and local network organization yield a structure capable of some of the most robust and organized states of synchronous activity to be observed in the central nervous system.

Despite extensive studies into the mechanisms of generation, the behavioral dependence, and the spatiotemporal distribution of hippocampal rhythmic activity we still do not know the function of these coherent activity patterns. A number of attractive proposals have been put forth.[35,36,88] Several of these relate closely to the well-established role of the hippocampus in learning and memory[114] and the propensity of hippocampal neurons to exhibit synaptic plasticity such as long-term potentiation and depression.[21,150] Studies investigating the latter phenomena have established that correlation of activity among synaptically connected neurons is a critical factor in the induction of changes in synaptic strength. Correlated activity is of course the essence of hippocampal rhythms. Thus, these coherent rhythmic states may provide the substrate for synaptic modifications and the storage of information.[88,126]

THE SOMATOMOTOR CORTEX AND ATTENTIVE BEHAVIOR

Another, less understood, but equally impressive form of synchronous rhythmic activity occurs in the somatosensory and motor cortices and thalamus of cats and monkeys. This activity has been observed primarily in intracortical field potential recordings but new data are rapidly emerging on the underlying neuronal mechanisms of generation. The signals range in frequency from 15- to 45-Hz and are most prevalent and highest in amplitude under conditions of alert attentive behavior. Early studies of this phenomenon revealed that during periods of attentive immobility the EEG recorded from the somatosensory cortex often shifted from a disorganized pattern of low frequencies to a clearly oscillatory pattern having a dominant frequency around 15 Hz in the monkey[136] and 35- to 45-Hz in the cat.[23,24] These rhythmic patterns of activity persisted for as long as the animal maintained a state of attentive behavior.

This effect was found to be particularly striking in the cat when the animal was able to view a mouse housed in another cage. Under these circumstances the cat displayed a stereotypical pattern of attentive behavior in which it remained completely motionless, its ears pointed forward and its gaze directed toward the mouse. During these periods the EEG in both the somatosensory cortex and ventrobasal thalamus showed a pronounced spectral peak at 35- to 40-Hz that was coherent between the two structures.[23] These data provided one of the first clear demonstrations of thalamocortical synchrony at a frequency range above that observed in spindle activity and the alpha rhythm. Subsequently these high-frequency rhythms in the cat have been localized to several areas of the somatosensory cortex[24] and have been found to depend on dopaminergic input from the ventral midbrain for their occurrence.[117]

A similar form of synchronous rhythmic activity has recently been observed in the motor cortex of alert macaque monkeys. In studies by two groups, field potentials[42,70,119,120,137] and multiple unit activity[43,119,120] were recorded from a number of electrodes in different regions of the motor cortical map. In one study activity was also recorded from the adjacent somatosensory cortex.[120] The animals were trained to perform a stereotyped motor task[137] as well as to perform voluntary movements of the hands in the absence of visual guidance.[120] During the periods prior to execution of the trained task, 25- to 35-Hz activity was readily apparent in both the field potential and less so in the unit activity. Correlation measurements revealed that the field potentials were synchronous over distances spanning up to 10 mm in motor cortex. These signals were found to abate during the onset and execution of the movement.[137] If, however, the animals were allowed to extract raisins from unseen locations by palpation, a task requiring significant attention, the 25- to 35-Hz activity increased in amplitude, and synchrony was observed within the motor cortex as well as between the motor and somatosensory cortex (Figure 12), a distance approaching 20 mm in the macaque.[119,120]

In a later study Murthy et al.[121] investigated the occurrence of interhemispheric synchrony in monkeys performing both trained and exploratory uni- and bi-manual tasks. Interhemispheric synchronization was found to exhibit many of the same properties of that observed within each hemisphere. During a trained task synchronous oscillations decreased at the onset of the movement. During exploratory movements, however, the magnitude of the oscillations and their synchrony increased above a no-movement control condition, and were present throughout the period of movement. Interestingly, the occurrence and magnitude of interhemispheric synchrony did not differ between uni- and bi-manual tasks. In the same study Murthy et al.[121] demonstrated that a similar phenomenon exists over regions of the somatomotor cortex in humans performing visuomotor tracking movements with the hands.

In an effort to investigate the neuronal mechanisms underlying these synchronous oscillations Chen and Fetz[40] performed simultaneous recordings of local field potentials, extracellular unit activity, and intracellular membrane potential in an awake monkey. This technical tour de force revealed a number of parallels to similar measurements made in the visual cortex and hippocampal formation (Figure 13). During episodes of oscillatory activity neurons tended to fire on the negative polarity of the field potential. Intracellular recordings of the membrane potential revealed that unit firing coincided with a rhythmic depolarization of the cell. It is likely that such oscillatory events are the result of synchronized volleys of excitatory synaptic inputs reflecting coordinated activation of the network. The

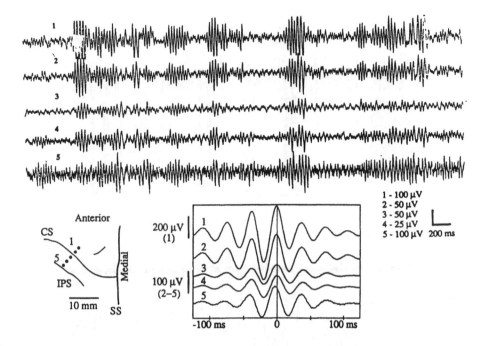

FIGURE 12 Local field potentials recorded simultaneously in five anterior-posterior tracks in the motor and somatosensory cortices of an alert monkey. Electrode sites, marked on the sketch of the cortical surface (lower left), straddled the central sulcus. Averages of the local field potentials aligned on triggers from oscillatory cycles in trace 1. The monkey was reaching for a raisin offered to the side of its head by the experimenter. From Murthy and Fetz.[120]

resulting extracellular current flow in a population of cells is likely to explain the presence and polarity of the local field potential. Taken together these data suggest a possible role of synchronous activation of motor and somatosensory cortex during the preparation of planned movement. The clear abatement of the rhythmic activity during the execution of a trained motor act suggests that such activity plays little or no role in learned or overtrained movements. However, during novel tasks requiring a degree of attention, the rhythmic activity is highly synchronous over large regions of somatomotor cortex including the two hemispheres. As in the other systems described above these coordinated states of activity reflect widespread and synchronous interactions among distributed populations of cells.

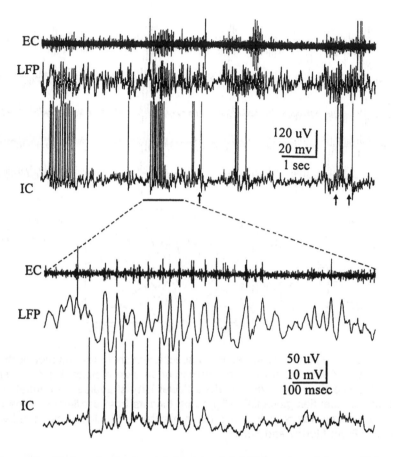

FIGURE 13   Oscillations in the local field potential (LFP), extracellular multiple unit activity (EC) and intracellular membrane potential from a single cell (IC) all recorded simultaneously in the motor cortex of a monkey. The LFP and multiple unit activity are recorded from the same electrode (negativity up). Arrows indicate subthreshold membrane potential depolarizations in phase with LFP oscillation cycles. Note that during oscillatory episodes groups of cells fire synchronously during the negative phase of the rhythm and these events are associated with rhythmic cycles of membrane depolarization. From Chen and Fetz.[40]

CONCLUSIONS

It is clear from the foregoing discussions that synchronized rhythmic activity is a general property of neuronal systems in the mammalian brain. Such activity occurs in a number of different systems, over a range of spatial and temporal scales, during different behavioral states and can be generated by a variety of different mechanisms. In the four systems described above, coherent rhythmic states of synchronous activity have been described in a number of different species including rats, rabbits, cats, and monkeys. Confirmation of these findings in humans has been more difficult. Nonetheless a number of reports have appeared over the years demonstrating similar phenomena in the human cerebral cortex.[39,71,121,125,128,131,134,140] At present, however, the functional significance of such coherent macroscopic states of activity is largely a matter of speculation. And the degree to which these states of activity can be related to particular functions depends on the system under study.

It does appear, however, that the generation of rhythmic activity patterns in groups of neurons can often arise from what otherwise seem to be mechanisms for maintaining stability. For instance, in many neural structures both short- and long-range excitatory connections exist in abundance. This powerful source of positive feedback can easily lead to runaway excitation and instability but is balanced by the ubiquitous presence of local inhibition. At the cellular level there is an analogous interplay between excitation and inhibition. The depolarization produced by inward currents is balanced by outward currents to maintain stability. Both forms of organization often produce rhythmic behavior. At the network level it can be easily demonstrated that the basic configuration of short- and long-range excitation coupled with local inhibition (Figure 14) readily leads to patterns of synchronous oscillation.[33,58,60,63,93,148,169,167] Sustained excitation arising from afferent and intrinsic sources leads to transient and repeated inhibition having a periodic time strucutre. The time delays inherent in the recurrent inhibitory circuit determine the frequency of oscillation. Similarly, the voltage and time dependence of inward and outward conductances confer a variety of intrinsic oscillatory behaviors on many different cell types.[8,104,105,112,123,144] When these network and membrane mechanisms function together the circuits in which they are embedded are tuned to synchronously oscillate within a particular frequency range.[151,152]

It comes as no surprise then that the nervous system should exhibit a wide range of oscillatory phenomena expressed at the level of single cells and networks of cells. The question arises as to whether such oscillatory activities are simply epiphenomena reflecting the physiological characteristics of neural structures. Or, alternatively, has the nervous system taken advantage of a class of ubiquitous neural mechanisms for specific functional purposes? The fact that oscillations are often associated with or give rise to macroscopic states of synchronous activity related to specific behaviors lends support to this alternative.[30] Perhaps then synchronous rhythms have evolved to dynamically control the grouping of populations of cells into organized assemblies. Such grouping may be essential for the coordination and integration of functions that are anatomically distributed.

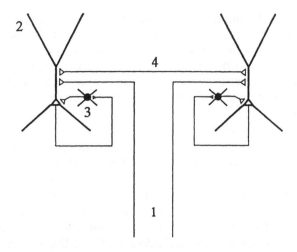

FIGURE 14 Schematic diagram of a simple neuronal circuit illustrating the basic elements thought to underlie the generation of synchronous oscillations. In such a network oscillations may arise by mechanisms intrinsic to the circuit, by oscillatory afferent input, or by a combination of the two. An oscillation arising through network interactions would begin when non-rhythmic afferent excitation (1) drives the excitatory cells (2), leading to the activation of a population of inhibitory interneurons (3). The ensuing inhibition of the excitatory cell population would produce a net disexcitation in the interneuron population leading to disinhibition of the excitatory cells. The sequence would repeat if the excitation arising from afferent and intrinsic sources were sufficiently sustained. Both short- and long-range collaterals of the excitatory cells (4) would serve to synchronize local and spatially distributed populations of cells. For oscillations to be in phase it is thought that a small percentage of these excitatory connections must synapse onto the inhibitory interneurons thereby conveying inhibtion over larger distances (not shown). Oscillations in the afferent input of such a network would produce a resonant state between the two systems as seen in the olfactory bulb-piriform cortex induced wave, the entorhinal cortex-hippocampal theta rhythm and the fast beta rhythms of the somatosensory system. In addition synchronous rhythms may arise through interactions among populations of excitatory and/or inhibitory cells that are intrinsically oscillatory. In the network illustrated, the oscillation frequency is controlled in large part by the time delays inherent in the recurrent inhibitory loop. This timing can be markedly influenced by the intrinsic membrane properties of the constituent cells.

ACKNOWLEDGMENTS

This work was supported by grants from the Office of Naval Research, The National Science Foundation, The National Eye Institute and fellowships from the Klingenstein Foundation, and the Sloan Foundation. I thank Gyorgy Buzsaki, Eberhard Fetz, Venkatesh Murthy, and Walter Freeman for their valuable comments and willingness to share their hard-earned experimental results. I also thank Yve Marder, Pedro Maldonado, and Steve Bressler for their helpful editorial comments on an earlier version of the manuscript.

REFERENCES

1. Adey, W. R., C. W. Dunlop, and C. E. Hendrix. "Hippocampal Slow Waves: Distribution and Phase Relations in the Course of Approach Learning." *Arch. Neurol.* **3** (1960): 74–90.

2. Adrian, E. D. "Olfactory Reactions in the Brain of the Hedgehog." *J. Physiol.* **100** (1942): 459–473.

3. Adrian, E. D. "The Electrical Activity of the Mammalian Olfactory Bulb." *Electroenceph. Clin. Neurophysiol.* **2** (1950): 377–388.

4. Adrian, E. D. "Sensory Discrimination: With Some Recent Evidence from the Olfactory Organ." *Brit. Med. Bull.* **6(4)** (1950): 330–333.

5. Adrian, E. D. "The Basis of Sensation: Some Recent Studies of Olfaction." *Brit. Med. J.* **Feb. 6** (1954).

6. Alonso, A., and E. Garcia-Austt. "Neuronal Sources of Theta Rhythm in the Entorhinal Cortex of the Rat I. Laminar Distribution of Theta Field Potentials." *Exp. Brain Res.* **67** (1987): 493–501.

7. Alonso, A., and E. Garcia-Austt. "Neuronal Sources of Theta Rhythm in the Entorhinal Cortex of the Rat II. Phase Relations Between Unit Discharges and Theta Field Potentials." *Exp. Brain Res.* **67** (1987): 502–509.

8. Alonso, A., and R. R. Llinas. "Subthreshold Na+-Dependent Theta-Like Rhythmicity in Stellate Cells of Entorhinal Cortex Layer II." *Nature* **342** (1989): 175–177.

9. Ariel, M., N. W. Daw, and R. K. Rader. "Rhythmicity in Rabbit Retinal Ganglion Cell Responses." *Vision Res.* **23(12)** (1983): 1485–1493.

10. Arnett, D. W. "Correlation Analysis of Units Recorded in the Cat Dorsal Lateral Geniculate Nucleus." *Exp. Brain Res.* **24** (1975): 111–130.

11. Bair, W., C. Koch, W. Newsome, K. Britten, and E. Niebur. "Power Spectrum Analysis of MT Neurons from Awake Monkey." *Soc. Neurosci. Abs.* **18** (1992): 11.12.

12. Bal, T., and D. McCormick. "Ionic Mechanisms of Rhythmic Burst Firing and Tonic Activity in the Nucleus Reticularis Thalami: A Mammalian Pacemaker." *J. Physiol.*, 1992.

13. Barlow, H. B., R. M. Hill, W. R. Levick. "Retinal Ganglion Cells Responding Selectively to Direction and Speed of Image Motion in the Rabbit." *J. Physiol. Lond.* **173** (1964):377–407.

14. Becker, C. J., and W. J. Freeman. "Prepyriform Electrical Activity After Loss of Peripheral or Central Input, or Both." *Physiol. & Behav.* **3** (1968): 597–599.

15. Bishop, P. O., W. R. Levick, and W. O. Williams. "Statitical Analyses of the Dark Discharge of Lateral Geniculate Neurons." *J. Physiol.* **170** (1964): 598–612.

16. Bland, B. H., P. Andersen, and T. Ganes. "Two Generators of Hippocampal Theta Activity in Rabbits." *Brain Res.* **94** (1975): 199–218.

17. Bland, B. H., and I. Q. Wishaw. "Generators and Topography of Hippocampal Theta (RSA) in the Anesthetized and Freely Moving Rat." *Brain Res.* **118** (1976): 259–280.

18. Bland, B. H., P. Andersen, T. Ganes, and O. Sveen. "Automated Analysis of Rhythmicity of Physiologically Identified Hippocampal Formation Neurons." *Exp. Brain Res.* **38** (1980): 205–219.

19. Bland, B. H. "The Physiology and Pharmacology of Hippocampal Formation Theta Rhythms." *Prog. in Neurobiol.* **26** (1986): 1–54.

20. Blasdel, G. G., and G. Salama. "Voltage-Sensitive Dyes Reveal a Modular Organization in Monkey Striate Cortex." *Nature* **321** (1986): 579–585.

21. Bliss, T. V. P., and T. Lomo. "Long-Lasting Potentiation of Synaptic Transmission in the Dentate Area of the Anesthetized Rabbit Following Stimulation of the Perforant Path." *J. Physiol. (Lond.)* **232** (1973): 331–356.

22. Boeijinga, P. H., and F. H. Lopes da Silva. "Modulations of EEG Activity in the Entorhinal Cortex and Forebrain Olfactory Areas During Odour Sampling." *Brain Res.* **478** (1989): 257–268.

23. Bouyer, J. J., M. F. Montaron, and A. Rougeul. "Fast Fronto-Parietal Rhythms During Combined Focused Attentive Behavior and Immobility in Cat: Cortical and Thalamic Localizations." *Electroencephal. & Clin. Neurophysiol.* **51** (1981): 244–252.

24. Bouyer, J. J., M. F. Montaron, J. M. Vahnee, M. P. Albert, and A. Rougeul. "Anatomical Localization of Cortical Beta Rhythms in Cat." *Neurosci.* **22(3)** (1987): 863–869

25. Bragin, A., G. Jando, Z. Nadasdy, J. Hetke, K. Wise, and G. Buzsaki. "Beta Frequency (40–100 Hz) Patterns in the Hippocampus: Modulation by Theta Activity." *Soc. Neurosci. Abs.* **19** (1993): 148.3.

26. Bressler, S. L., and W. J. Freeman. "Frequency Analysis of Olfactory System EEG in Cat, Rabbit, and Rat." *Electroencephal. & Clin. Neurophysiol.* **50** (1980): 19–24.

27. Bressler, S. L. "Spatial Organization of EEGs from Olfactory Bulb and Cortex." *Electroencephal. & Clin. Neurophysiol.* **57** (1984): 270–276.

28. Bressler, S. L. "Relation of Olfactory Bulb and Cortex: I. Spatial Variation of Bulbocortical Interdependence." *Brain Res.* **409** (1987): 285–293.
29. Bressler, S. L. "Relation of Olfactory Bulb and Cortex: II. Model for Driving of Cortex by Bulb." *Brain Res.* **409** (1987): 294–301.
30. Bressler, S. L., R. Coppola, and R. Nakamura. "Episodic Multiregional Cortical Coherence at Multiple Frequencies During Visual Task Performance." *Nature* **366** (1993): 153–156.
31. Bringuier, V., Y. Fregnac, D. Debanne, D. Shulz, and A. Baranyi. "Synaptic Origin of Rhythmic Visually Evoked Activity in Kitten Area 17 Neurones." *NeuroReport* **3** (1992): 1065–1068.
32. Bullier, J., M. H. J. Munk, and L. G. Nowak. "Synchronization of Neuronal Firing in Areas V1 and V2 of the Monkey." *Soc. Neurosci. Abstr.* **18** (1992): 11.7.
33. Bush, P. C., and R. J. Douglas. "Synchronization of Bursting Action Potential Discharge in a Model Network of Neocortical Neurons." *Neural Comp.* **3** (1991): 19–30.
34. Buzsaki, G., L. S. Leung, and C. H. Vanderwolf. "Cellular Bases of Hippocampal EEG in the Behaving Rat." *Brain Res. Rev.* **6** (1983): 139–171.
35. Buzsaki, G. "Hippocampal Sharp Waves: Their Origin and Significance." *Brain Res.* **398** (1986): 242–252.
36. Buzsaki, G. "Two-Stage Model of Memory Trace Formation: A Role for 'Noisy' Brain States." *Neurosci.* **31(3)** (1989): 551–570.
37. Buzsaki, G., Z. Horvath, R. Urioste, J. Hetke, and K. Wise. "High-Frequency Network Oscillation in the Hippocampus." *Science* **256** (1992): 1025–1027.
38. Cajal, S. R. *Studies on the Cerebral Cortex (limbic structures).* Translated by L. M. Kraft. London: Lloyd-Luke, 1955.
39. Chatrian, G. E., R. G. Bickford, and A. Uilein. "Depth Electrographic Study of a Fast Rhythm Evoked from the Human Calcarine Region by Steady Illumination." *Electroencephal. & Clin. Neurophysiol.* **12** (1960): 167–176.
40. Chen, D. F., and E. E. Fetz. "Effect of Synchronous Neural Activity on Synaptic Transmission in Primate Cortex." *Soc. Neurosci. Abstr.* **19** (1993): 319.7.
41. Desimone, R., and L. G. Ungerleider. "Neural Mechanisms of Visual Processing in Monkeys." In *Handbook of Neuropsychology*, edited by F. Boller and J. Grafman, Vol. 2, Ch. 14. Elsevier Science Publishers, 1989.
42. Donoghue, J. P., and J. N. Sanes. "Dynamic Modulation of Primate Motor Cortex Output During Movement." *Neurosci. Soc. Abstr.* **17** (1991): 407.5.
43. Donoghue, J. P., G. Gaal, M. Niethammer, and J. N. Sanes. "Oscillations in Local Field Potentials and Neural Discharge in Monkey Motor Cortex." *Soc. Neurosci. Abs.* **19** (1993): 319.5.
44. Doty, R. W., and D. S. Kimura. "Oscillatory Potentials in the Visual System of Cats and Monkeys." *J. Physiol.* **168** (1963): 205–218.
45. Dreher, B. "Thalamocortical and Corticocortical Interconnections in the Cat Visual System: Relation to the Mechanisms of Information Processing." In

Visual Neuroscience, edited by J. D. Pettigrew, K. J. Sanderson, and W. R. Levick, 290–314. New York: Cambridge University Press, 1986.

46. Eckhorn, R., R. Bauer, W. Jordan, M. Brosch, W. Kruse, M. Munk, and H. J. Reitboeck. "Coherent Oscillations: A Mechanism of Feature Linking in the Visual Cortex?" *Biol. Cybern.* **60** (1988): 121–130.

47. Eckhorn, R., A. Frien, R. Bauer, T. Woelbern, and H. Kehr. "High-Frequency (60–90 Hz) Oscillations in Primary Visual Cortex of Awake Monkey." *NeuroReport* **4** (1993): 243–246.

48. Engel, A. K., P. Koenig, C. M. Gray, and W. Singer. "Stimulus-Dependent Neuronal Oscillations in Cat Visual Cortex: Inter-Columnar Interaction as Determined by Cross-Correlation Analysis." *Eur. J. Neurosci.* **2** (1990): 588–606.

49. Engel, A. K., A. K. Kreiter, P. Koenig, and W. Singer. "Synchronization of Oscillatory Neuronal Responses Between Striate and Extrastriate Visual Cortical Areas of the Cat." *Proc. Natl. Acad. Sci.* **88** (1991): 6048–6052.

50. Engel, A. K., P. Koenig, A. K. Kreiter, and W. Singer. "Interhemispheric Sychronization of Oscillatory Responses in Cat Visual Cortex." *Science* **252** (1991): 1177–1179.

51. Engel, A. K., P. Koenig, and W. Singer. "Direct Physiological Evidence for Scene Segmentation by Temporal Coding." *Proc. Natl. Acad. Sci.* **88** (1991): 9136–9140.

52. Felleman, D. J., and D. C. Van Essen. "Distributed Hierarchical Processing in the Primate Cerebral Cortex." *Cerebral Cortex* **1(1)** (1991): 1–47.

53. Ferster, D., and S. LeVay. "The Axonal Arborizations of Lateral Geniculate Neurons in the Striate Cortex of the Cat." *J. Comp. Neurol.* **182** (1978): 923–944.

54. Ferster, D. "Orientation Selectivity of Synaptic Potentials in Neurons of Cat Primary Visual Cortex." *J. Neurosci.* **6(5)** (1986): 1284–1301.

55. Fox, S. E., and J. B. Ranck. "Electrophysiological Characteristics of Hippocampal Complex-Spike Cells and Theta Cells." *Exp. Brain Res.* **41** : 399–410.

56. Freeman, W. J. "Distribution in Space and Time of Prepyriform Electrical Activity." *J. Neurophysiol.* **22** (1959): 644–666.

57. Freeman, W. J. "Correlation of Electrical Activity of Prepiriform Cortex and Behavior in Cat." *J. Neurophysiol.* **23** (1960): 111–131.

58. Freeman, W. J. "Analog Simulation of Prepiriform Cortex in the Cat." *Math. BioSci.* **2** (1968): 181–190.

59. Freeman, W. J. "Average Transmission Distance from Mitral-Tufted to Granule Cells in Olfactory Bulb." *Electroencephal. & Clin. Neurophysiol.* **36** (1974): 609–618.

60. Freeman, W. J. *Mass Action in the Nervous System.* New York: Academic Press, 1975.

61. Freeman, W. J. "Spatial Properties of an EEG Event in the Olfactory Bulb and Cortex." *Electroencephal. & Clin. Neurophysiol.* **44** (1978): 586–605.

62. Freeman, W. J. "Spatial Frequency Analysis of an EEG Event in the Olfactory Bulb." In *Multidisciplinary Perspectives in Event-Related Brain Potential Research*, edited by D. A. Otto, EPA-600/9-77-043, 531–546. Washington, DC: U.S. Government Printing Office, 1978.

63. Freeman, W. J. "Nonlinear Dynamics of Paleocortex Manifested in the Olfactory EEG." *Biol. Cybern.* **35** (1979): 21–37.

64. Freeman, W. J., and W. Schneider. "Changes in Spatial Patterns of Rabbit Olfactory EEG with Conditioning to Odors." *Psychophysiol.* **19(1)** (1982): 44–56.

65. Freeman, W. J. "Analytic Techniques Used in the Search for the Physiological Basis for the EEG." In *Handbook of Electroencephalography and Clinical Neurophysiology*, edited by A. Gevins and A. Remond, Vol 3A, Part 2, Ch. 18. Amsterdam: Elsevier, 1985.

66. Freeman, W. J., and C. A. Skarda. "Spatial EEG Patterns, Nonlinear Dynamics and Perception: The Neo-Sherringtonian View." *Brain Res. Rev.* **10** (1985): 147–175.

67. Freeman, W. J., and G. Viana Di Prisco. "EEG Spatial Pattern Differences with Discriminated Odors Manifest Chaotic and Limit Cycle Attractors in Olfactory Bulb of Rabbits." In *Brain Theory*, edited by G. Palm and A. Aertsen, 97–119. Heidelberg, Berlin: Springer-Verlag, 1986.

68. Freeman, W. J. "The Physiology of Perception." *Sci. Am.* **264(2)** (1991): 78–85.

69. Fuster, J. M., A. Herz, and O. D. Creutzfeldt. "Interval Analysis of Cell Discharge in Spontaneous and Optically Modulated Activity in the Visual System." *Arch. Ital. Biol.* **103** (1965): 159–177.

70. Gaal, G., J. N. Sanes, and J. P. Donoghue. "Motor Cortex Oscillatory Neural Activity During Voluntary Movement in Macaca Fascicularis." *Soc. Neurosci. Abstr.* **18** (1992): 355.14.

71. Galambos, R., S. Makeig, and P. Talmachoff. "A 40-Hz Auditory Potential Recorded from the Human Scalp." *Proc. Nat. Acad. Sci.* **78** (1981): 2643–2647.

72. Ghose, G. M., and R. D. Freeman. "Oscillatory Discharge in the Visual System: Does It Have a Functional Role?" *J. Neurophysiol.* **68** (1992): 1558–1574.

73. Gochin, P. M., E. K. Miller, C. G. Cross, and G. L. Gerstein. "Functional Interactions Among Neurons in Inferior Temporal Cortex of the Awake Macaque." *Exp. Brain Res.* **84** (1991): 505–516.

74. Gray, C., and W. Singer. "Stimulus-Specific Neuronal Oscillations in the Cat Visual Cortex: A Cortical Functional Unit." *Soc. Neurosci. Abstr.* **13** (1987): 404.3

75. Gray, C. M., and J. E. Skinner. "Centrifugal Regulation of Neuronal Activity in the Olfactory Bulb of the Waking Rabbit as Revealed by Reversible Cryogenic Blockade." *Exp. Brain Res.* **69** (1988): 378–386.

76. Gray, C. M., and J. E. Skinner. "Field Potential Response Changes in the Rabbit Olfactory Bulb Accompany Behavioral Habituation During the Repeated Presentation of Unreinforced Odors." *Exp. Brain Res.* **73** (1988): 189–197.

77. Gray, C. M., and W. Singer. "Stimulus-Specific Neuronal Oscillations in Orientation Columns of Cat Visual Cortex." *Proc. Nat. Acad. Sci.* **86** (1989): 1698–1702.

78. Gray, C. M., P. Koenig, A. K. Engel, and W. Singer. "Stimulus-Specific Neuronal Oscillations in Cat Visual Cortex Exhibit Inter-Columnar Synchronization Which Reflects Global Stimulus Properties." *Nature* **338** (1989): 334–337.

79. Gray, C., A. K. Engel, P. Koenig, and W. Singer. "Stimulus-Dependent Neuronal Oscillations in Cat Visual Cortex: Receptive Field Properties and Feature Dependence." *Eur. J. Neurosci.* **2** (1990): 607–619.

80. Gray, C. M., A. K. Engel, P. Koenig, and W. Singer. "Synchronization of Oscillatory Neuronal Responses in Cat Striate Cortex: Temporal Properties." *Vis. Neurosci.* **8** (1992): 337–347.

81. Gray, C. M., A. K. Engel, P. Koenig, and W. Singer. "Mechanisms Underlying the Generation of Neuronal Oscillations in Cat Visual Cortex." In *Induced Rhythmicities in the Brain*, edited by T. Bullock and E. Basar, 1992.

82. Gray, C. M., and G. Viana Di Prisco. "Properties of Stimulus-Dependent Rhythmic Activity of Visual Cortical Neurons in the Alert Cat." *Soc. Neurosci. Abstr.* **19** (1993): 359.8.

83. Green, J. D., and A. Arduini. "Hippocampal Electrical Activity in Arousal." *J. Neurophysiol.* **17** (1954): 533–557.

84. Haberly, L. B., and J. L. Price. "Association and Commissural Fiber Systems of the Olfactory Cortex of the Rat: Systems Originating in the Piriform Cortex and Adjacent Areas." *J. Comp. Neurol.* **178** (1978): 711–740.

85. Haberly, L. B., and J. M. Bower. "Analysis of Association Fiber System in Piriform Cortex with Intracellular Recording and Staining Techniques." *J. Neurophysiol.* **51(1)** (1984): 90–112.

86. Haberly, L. B., and J. M. Bower. "Olfactory Cortex: Model Circuit for Study of Associative Memory?" *TINS* **12(7)** (1989): 258–264.

87. Hubel, D. H., and T. N. Wiesel. "Receptive Fields and Functional Architecture in Two Nonstriate Visual Areas (18 and 19) of the Cat." *J. Neurophysiol.* **28** (1965): 229–289.

88. Huerta, P. T., and J. E. Lisman. "Heightened Synaptic Plasticity of Hippocampal CA1 Neurons During a Cholinergically Induced Rhythmic State." *Nature* **364** (1993): 723–725.

89. Hughes, J. R., D. E. Hendrix, N. S. Wetzel, and J. W. Johnston. "Correlations Between Electrophysiological Activity from the Human Olfactory Bulb and the Subjective Response to Odoriferous Stimuli." In *Olfaction and Taste III*, edited by C. Pfaffman. New York: Rockefeller, 1969.

90. Jagadeesh, B., C. M. Gray, and D. Ferster. "Visually-Evoked Oscillations of Membrane Potential in Neurons of Cat Striate Cortex Studied with *In Vivo* Whole Cell Patch Recording." *Science* **257** (1992): 552–554.

91. Jung, R., and A. Kornmuller. "Eine Methodik der Abteilung Lokalisierter Potential Schwankingen aus Subcorticalen Hirnyebieten." *Arch. Psychiat. Neruenkr.* **109** (1938): 1–30.

92. Kishi, K., K. Mori, and H. Ojima. "Distribution of Local Axon Collaterals of Mitral, Displaced Mitral, and Tufted Cells in the Rabbit Olfactory Bulb." *J. Comp. Neurol.* **225** (1984): 511–526.

93. Koenig, P., and T. B. Schillen. "Stimulus-Dependent Assembly Formation of Oscillatory Responses: I. Synchronization." *Neural Comp.* **3** (1991): 155–166.

94. Koenig, P., A. K. Engel, and W. Singer. "The Relation Between Oscillatory Activity and Long-Range Synchronization in Cat Visual Cortex." *Proc. Natl. Acad. Sci.* (1994): in press.

95. Kreiter, A. K., and W. Singer. "Oscillatory Neuronal Responses in the Visual Cortex of the Awake Macaque Monkey." *Eur. J. Neurosci.* **4** (1992): 369–375.

96. Kreiter, A. K., A. K. Engel, and W. Singer. "Stimulus Dependent Synchronization in the Caudal Superior Temporal Sulcus of Macaque Monkeys." *Soc. Neurosci. Abstr.* **18** (1992): 11.11.

97. Kuperstein, M., H. Eichenbaum, and T. VanDeMark. "Neural Group Properties in the Rat Hippocampus During the Theta Rhythm." *Exp. Brain Res.* **61** (1986): 438–442.

98. Laufer, M., and M. Verzeano. "Periodic Activity in the Visual System of the Cat." *Vision Res.* **7** (1967): 215–229.

99. Leung, L. S., and C. C. Yim. "Intrinsic Membrane Potential Oscillations in Hippocampal Neurons *in vitro*." *Brain Res.* **553** (1991): 261–274.

100. Leung, L. S. "Fast (Beta) Rhythms in the Hippocampus: A Review." *Hipocampus* **2(2)** (1992): 93–98.

101. Libet, B., and R. W. Gerard. "Control of the Potential Rhythm of the Isolated Frog Brain." *J. Neurophysiol.* **2** (1939): 153–169.

102. Livingstone, M. S. "Visually Evoked Oscillations in Monkey Striate Cortex." *Soc. Neurosci. Abstr.* **17** (1991): 73.3.

103. Livingstone, M. S., and D. H. Hubel. "Segregation of Form, Color, Movement, and Depth: Anatomy, Physiology, and Perception." *Science* **240** (1988): 740–749.

104. Llinas, R., and Y. Yarom. "Oscillatory Properties of Guinea-Pig Inferior Olivary Neurons and Their Pharmacological Modulation: An *In Vitro* Study." *J. Physiol.*, **376** (1986): 163–182.

105. Llinas, R. R. "The Intrinsic Electrophysiological Properties of Mammalian Neurons: Insights into Central Nervous System Function." *Science* **242** (1988): 1654–1664.

106. Llinas, R. R., A. A. Grace, and Y. Yarom. "*In Vitro* Neurons in Mammalian Cortical Layer 4 Exhibit Intrinsic Oscillatory Activity in the 10- to 50-Hz Frequency Range." *Proc. Natl. Acad. Sci.* **88** (1991): 897–901.

107. Lund, J. S., G. H. Henry, C. L. MacQueen, and A. R. Harvey. "Anatomical Organization of the Primary Visual Cortex (Area 17) of the Cat: A Comparison with Area 17 of the Macaque Monkey." *J. Comp. Neurol.* **184** (1979): 599–618.

108. Malsburg, C. von der "The Correlation Theory of Brain Function." Internal Report, Max-Planck-Institute for Biophysical Chemistry, Gottingen, West Germany, 1981.

109. Malsburg, C. von der, "Nervous Structures with Dynamical Links." *Ber. Bunsenges. Phys. Chem.* **89** (1985): 703–710.

110. Malsburg, C. von der, and W. Schneider. "A Neural Cocktail-Party Processor." *Biol. Cybern.* **54** (1986): 29–40.

111. Martin, K. A. C. "Neuronal Circuits in Cat Striate Cortex." In *Cerebral Cortex, volume 2, Functional Properties of Cortical Cells*, edited by E. G. Jones and A. Peters. New York: Plenum Press, 1984.

112. McCormick, D. A., and H. Pape. "Properties of a Hyperpolarization-Activated Cation Current and its role in Rhythmic Oscillation in Thalamic Relay Neurones." *J. Physiol.* **431** (1990): 291–318

113. McCormick, D. A., C. M. Gray, and Z. Wang. "Chattering Cells: A New Physiological Subtype Which May Contribute to 20–60 Hz Oscillations in Cat Visual Cortex." *Soc. Neurosci. Abstr.* **19** (1993): 359.9.

114. Milner, B. "Amnesia Following Operation on the Temporal Lobes." In *Amnesia*, edited by C. W. M. Whitty and O. L. Zangwill, 109–133. London: Butterworths, 1966.

115. Milner, P. "A Model for Visual Shape Recognition." *Psychol. Rev.* **81(6)** (1974): 521–535.

116. Mitchell, S. J., and J. B. Ranck. "Generation of Theta Rhythm in Medial Entorhinal Cortex of Freely Moving Rats." *Brain Res.* **178** (1980): 49–66.

117. Montaron, M., J. Bouyer, A. Rougeul, and P. Buser. "Ventral Mesencephalic Tegmentum (VMT) Controls Electrocortical Beta Rhythms and Associated Attentive Behaviour in the Cat." *Behav. Brain Res.* **6** (1982): 129–145

118. Munemori, J., K. Hara, M. Kimura, and R. Sato. "Statistical Features of Impulse Trains in Cat's Lateral Geniculate Neurons." *Biol. Cybern.* **50** (1984): 167–172.

119. Murthy, V. N., D. G. Chen, and E. E. Fetz. "Spatial Extent and Behavioral Dependence of Coherence of 25–35 Hz Oscillations in Primate Sensorimotor Cortex." *Soc. Neurosci. Abstr.* **18** (1992): 355.12.

120. Murthy, V. N., and E. E. Fetz. "Coherent 25–35 Hz Oscillations in the Sensorimotor Cortex of the Awake Behaving Monkey." *Proc. Natl. Acad. Sci.* **89** (1992): 5670–5674.

121. Murthy, V. N., F. Aoki, and E. E. Fetz. "Synchronous Oscillations in Sensorimotor Cortex of Awake Monkeys and Humans." In *Oscillatory Event-Related Brain Dynamics*, edited by C. Pantev, T. Elbert, and B. Lutkenhoener. New York: Plenum, 1994.

122. Nelson, J. I., P. A. Salin, M. H.-J. Munk, M. Arzi, and J. Bullier. "Spatial and Temporal Coherence in Cortico-Cortical Connections: A Cross-Correlation Study in Areas 17 and 18 in the Cat." *Vis. Neurosci.* **9** (1992): 001–017.

123. Nunez, A., E. Garcia-Austt, and W. Buno. "Intracellular Theta-Rhythm Generation in Identified Hippocampal Pyramids." *Brain Res.* **416** (1987): 289–300.

124. Ottoson, D. "Studies on Slow Potentials in the Rabbit's Olfactory Bulb and Nasal Mucosa." *Acta. Physiol. Scand.* **47** (1959): 136–148.

125. Pantev, C., S. Makeig, M. Hoke, R. Galambos, S. Hampson, and C. Galen. "Human Auditory Evoked Gamma-Band Magnetic Fields." *Proc. Natl. Acad. Sci.* **88** (1991): 8996–9000.

126. Pavlides, C., Y. J. Greenstein, M. Grudman, and J. Winson. "Long-Term Potentiation in the Dentate Gyrus is Induced Preferentially on the Positive Phase of Theta Rhythm." *Brain Res.* **439** (1988): 383–387.

127. Payne, B. R. "Evidence for Visual Cortical Area Homologs in Cat and Macaque Monkey." *Cerebral Cortex* **3** (1993): 1–25.

128. Perez-Borja, C., F. A. Tyce, C. McDonald, and A. Uihlein. "Depth Electrographic Studies of a Focal Fast Response to Sensory Stimulation in the Human." *Electroencephal. & Clin. Neurophysiol.* **13** (1961): 695–702.

129. Petsche, H., G. Stumpf, and G. Gogolak. "The Significance of the Rabbit's Septum as a Relay Station Between the Midbrain and the Hippocampus." *Electroencephal. & Clin. Neurophysiol.* **19** (1962): 25–33.

130. Petsche, H., G. Gogolak, and P. A. Van Zwieten. "Rhythmicity of Septal Cell Discharges at Various Levels of Reticular Excitation." *Electroencephal. & Clin. Neurophysiol.* **19** (1965): 25–33.

131. Pfurtscheller, G., and C. Neuper. "Simultaneous EEG 10 Hz Desynchronization and 40 Hz Synchronization During Finger Movements." *NeuroReport* **3** (1992): 1057–1060.

132. Rall, W., G. M. Shepherd, T. S. Reese, and M. W. Brightman. "Dendrodendritic Synaptic Pathway for Inhibition in the Olfactory Bulb." *Exp. Neurol.* **14** (1966): 44–56.

133. Rall, W., and G. M. Shepherd. "Theoretical Reconstruction of Field Potentials and Dendrodendritic Synaptic Interactions in Olfactory Bulb." *J. Neurophysiol.* **31** (1968): 884–915.

134. Ribary, U., A. A. Joannides, K. D. Singh, R. Hasson, J. P. R. Bolton, F. Lado, A. Mogilner, and R. Llinas. "Magnetic Field Tomography of Coherent Thalamocortical 40 Hz Oscillations in Humans." *Proc. Natl. Acad. Sci.* **88** (1991): 11037–11041.

135. Rosenquist, A. C. "Connections of Visual Cortical Areas in the Cat." In *Cerebral Cortex*, edited by A. Peters and E. G. Jones, 81–117. New York: Plenum, 1985.

136. Rougeul, A., J. J. Bouyer, L. Dedet, and O. Debray. "Fast Somato-Parietal Rhythms During Combined Focal Attention and Immobility in Baboon and Squirrel Monkey." *Electroencephal. & Clin. Neurophysiol.* **46** (1979): 310–319.

137. Sanes, J. N., and J. P. Donoghue. "Oscillations in Local Field Potentials of the Primate Motor Cortex During Voluntary Movement." *Proc. Natl. Acad. Sci.* **90** (1993): 4470–4474.

138. Schwarz, C., and J. Bolz. "Functional Specificity of a Long-range Horizontal Connection in Cat Visual Cortex: A Cross-Correlation Study." *J. Neurosci.* **11(10)** (1991): 2995–3007

139. Schoenfeld, T. A., J. E. Marchand, and F. Macrides. "Topographic Organization of Tufted Cell Axonal Projections in the Hamster Main Olfactory Bulb: An Intrabulbar Associational System." *J. Comp. Neurol.* **235** (1985): 503–518.

140. Sem-Jacobsen, C. W., M. C. Petersen, H. W. Dodge, J. A. Lazarte, and C. B. Holman. "Electroencephalographic Rhythms from the Depths of the Parietal, Occipital and Temporal Lobes in Man." *Electroencephal. & Clin. Neurophysiol.* **8** (1956): 263–278.

141. Sereno, M. I., and J. M. Allman. "Cortical Visual Areas in Mammals." In *The Neural Basis of Visual Function*, edited by A. Leventhal, 160–172. New York: MacMillan, 1991.

142. Shepherd, G. M. "Synaptic Organization of the Mammalian Olfactory Bulb." *Physiol. Rev.* **52** (1972): 864–917.

143. Sillito, A. M., H. E. Jones, and J. Davis. "Corticofugal Feedback and Stimulus-Dependent Correlations in the Firing of Simultaneously Recorded Cells in the Dorsal Lateral Geniculate." *Soc. Neurosci. Abstr.* **19** (1993): 218.5.

144. Silva, L. R., Y. Amitai, and B. W. Connors. "Intrinsic Oscillations of Neocortex Generated by Layer 5 Pyramidal Neurons." *Science* **251** (1990): 432–435.

145. Singer, W. "Search for Coherence: A Basic Principle of Cortical Self-Organization." *Concepts in Neurosci.* **1(1)** (1990): 1–26.

146. Singer, W. "Synchronization of Cortical Activity and Its Putative Role in Information Processing and Learning." *Ann. Rev. Physiol.* **55** (1993): 349–374.

147. Spear, P. "Functions of Extrastriate Visual Cortex in Non-Primate Species." In *The Neural Basis of Visual Function*, edited by A. Leventhal, 339–370. New York: MacMillan, 1991.

148. Sporns, O., J. A. Gally, G. N. Reeke, and G. M. Edelman. "Reentrant Signaling Among Simulated Neuronal Groups Leads to Coherency in Their Oscillatory Activity." *Proc. Natl. Acad. Sci.* **86** (1989): 7265–7269.

149. Sporns, O., G. Tononi, and G. M. Edelman. "Modeling Perceptual Grouping and Figure-Ground Segregation by Means of Active Reentrant Connections." *Proc. Natl. Acad. Sci.* **88** (1991): 129–133.

150. Stanton, P. K., and T. J. Sejnowski. "Associative Long-term Depression in the Hippocampus Induced by Hebbian Covariance." *Nature* **339** (1989): 215–218.

151. Steriade, M., E. G. Jones, and R. R. Llinas. *Thalamic Oscillations and Signaling.* New York: John Wiley, 1990.
152. Steriade, M., D. A. McCormick, and T. J. Sejnowski. "Thalamocortical Oscillations in the Sleeping and Aroused Brain." *Science* **262** (1993): 679–685.
153. Tamamaki, N., K. Abe, and Y. Nojyo. "Three-Dimensional Analysis of the Whole Axonal Arbors Originating from Single CA2 Pyramidal Neurons in the Rat Hippocampus with the Aid of a Computer Graphic Technique." *Brain Res.* **452** (1988): 255–272.
154. Thommesen, G. "The Spatial Distribution of Odor-Induced Potentials in the Olfactory Bulb of Char and Trout (Salmonidae)." *Acta. Physiol. Scand.* **102** (1978): 205–217.
155. Tombol, T., and H. Petsche. "The Histological Organization of the Pacemaker for the Hippocampal Theta Rhythm in the Rabbit." *Brain Res.* **12** (1969): 414–426.
156. Tovee, M. J., and E. T. Rolls. "Oscilaltory Activity is Not Evident in the Primate Temporal Visual Cortex with Static Stimuli." *NeuroReport* **3** (1992): 369–372.
157. Toyama, K., M. Kimura, and K. Tanaka. "Cross-Correlation Analysis of Interneuronal Connectivity in Cat Visual Cortex." *J. Neurophysiol.* **46(2)** (1981): 191–201.
158. Toyama, K., M. Kimura, and K. Tanaka. "Organization of Cat Visual Cortex as Investigated by Cross-Correlation Technique." *J. Neurophysiol.* **46(2)** (1981): 202–213.
159. Traub, R. D., R. Miles, and R. K. S. Wong. "Model of the Origin of Rhythmic Population Oscillations in the Hippocampal Slice." *Science* **243** (1989): 1319–1325.
160. Ts'o, D. Y., C. D. Gilbert, and T. N. Wiesel. "Relationships Between Horizontal Interactions and Functional Architecture in Cat Striate Cortex as Revealed by Cross-Correlation Analysis." *J. Neurosci.* **6(4)** (1986): 1160–1170.
161. Ts'o, D. Y., and C. G. Gilbert. "The Organization of Chromatic and Spatial Interactions in the Primate Striate Cortex." *J. Neurosci.* **8(5)** (1988): 1712–1727.
162. T'so, D. Y., R. D. Frostig, E. E. Lieke, and A. Grinvald. "Functional Organization of Primate Visual Cortex Revealed by High-Resolution Optical Imaging." *Science* **249** (1990): 417–420.
163. Vanderwolf, C. H. "Hippocampal Electrical Activity and Voluntary Movement in the Rat." *Electroencephal. & Clin. Neurophysiol.* **26** (1969): 407–418.
164. Viana Di Prisco, G., and W. J. Freeman. "Odor-Related Bulbar EEG Spatial Pattern Analysis During Appetitive Conditioning in Rabbits." *Behav. Neurosci.* **99(5)** (1985): 964–978.
165. von Krosigk, M., T. Bal, and D. A. McCormick. "Cellular Mechanisms of a Synchronized Oscillation in the Thalamus." *Science* **261** (1993): 361–364.
166. Willey, T. J. "The Ultrastructure of the Cat Olfactory Bulb." *J. Comp. Neurol.* **152** (1973): 211–232.

167. Wilson, M., and J. M. Bower. "Cortical Oscillations and Temporal Interactions in a Computer Simulation of Piriform Cortex." *J. Neurophysiol.* **67(4)** (1992): 981–995.

168. Wilson, M. A., and B. L. McNaughton. "Dynamics of the Hippocampal Ensemble Code for Space." *Science* **261** (1993): 1055–1058.

169. Wilson, H. R., and J. D. Cowan. "Excitatory and Inhibitory Interactions in Localized Populations of Model Neurons." *Biophys. J.* **12** (1972): 1–24.

170. Ylinen, A., A. Sik, A. Bragin, G. Jando, and G. Buzsaki. "Intracellular Correlates of Hippocampal Sharp Wave Bursts *In Vivo*." *Soc. Neurosci. Abs.* **19** (1993): 148.4.

171. Young, M. P., K. Tanaka, and S. Yamane. "On Oscillating Neuronal Responses in the Visual Cortex of the Monkey." *J. Neurophysiol.* **67(6)** (1992): 1464–1474.

Bernardo A. Huberman and Tad Hogg
Xerox Palo Alto Research Center, Palo Alto, California 94304 USA

The Emergence of Computational Ecologies

This chapter originally appeared in *1992 Lectures in Complex Systems*, edited by L. Nadel and D. Stein, 185–205. Santa Fe Institute Studies in the Sciences of Complexity, Lect. Vol. V. Reading, MA: Addison-Wesley, 1993. Reprinted by permission.

We describe a form of distributed computation in which agents have incomplete knowledge and imperfect information on the state of the system, and an instantiation of such systems based on market mechanisms. When agents can choose among several resources, the dynamics of the system can be oscillatory and even chaotic. A mechanism is described for achieving global stability through local controls.

1. INTRODUCTION

Propelled by advances in software design and increasing connectivity of computer networks, distributed computational systems are starting to spread throughout offices, laboratories, countries, and continents. In these systems, computational processes consisting of the active execution of programs can spawn new ones in other machines as they make use of printers, file servers, and other machines of the network as the need arises. In the most complex applications, various processes can collaborate to solve problems, while competing for the available computational resources, and may also directly interact with the physical world. This contrasts with the more familiar stand-alone computers, with traditional methods of centralized scheduling for resource allocation and programming methods based on serial /hbox-processing.

The effective use of distributed computation is a challenging task, since the processes must obtain resources in a dynamically changing environment and must be designed to collaborate despite a variety of asynchronous and unpredictable changes. For instance, the lack of global perspectives for determining resource allocation requires a very different approach to system-level programming and the creation of suitable languages. Even implementing reliable methods whereby processes can compute in machines with diverse characteristics is difficult.

As these distributed systems grow, they become a community of concurrent processes, or a *computational ecosystem*,[5] which, in their interactions, strategies, and lack of perfect knowledge, are analogous to biological ecosystems and human economies. Since all of these systems consist of a large number of independent actors competing for resources, this analogy can suggest new ways to design and understand the behavior of these emerging computational systems. In particular, these existing systems have methods to deal successfully with coordinating asynchronous operations in the face of imperfect knowledge. These methods allow the system as a whole to adapt to changes in the environment or disturbances to individual members, in marked contrast to the brittle nature of most current computer programs which often fail completely if there is even a small change in their inputs or an error in the program itself. To improve the reliability and usefulness of distributed computation, it is therefore interesting to examine the extent to which this analogy can be exploited.

Based on the law of large numbers, statistical mechanics has taught us that many universal and generic features of large systems can be quantitatively understood as approximations to the average behavior of infinite systems. Although such infinite models can be difficult to solve in detail, their overall qualitative features can be determined with a surprising degree of accuracy. Since these features are universal in character and depend only on a few general properties of the system, they can be expected to apply to a wide range of actual configurations. This is the case when the number of relevant degrees of freedom in the system, as well as the number of interesting parameters, is small. In this situation, it becomes useful to treat the unspecified internal degrees of freedom as if they are given by a probability distribution. This implies assuming a lack of correlations between the unspecified and specified degrees of freedom. This assumption has been extremely successful in statistical mechanics. It implies that although degrees of freedom may change according to purely deterministic algorithms, because they are unspecified, they appear to be effectively random to an outside observer.

Consider, for instance, massively parallel systems which are desired to be robust and adaptable. They should work in the presence of unexpected errors and with changes in the environment in which they are embedded (i.e., fail soft). This implies that many of the system's internal degrees of freedom will be adjustable by taking on a range of possible configurations. Furthermore, their large size will necessarily enforce a perspective that concentrates on a few relevant variables. Although these considerations suggest that the assumptions necessary for a statistical description hold for these systems, experiments will be necessary for deciding their applicability.

While computational and biological ecosystems share a number of features, we should also note there are a number of important differences. For instance, in contrast to biological individuals, computational agents are programmed to complete their tasks as soon as possible, which in turn implies a desirability for their earliest death. This task completion may also involve terminating other processes spawned to work on different aspects of the same problem, as in parallel search, where the first process to find a solution terminates the others. This rapid turnover of agents can be expected to lead to dynamics at much shorter time scales than seen in biological or economic counterparts.

Another interesting difference between biological and computational ecologies lies in the fact that for the latter the local rules (or programs for the processes) can be arbitrarily defined, whereas in biology those rules are quite fixed. Moreover, in distributed computational systems the interactions are not constrained by a Euclidean metric, so that processes separated by large physical distances can strongly affect each other by passing messages of arbitrary complexity between them. And last but not least, in computational ecologies the rationality assumption of game theory can be explicitly imposed on their agents, thereby making these systems amenable to game dynamic analyses suitably adjusted for their intrinsic characteristics. On the other hand, computational agents are considerably less sophisticated in their decision-making capacity than people, which could prevent expectations based on observed human performance from being realized.

By now there are a number of distributed computational systems that exhibit many of the above characteristics and offer increased performance when compared with traditional operating systems. *Enterprise*[8] is a marketlike scheduler in which independent processes or agents are allocated at run time among remote idle workstations through a bidding mechanism. A more evolved system, *Spawn*,[12] is organized as a market economy composed of interacting buyers and sellers. The commodities in this economy are computer-processing resources, specifically, slices of CPU time on various types of computers in a distributed computational environment. The system has been shown to provide substantial improvements over more conventional systems, while providing dynamic response to changes and resource sharing.

From a scientific point of view, the analogy between distributed computation and natural ecologies brings to mind the spontaneous appearance of organized behavior in biological and social systems, where agents can engage in cooperating strategies while working on the solution of particular problems. In some cases, the strategy mix used by these agents evolves toward an asymptotic ratio that is constant in time and stable against perturbations. This phenomenon sometimes goes under the name of evolutionarily stable strategy (ESS). Recently, it has been shown that spontaneous organization can also exist in open computational systems when agents can choose among many possible strategies while collaborating in the solution of computational tasks. In this case, however, imperfect knowledge and delays in information introduce asymptotic oscillatory and chaotic states that exclude the existence of simple ESS's. This is an important finding in light of studies which resort to notions of evolutionarily stable strategies in the design and prediction of an open system's performance.

In what follows we will describe a market-based computational ecosystem and a theory of distributed computation. The theory describes the collective dynamics of computational agents, while incorporating many of the features endemic to such systems, including distributed control, asynchrony, resource contention, and extensive communication among agents. When processes can choose among many possible strategies while collaborating in the solution of computational tasks, the dynamics leads to asymptotic regimes characterized by complex attractors. Detailed experiments have confirmed many of the theoretical predictions while uncovering new phenomena, such as chaos induced by overly clever decision-making procedures.

Next, we deal with the problem of controlling chaos in such systems, for we have discovered ways of achieving global stability through local controls inspired by fitness mechanisms found in nature. Furthermore, we show how diversity enters into the picture, along with the minimal amount of such diversity that is required to achieve stable behavior in a distributed computational system.

2. COMPUTATIONAL MARKETS FOR RESOURCE ALLOCATION

Allocating resources to competing tasks is one of the key issues for making effective use of computer networks. Examples include deciding whether to run a task in parallel on many machines or serially on one, and whether to save intermediate results or recompute them as needed. The similarity of this problem to resource allocation in market economies has prompted considerable interest in using analogous techniques to schedule tasks in a network environment. In effect, a coordinated solution to the allocation problem is obtained using Adam Smith's "invisible hand."[10] Although unlikely to produce the same optimal allocation made by an omniscient controller with unlimited computational capability, it can perform well compared to other feasible alternatives.[1,7] As in economics,[3] the use of prices provides a flexible mechanism for allocating resources, with relatively low information requirements: a single price summarizes the current demand for each resource, whether processor time, memory, communication bandwidth, use of a database, or control of a particular sensor. This flexibility is especially desirable when resource preferences and performance measures differ among tasks. For instance, an intensive numerical simulation's need for fast floating-point hardware is quite different from an interactive text editor's requirement for rapid response to user commands or a database's search requirement for rapid access to the data and fast query matching.

As a conceptual example of how this could work in a computational setting, suppose that a number of database search tasks are using networked computers to find items of interest to various users. Furthermore, suppose that some of the machines have fast floating-point hardware but are otherwise identical. Assuming the search tasks make little use of floating-point operations, their performance will not depend on whether they run on a machine with fast floating-point hardware. In a market-based system, these programs will tend to value each machine based on how many other tasks it is running, leading to a uniform load on the machines. Now suppose some floating-point intensive tasks arrive in the system. These will definitely prefer the specialized machines and consequently bid up the price of those particular resources. Observing that the price for some machines has gone up, the database tasks will then tend to migrate toward those machines without the fast floating-point hardware. Importantly, because of the high cost of modifying large existing programs, the database tasks will not need to be rewritten to adjust for the presence of the new tasks. Similarly, there is no need to reprogram the scheduling method of a traditional central controller, which is often very time consuming.

This example illustrates how a reasonable allocation of resources could be brought about by simply having the tasks be sensitive to current resource price. Moreover, adjustments can take place continually as new uses are found for particular network resources (which could include specialized databases or proprietary algorithms as well as the more obvious hardware resources) that do not require all users to agree on, or even know about, these new uses thus encouraging an incremental and experimental approach to resource allocation.

While this example motivates the use of market-based resource allocation, a study of actual implementations is required to see how large the system must be for its benefits to appear and whether any of the differences between simple computer programs and human agents pose additional problems. In particular, a successful use of markets requires a number of changes to traditional computer systems. First, the system must provide an easily accessible, reliable market so that buyers and sellers can quickly find each other. Second, individual programs must be price sensitive so they will respond to changes in relative prices among resources. This implies that the programs must, in some sense at least, be able to make choices among various resources based on how well suited they are for the task at hand.

A number of marketlike systems have been implemented over the years.[8,11,12] Most instances focus on finding an appropriate machine for running a single task. While this is important, further flexibility is provided by systems that use market mechanisms to also manage a collection of parallel processes contributing to the solution of a single task. In this latter case, prices give a flexible method for allocating resources among multiple competing heuristics for the same problem based on their perceived progress. Thus it greatly simplifies the development of programs that adjust to unpredictable changes in resource demand or availability. So we have a second reason to consider markets: not only may they be useful for flexible allocation of computational resources among competing tasks, but the simplicity of the price mechanism could provide help with designing cooperative parallel programs.

One such system is Spawn,[12] in which each task, starting with a certain amount of money corresponding to its relative priority, bids for the use of machines on the network. In this way, each task can allocate its budget toward those resources most important for it. In addition, when prices are low enough, some tasks can split into several parts which run in parallel, as shown in Figure 1, thereby adjusting the number of machines devoted to each task based on the demand from other users. From a user's point of view, starting a task with the Spawn system amounts to giving a command to execute it and the necessary funding for it to buy resources.

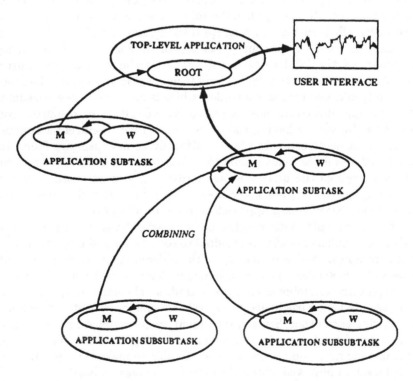

FIGURE 1 Managing parallel execution of subtasks in Spawn. Worker processes (W) report progress to their local managers (M) who in turn make reports to the next higher level of management. Upper management combines data into aggregate reports. Finally, the root manager presents results to the user. Managers also bid for the use of additional machines and, if successful, spawn additional subtasks on them.

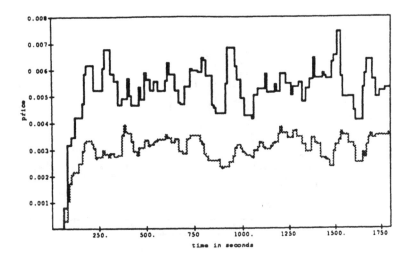

FIGURE 2   Price as a function of time (in seconds) in an inhomogeneous Spawn network consisting of three Sun 4/260's and six Sun 4/110's running four independent tasks. The average price of the 260's is in black, the less powerful 110's in gray.

The Spawn system manages auctions on each of the participating machines and the use of resources by each participating task, and provides communication paths among the spawned processes. It remains for the programmer to determine the specific algorithms to be used and the meaningful subtasks into which to partition the problem. That is, the Spawn system provides the price information and a market, but the individual programs must be written to make their own price decisions to effectively participate in the market. To allow existing, nonprice-sensitive programs to run within the Spawn system without modification, we provided a simple default manager that simply attempted to buy time on a single machine for that task. Users could then gradually modify this manager for their particular task, if desired, to spawn subtasks or to use market strategies more appropriate for the particular task.

Studies with this system show that an equilibrium price can be meaningfully defined with even a few machines participating. A specific instance is shown in Figure 2. Despite the continuing fluctuations, this small network reaches a rough price equilibrium. Moreover, the ratio of prices between the two machines closely matches their relative speeds, which was the only important difference between the two types of machine for these tasks. An additional experiment studied a network with some lengthy, low-priority tasks to which was added a short, high-priority task. The new task rapidly expands throughout the network by outbidding the existing tasks and driving the price of CPU time up, as shown in Figure 3. It is able therefore

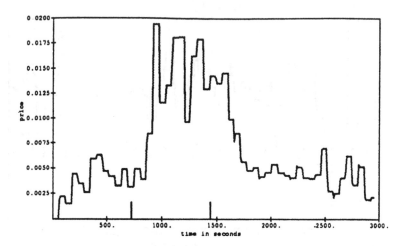

FIGURE 3 Price as a function of time (in seconds) when a high-priority task is introduced into a Spawn network running low-priority jobs. The first vertical line segment on the time axis marks the introduction of the high-priority task, and the second one the termination of its funding.

to utilize briefly a large number of networked machines and to illustrate the inherent flexibility of market-based resource allocation. Although the very small networks used in these experiments could be adequately managed centrally, these results do show that expected market behavior can emerge even in small cases.

Computer market systems can be used to experimentally address a number of additional issues. For instance, they can help in understanding what happens when more sophisticated programs begin to use the network, e.g., processes that attempt to anticipate future loads so as to maximize their own resource usage. Such behavior can destabilize the overall system. Another area of interest is the emergence of diversity or specialization from a group of initially similar machines. For example, a machine might cache some of the routines or data commonly used by its processes, giving it a comparative advantage in bids for similar tasks in the future. Ultimately this could result in complex organizational structures embedded within a larger market framework.[9] Within these groups, some machines could keep track of the kinds of problems for which others perform best and use this information to guide new tasks to appropriate machines. In this way the system could gradually learn to perform common tasks more effectively.

These experiments also highlighted a number of more immediate practical issues. In setting up Spawn, it was necessary to find individuals willing to allow their machines to be part of the market. While it would seem simple enough to do so, in practice a number of incentives were needed to overcome the natural reluctance of people to have other tasks running on their machines. This reluctance is partly based on perceived limitations on the security of the network and the individual operating systems, for it was possible that a remote procedure could crash an individual machine or consume more resources than anticipated. In particular, users with little need for computer-intensive tasks saw little benefit from participating since they had no use for the money collected by their machines. This indicates the need to use real money in such situations so that these users could use their revenues for their own needs. This in turn brings the issue of computer security to the forefront so users will feel confident that no counterfeiting of money takes place and tasks in fact will be limited to use only resources they have paid for.

Similarly, for those users participating in the system as buyers, they need to have some idea of what amount of money is appropriate to give a task. In a fully developed market, there could easily be tools to monitor the results of various auctions and, hence, give a current market price for resources. However, when using a newly created market with only a few users, tools are not always available to give easy access to prices and, even if they are, the prices have large fluctuations. Effective use of such a system also requires users to have some idea of what resources are required for their programs or, better yet, to encode that information in the program itself so it will be able to respond to available resources—e.g., by spawning subtasks—more rapidly than the users can. Conversely, there must be a mechanism whereby sellers can make available information about the characteristics of their resources (e.g., clock speed, available disk space, or special hardware). This can eventually allow for more complex market mechanisms, such as auctions that attempt to sell simultaneous use of different resources (e.g., CPU time and fast memory) or future use of currently unavailable resources to give tasks a more predictable use of resources. Developing and evaluating a variety of auction and price mechanisms that are particularly well suited to these computational tasks is an interesting open problem.

Finally, these experimental systems help clarify the differences between human and computer markets. For instance, computational processes can respond to events much more rapidly than people, but they are far less sophisticated. Moreover, unlike the situation with people, particular incentive structures, rationality assumptions, etc. can be explicitly built into computational processes, allowing for the possibility of designing particular market mechanisms. This could lead to the ironic situation in which economic theory has greater predictability for the behavior of computational markets than for that of the larger, and more complex, human economy.

3. CHAOS IN COMPUTATIONAL ECOSYSTEMS

The systems we have been discussing are basically made up of simple agents with fast response times, compared to human agents in more complex and slower economic settings. This implies that an understanding of the behavior of computational ecosystems requires focusing on the dynamics of collections of agents capable of a set of simple decisions.

Since decisions in a computational ecosystem are not centrally controlled, agents independently and asynchronously select among the available choices based on their perceived payoff. These payoffs are actual computational measures of performance, such as the time required to complete a task, accuracy of the solution, amount of memory required, etc. In general, the payoff G_r for using resource r depends on the number of agents already using it. In a purely competitive environment, the payoff for using a particular resource tends to decrease as more agents make use of it. Alternatively, the agents using a resource could assist one another in their computations, as might be the case if the overall task could be decomposed into a number of subtasks. If these subtasks communicate extensively to share partial results, the agents will be better off using the same computer rather than running more rapidly on separate machines and then being limited by slow communications. As another example, agents using a particular database could leave index links that are useful to others. In such cooperative situations, the payoff of a resource then would increase as more agents use it, until it became sufficiently crowded.

Imperfect information about the state of the system causes each agent's perceived payoff to differ from the actual value, with the difference increasing when there is more uncertainty in the information available to the agents. This type of uncertainty concisely captures the effect of many sources of errors such as some program bugs, heuristics incorrectly evaluating choices, errors in communicating the load on various machines, and mistakes in interpreting sensory data. Specifically, the perceived payoffs are taken to be normally distributed, with standard deviation σ, around their correct values. In addition, information delays cause each agent's knowledge of the state of the system to be somewhat out of date. Although for simplicity we will consider the case in which all agents have the same effective delay, uncertainty, and preferences for resource use, we should mention that the same range of behaviors is also found in more general situations.[4]

As a specific illustration of this approach, we consider the case of two resources, so the system can be described by the fraction f of agents which are using resource 1 at any given time. Its dynamics is then governed by[5]

$$\frac{df}{dt} = \alpha(\rho - f) \qquad (1)$$

where α is the rate at which agents reevaluate their resource choice and ρ is the probability that an agent will prefer resource 1 over 2 when it makes a choice. Generally, ρ is a function of f through the density-dependent payoffs. In terms of the payoffs and uncertainty, we have

$$\rho = \frac{1}{2}\left(1 + \operatorname{erf}\left(\frac{G_1(f) - G_2(f)}{2\sigma}\right)\right) \qquad (2)$$

where σ quantifies the uncertainty. Notice that this definition captures the simple requirement that an agent is more likely to prefer a resource when its payoff is relatively large. Finally, delays in information are modeled by supposing that the payoffs that enter into ρ at time t are the values they had at a delayed time $t - \tau$.

For a typical system of many agents with a mixture of cooperative and competitive payoffs, the kinds of dynamical behaviors exhibited by the model are shown in Figure 4. When the delays and uncertainty are fairly small, the system converges to an equilibrium point close to the optimal obtainable by an omniscient, central controller. As the information available to the agents becomes more corrupted, the equilibrium point moves further from the optimal value. With increasing delays,

(a)

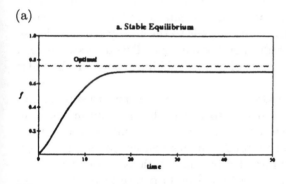

(b)

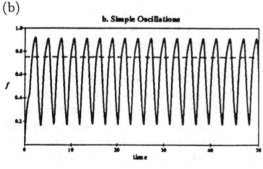

(c)

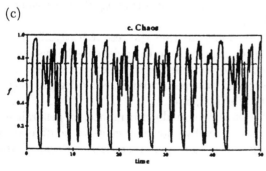

FIGURE 4 Typical behaviors for the fraction f of agents using resource 1 as a function of time for successively longer delays: (a) relaxation toward stable equilibrium, (b) simple persistent oscillations, and (c) chaotic oscillations. The payoffs are $G_1 = 4 + 7f - 5.333f^2$ for resource 1 and $G_2 = 4 + 3f$ for resource 2. The time scale is in units of the delay time τ, $\sigma = 1/4$, and the dashed line shows the optimal allocation for these payoffs.

the equilibrium eventually becomes unstable, leading to the oscillatory and chaotic behavior shown in the figure. In these cases, the number of agents using particular resources continues to vary so that the system spends relatively little time near the optimal value, with a consequent drop in its overall performance. This can be due to the fact that chaotic systems are unpredictable, hence making it difficult for individual agents to automatically select the best resources at any given time.

4. THE USES OF FITNESS

We will now describe an effective procedure for controlling chaos in distributed systems.[4] It is based on a mechanism that rewards agents according to their actual performance. As we shall see, such an algorithm leads to the emergence of a diverse community of agents out of an essentially homogenous one. This diversity in turn eliminates chaotic behavior through a series of dynamical bifurcations which render chaos a transient phenomenon.

The actual performance of computational processes can be rewarded in a number of ways. A particularly appealing one is to mimic the mechanism found in biological evolution, where fitness determines the number of survivors of a given species in a changing environment. In computation this mechanism is called a *genetic algorithm*.[2] Another example is provided by computational systems modeled on ideal economic markets,[9,12] which reward good performance in terms of profits. In this case, agents pay for the use of resources, and they in turn are paid for completing their tasks. Those making the best choices collect the most currency and are able to outbid others for the use of resources. Consequently they come to dominate the system.

While there is a range of possible reward mechanisms, their net effect is to increase the proportion of agents that are performing successfully, thereby decreasing the number of those who are less successful. It is with this insight in mind that we developed a general theory of effective reward mechanisms without resorting to the details of their implementations. Since this change in agent mix in turn will change the choices made by every agent and their payoffs, those that were initially most successful need not be so in the future. This leads to an evolving diversity whose eventual stability is by no means obvious.

Before proceeding with the theory, we point out that the resource payoffs that we will consider are instantaneous ones (i.e., shorter than the delays in the system), e.g., work actually done by a machine, currency actually received, etc. Other reward mechanisms, such as those based on averaged past performance, could lead to very different behavior from the one exhibited in this paper.

In order to investigate the effects of rewarding actual performance, we generalize the previous model of computational ecosystems by allowing agents to be different types, a fact which gives them different performance characteristics. Recall that the agents need to estimate the current state of the system based on imperfect and delayed information in order to make good choices. This can be done in a number of ways, ranging from extremely simple extrapolations from previous data to complex forecasting techniques. The different types of agents then correspond to the various ways in which they can make these extrapolations.

Within this context, a computational ecosystem can be described by specifying the fraction of agents, f_{rs} of a given type s using a given resource r at a particular time. We will also define the total fraction of agents using a resource of a particular type as

$$f_r^{\text{res}} = \sum_s f_{rs}$$

$$f_s^{\text{type}} = \sum_r f_{rs} \qquad (3)$$

respectively.

As mentioned previously, the net effect of rewarding performance is to increase the fraction of highly performing agents. If γ is the rate at which performance is rewarded, then Eq. (1) is enhanced with an extra term which corresponds to this reward mechanism. This gives

$$\frac{df_{rs}}{dt} = \alpha(f_s^{\text{type}}\rho_{rs} - f_{rs}) + \gamma(f_r^{\text{res}}\eta_s - f_{rs}) \qquad (4)$$

where the first term is analogous to that of the previous theory and the second term incorporates the effect of rewards on the population. In this equation, ρ_{rs} is the probability that an agent of type s will prefer resource r when it makes a choice and η_s is the probability that new agents will be of type s, which we take to be proportional to the actual payoff associated with agents of type s. As before, α denotes the rate at which agents make resource choices and the detailed interpretation of γ depends on the particular reward mechanism involved. For example, if they are replaced on the basis of their fitness, it is the rate at which this happens. On the other hand, in a market system, γ corresponds to the rate at which agents are paid. Notice that in this case, the fraction of each type is proportional to the wealth of agents of that type.

Since the total fraction of agents of all types must be one, a simple form of the normalization condition can be obtained if one considers the relative payoff, which is given by

$$\eta_s = \frac{\sum_r f_{rs}G_r}{\sum_r f_r^{\text{res}}G_r}. \qquad (5)$$

Note that the numerator is the actual payoff received by agents of type s given their current resource usage and the denominator is the total payoff for all agents in the system, both normalized to the total number of agents in the system. This form assumes positive payoffs: e.g., they could be growth rates. If the payoffs can be negative (e.g., they are currency changes in an economic system), one can use instead the difference between the actual payoffs and their minimum value m. Since the η_s must sum to 1, this will give

$$\eta_s = \frac{\sum_r f_{rs} G_r - m}{\sum_r f_r^{\text{res}} G_r - Sm} \tag{6}$$

where S is the number of agent types, and which reduces to the previous case when $m = 0$.

Summing Eq. (4) over all resources and types gives

$$\frac{df_r^{\text{res}}}{dt} = \alpha \left(\sum_s f_s^{\text{type}} \rho_{rs} - f_r^{\text{res}} \right)$$

$$\frac{df_s^{\text{type}}}{dt} = \gamma \left(\eta_s - f_s^{\text{type}} \right) \tag{7}$$

which describe the dynamics of overall resource use and the distribution of agent types, respectively. Note that this implies that those agent types which receive greater than average payoff (i.e., types for which $\eta_s > f_s^{\text{type}}$ will increase in the system at the expense of the low performing types).

Note that the actual payoffs can only reward existing types of agents. Thus, in order to introduce new variations into the population, an additional mechanism is needed (e.g., corresponding to mutation in genetic algorithms or learning).

5. RESULTS

In order to illustrate the effectiveness of rewarding actual payoffs in controlling chaos, we examine the dynamics generated by Eq. (4) for the case in which agents choose among two resources with cooperative payoffs, a case which, as we have shown, generates chaotic behavior in the absence of rewards.[5,6] As in the particular example of Figure 4(c), we use $\tau = 10$; $G_1 = 4 + 7f_1 - 5.333f_1^2$; $G_2 = 7 - 3f_2$; $\sigma = 1/4$; and an initial condition in which all agents start by using resource 2.

One kind of diversity among agents is motivated by the simple case in which the system oscillates with a fixed period. In this case, those agents that are able to discover the period of the oscillation can then use this knowledge to reliably estimate the current system state in spite of delays in information. Notice that this estimate does not necessarily guarantee that they will keep performing well in the future, for their choice can change the basic frequency of oscillation of the system.

In what follows, the diversity of agent types corresponds to the different past horizons, or extra delays, that they use to extrapolate to the current state of the system. These differences in estimation could be due to the variety of procedures for analyzing the system's behavior. Specifically, we identify different agent types with the different assumed periods that range over a given interval. Thus, agents of type s use an effective delay of $\tau + s$ while evaluating their choices.

The resulting behavior, shown in Figure 5, should be contrasted with Figure 4(c). We used an interval of extra delays ranging from 0 to 40. As shown, the introduction of actual payoffs induces a chaotic transient that, after a series of dynamical bifurcations, settles into a fixed point that signals stable behavior. Furthermore, this fixed point is exactly that obtained in the case of no delays. This equilibrium is stable against perturbations because, if the system were perturbed again (as shown in Figure 6), it rapidly returns to its previous value. In additional

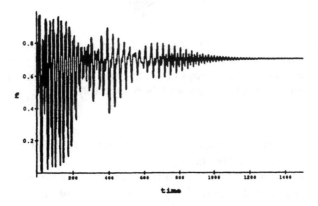

FIGURE 5 Fraction of agents using resource 1 as a function of time with adjustment based on actual payoff. These parameters correspond to Figure 4(c), so without the adjustment, the system would remain chaotic.

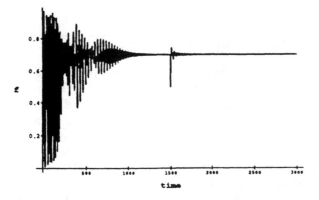

FIGURE 6 Behavior of the system shown in Figure 5 with a perturbation introduced at time 1500.

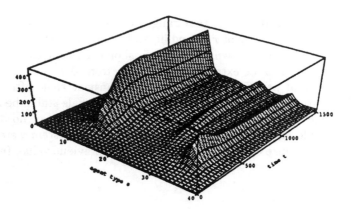

FIGURE 7   Ratio $f_s^{\text{type}}(t)/f_s^{\text{type}}(0)$ of the fraction of agents of each type, normalized to their initial values, as a function of time. Note there are several peaks, which correspond to agents with extra delays of 12, 26, and 34 time units. Since $\tau = 10$, these match periods of length 22, 36, and 44 respectively.

experiments, with a smaller range of delays, we found that the system continued to oscillate without achieving the fixed point.

This transient chaos and its eventual stability can be understood from the distribution of agents with extra delays as a function of time. As can be seen in Figure 7, actual payoffs lead to a highly heterogeneous system characterized by a diverse population of agents of different types. It also shows that the fraction of agents with certain extra delays increases greatly. These delays correspond to the major periodicities in the system.

6. STABILITY AND MINIMAL DIVERSITY

As we showed in the previous section, rewarding the performance of large collections of agents engaging in resource choices leads to a highly diverse mix of agents that stabilize the system. This suggests that the real cause of stability in a distributed system is sufficient diversity, and the reward mechanism is an efficient way of automatically finding a good mix. This raises the interesting question of the minimal amount of diversity needed in order to have a stable system.

The stability of a system is determined by the behavior of a perturbation around equilibrium, which can be found from the linearized version of Eq. (4). In our case, the diversity is related to the range of different delays that agents can have. For a continuous distribution of extra delays, the characteristic equation is obtained by assuming a solution of the type $e^{\lambda t}$ in the linearized equation, giving

$$\lambda + \alpha - \alpha\rho' \int ds f(s) e^{-\lambda(s+r)} = 0. \tag{8}$$

Stability requires that all the values of λ have negative real parts, so that perturbations will relax back to equilibrium. As an example, suppose agent types are uniformly distributed in $(0, S)$. Then $f(s) = 1/S$, and the characteristic equation becomes

$$\lambda + \alpha - \alpha\rho' \frac{1 - e^{-\lambda S}}{\lambda S} e^{-\lambda r} = 0. \tag{9}$$

Defining a normalized measure of the diversity of the system for this case by $\eta \equiv S/\tau$, introducing the new variable $z \equiv \lambda\tau(1 - \eta)$, and multiplying Eq. (9) by $\tau(1 + \eta)ze^z$ introduces an extra root at $z = 0$ and gives

$$(z^2 + az)e^z - b + be^{rz} = 0 \tag{10}$$

where

$$a = \alpha\tau(1 + \eta) > 0,$$
$$b = -\rho' \frac{\alpha\tau(1 + \eta)^2}{\eta} > 0, \tag{11}$$
$$r = \frac{\eta}{1 + \eta} \in (0, 1).$$

The stability of the system with a uniform distribution of agents with extra delays thus reduces to finding the condition under which all roots of Eq. (10), other than $z = 0$, have negative real parts. This equation is a particular instance of an *exponential polynomial*, having terms that consist of powers multiplied by exponentials. Unlike regular polynomials, these objects generally have an infinite number of roots and are important in the study of the stability properties of differential-delay equations. Established methods can be used then to determine when they have roots with positive real parts. This in turn defines the stability boundary of the equation. The result for the particular case in which $\rho' = -3.41044$, corresponding to the parameters used in Section 5, is shown in Figure 8(a).

Similarly, if we choose an exponential distribution of delays, i.e., $f(s) = 1/Se^{-s/S}$ with positive S, the characteristic equation acquires the form

$$(z^2 + pz + q)e^z + r = 0 \tag{12}$$

where

$$p = \alpha\tau + \frac{1}{\eta} > 0,$$
$$q = \frac{\alpha\tau}{\eta} > 0, \tag{13}$$
$$r = \frac{-\alpha\tau\rho'}{\eta} > 0,$$

(a)

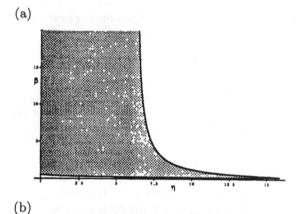

(b)

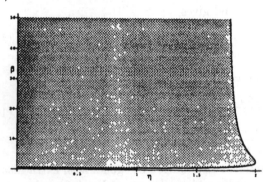

FIGURE 8   Stability as a function of $\beta = \alpha\tau$ and $\eta = S/\tau$ for two possible distributions of agent types: (a) $f(s) = 1/S$ in $(0, S)$ and (b) $f(s) = 1/Se^{-s/S}$. The system is unstable in the shaded regions and stable to the right and below the curves.

and $z \equiv \lambda\tau$. An analysis similar to that for the uniform distribution case leads to the stability diagram shown in Figure 8(b).

Although the actual distributions of agent types can differ from these two cases, the similarity between the stability diagrams suggests that, regardless of the magnitude of β, one can always find an appropriate mix that will make the system stable. This property follows from the vertical asymptote of the stability boundary. It also illustrates the need for a minimum diversity in the system to stablize it when the delays are not too small.

Having established the right mix that produces stability, one may wonder whether a static assignment of agent types at an initial time would not constitute a simpler and more direct procedure to stabilize the system without resorting to a dynamic reward mechanism. While this is indeed the case in a nonfluctuating environment, such a static mechanism cannot cope with changes in both the nature of the system (e.g., machines crashing) and the arrival of new tasks or fluctuating loads. A dynamic procedure is needed precisely to avoid this vulnerability by keeping the system adaptive.

Having seen how sufficient diversity stabilizes a distributed system, we now turn to the mechanisms that can generate such heterogeneity as well as the time that it takes for the system to stabilize. In particular, the details of the reward procedures determine whether the system can even find a stable mix of agents. In the cases described above, reward was proportional to actual performance, as measured by the payoffs associated with the resources used. One might also wonder whether stability would be achieved more rapidly by giving greater (than their fair share) increases to the top performers.

We have examined two such cases: (a) rewards proportional to the square of their actual performance and (b) one giving all the rewards to top performers (e.g., those performing at the ninetieth percentile or better in the population). In the former case we observed stability with a shorter transient whereas, in the latter case, the mix of agents continued to change through time, thus preventing stable behavior. This can be understood in terms of our earlier observation that, whereas a small percentage agents can identify oscillation periods and thereby reduce their amplitude, a large number of them no longer can perform well.

Note that the time to reach equilibrium is determined by two parameters of the system. The first is the time that it takes to find a stable mix of agent types, which is governed by γ, and the second is the rate at which perturbations relax, given the stable mix. The latter is determined by the largest real part of any of the roots, λ, of the characteristic equation.

7. DISCUSSION

In this paper we have presented a case for treating distributed computation as an ecosystem, an analogy that turns out to be quite fruitful in the analysis, design, and control of such systems. Resource contention, complex dynamics, and reward mechanisms seem to be ubiquitous in distributed computation, making it also a tool for the study of natural ecosystems in spite of the many differences between computational processes and organisms.

Since chaotic behavior seems to be the natural result of interacting processes with imperfect and delayed information, the problem of controlling such systems is of paramount importance. We discovered that rewards based on the actual performance of agents in a distributed computational system can stabilize an otherwise chaotic or oscillatory system. This leads in turn to greatly improved system performance.

In all these cases, stability is achieved by making chaos a transient phenomena. In the case of distributed systems, the addition of the reward mechanism has the effect of dynamically changing the control parameters of the resource allocation dynamics in such a way that a global fixed point of the system is achieved. This brings the issue of the length of the chaotic transient as compared to the time needed for most agents to complete their tasks. Even when the transients are long, the results of this study show that the range gradually decreases, thereby improving performance even before the fixed point is achieved.

A particularly relevant question for distributed systems is the extent to which these results generalize beyond the mechanism that we studied. We only considered the specific situation of a collection of agents with different delays in their appraisal of the system evolution. Hence it remains an open question whether using rewards to increase diversity works more generally than in the case of extra delays.

Since we only considered agents choosing between two resources, it is important to understand what happens when the agents have many resources to choose from. One may argue that since diversity is the key to stability, a plurality of resources provides enough channels to develop the necessary heterogeneity, which is what we observed in situations with three resources. Another note of caution: While we have shown that sufficient diversity can, on average, stabilize the system, in practice a fluctuation could wipe out those agent types that otherwise would be successful in stabilizing the system. Thus, we need either a large number of each kind of agent or a mechanism, such as mutation, to create new kinds of agents.

A final issue concerns the time scales over which rewards are assigned to agents. In our treatment, we assumed the rewards were always based on the performance at the time they were given. Since in many cases this procedure is delayed, there remains the question of the extent to which rewards based on past performance are also able to stabilize chaotic distributed systems.

The fact that these simple resource allocation mechanisms work and produce a stable environment provides a basis for developing more complex software systems that can be used for a wide range of computational problems.

REFERENCES

1. Ferguson, D., Y. Yemini, and C. Nikolaou. "Microeconomic Algorithms for Load Balancing in Distributed Computer Systems." In *International Conference on Distributed Computer Systems*, 491–499. Washington, DC: IEEE, 1988.
2. Goldberg, D. E. *Genetic Algorithms in Search, Optimization and Machine Learning*. Reading, MA: Addison-Wesley, 1989.
3. Hayek, F. A. "Competition as a Discovery Procedure." In *New Studies in Philosophy, Politics, Economics and the History of Ideas*, 179–190. Chicago: University of Chicago Press, 1978.
4. Hogg, T., and B. A. Huberman. "Controlling Chaos in Distributed Systems." *IEEE Trans. Sys., Man & Cyber.* **21(6)** (1991): 1325–1332.
5. Huberman, B. A., and T. Hogg. "The Behavior of Computational Ecologies." In *The Ecology of Computation*, edited by B. A. Huberman, 77–115. Amsterdam: North-Holland, 1988.
6. Kephart, J. O., T. Hogg, and B. A. Huberman. "Dynamics of Computational Ecosystems." *Phys. Rev. A* **40** (1989): 404–421.
7. Kurose, J. F., and R. Simha. "A Microeconomic Approach to Optimal Resource Allocation in Distributed Computer Systems." *IEEE Trans. Comp.* **38(5)** (1989): 705–717.
8. Malone, T. W., R. E. Fikes, K. R. Grant, and M. T. Howard. "Enterprise: A Market-Like Task Scheduler for Distributed Computing Environments." In *The Ecology of Computation*, edited by B. A. Huberman, 177–205. Amsterdam: North-Holland, 1988.
9. Miller, M. S., and K. E. Drexler. "Markets and Computation: Agoric Open Systems." In *The Ecology of Computation*, edited by B. A. Huberman, 133–176. Amsterdam: North-Holland, 1988.
10. Smith, A. *An Inquiry into the Nature and Causes of the Wealth of Nations*. Chicago: University of Chicago Press, 1976. Reprint of the 1776 edition.
11. Sutherland, I. E. "A Futures Market in Computer Time." *Comm. ACM* **11(6)** (1968): 449–451.
12. Waldspurger, C. A., T. Hogg, B. A. Huberman, J. O. Kephart, and W. S. Stornetta. "Spawn: A Distributed Computational Economy." *IEEE Trans. Software Engr.* **18(2)** (1992): 103–117.

REFERENCES

Erica Jen
Theoretical Division, Los Alamos National Laboratory, Los Alamos, NM 87545

Preimages and Forecasting for Cellular Automata

This chapter originally appeared in *1989 Lectures in Complex Systems*, edited by E. Jen, 371–388. Santa Fe Institute Studies in the Sciences of Complexity, Lect. Vol. II. Reading, MA: Addison-Wesley, 1990. Reprinted by permission.

Cellular automata are mathematical systems characterized by discreteness (in space, time, and state values), determinism, and local interaction. Few analytical techniques exist for such systems. The rule matrix and degree vectors of a cellular automaton—both of which are determined *a priori* from the function defining the automaton, rather than *a posteriori* from simulations of its evolution—are introduced here as tools for understanding certain qualitative features of automaton behavior. The rule matrix represents in convenient form the information contained in an automaton's rule table; the degree vectors are computed from the rule matrix and reflect the extent to which the system is "one-to-one" versus "many-to-one" on restricted subspaces of the mapping. The rule matrix and degree vectors determine, for example, several aspects of the enumeration and "prediction" of preimages for spatial sequences evolving under the rule, where the preimages of a sequence S are defined to be the set of sequences mapped by the automaton rule onto S.

INTRODUCTION

Cellular automata are a class of mathematical systems characterized by discreteness (in space, time, and state values), determinism, local interaction, and an inherently parallel form of evolution. As such, cellular automata are prototypical examples of discrete dynamical systems, and are of particular interest since they are simple in construction, and yet capable of generating complicated behavior.

There is a growing recognition that these simple systems are important both theoretically and practically. Theoretically, the distinctive characteristics of cellular automata pose a number of challenging problems in the analysis of discrete dynamical systems. Practically, as the evidence of lattice gases[1,2] indicates, cellular automata-based algorithms may provide alternative and efficient techniques for solving certain partial differential equations.

At present, few tools exist for the analysis of cellular automata behavior. Problems in cellular automata research pose special difficulties since they often fall outside the purview of traditional continuous mathematics; cellular automata problems typically reflect, in both their formulation and their solution, the features of discreteness and local interaction that make these systems distinctive. Aside from anything else, the evolution of a typical cellular automaton is governed not by an equation, such as, say, a partial differential equation, but by a "rule table" consisting of a list of the discrete states that occur in the automaton together with the values to which these states are to be mapped in one iteration of the rule. Not much of calculus applies to such systems.

One way to develop analytical insight into these systems is to restrict attention to simple cellular automata, and in this context, to identify interesting aspects of their behavior and to develop analytical techniques for understanding this behavior. The simplest cellular automata are those designated by Wolfram[12] as "elementary": defined on a one-dimensional spatial lattice, these automata consist of binary-valued sites that evolve in time according to a nearest-neighbor interaction rule. While higher-dimensional and/or multiple site-valued systems will of course differ in important respects, the phenomenology and analysis of one-dimensional systems can be expected to be fundamental in the understanding of cellular automata as a general class of mathematical systems.

Examples of questions that naturally arise in the study of elementary cellular automata include the complexity of the patterns they generate; the reversibility of their behavior; the dependence of behavior on boundary conditions; the number and size of their attractors; and the probability distribution of spatial sequences generated by these systems. (For discussion of these topics, see for example Hurd,[4] Jen,[7] Martin,[10] and Wolfram.[11,12]) This lecture focuses on problems related to probability distributions for elementary cellular automata—in particular, the enumeration and scaling of preimages for spatial sequences. The emphasis is on the development of matrix techniques for the analysis of these preimage problems.

The preimages of a sequence S are the set of sequences mapped by the rule onto S. By definition, the number of preimages of a sequence provides the frequency of its occurrence after one application of the automaton rule to a random initial condition. ("Random" implies that all finite tuples occur equally often in the initial condition.) The distribution of preimages thus is equivalent to the probability distribution obtained after one iteration of the rule applied to a uniform measure.

The analysis of preimages will depend upon tools (to be introduced here) known as the "rule matrix" and the "degree vectors" of an automaton. The rule matrix represents the rule table for the automaton in a convenient matrix form; increasing powers of the matrix capture the effect of evolution under the rule table for spatial sequences of increasing length. The matrix turns out to be convenient for clarifying the relationship between the detailed structure of the rule table and global properties of the sequences generated by the automaton rule. The degree vectors, which are constructed from the rule matrix, reflect the "dissipative" versus "conservative" nature of the rule.

Two applications of the rule matrix technique are considered here: first, using the matrix to obtain a formula for the number of preimages associated with any sequence S and thereby to understand certain qualitative features of cellular automata behavior; second, using the matrix to identify the amount of information on a sequence S needed to "predict" a specific feature that ostensibly depends upon the entire sequence. The feature of interest is also related to preimages, and is defined as the site value whose appendage to the given sequence S results in the minimization—or maximization of preimages for the new sequence.

RULE MATRIX OF AN ELEMENTARY CELLULAR AUTOMATON

An "elementary"[12] (that is, nearest-neighbor, binary site-valued) cellular automaton is defined by

$$x_i^{t+1} = f(x_{i-1}^t, x_i^t, x_{i+1}^t), \quad f : \{0,1\}^3 \to \{0,1\},$$

where x_i^t denotes the value of site i at time t, and f represents the "rule" defining the automaton.

Since the domain of f is the set of 2^3 possible 3-tuples, the rule function f is completely defined by specifying the "rule table" of values $a_i \in \{0,1\}$ with $i = 0, 1, \ldots, 7$ such that

$$000 \to a_0, \quad 001 \to a_1, \ldots, 111 \to a_7,$$

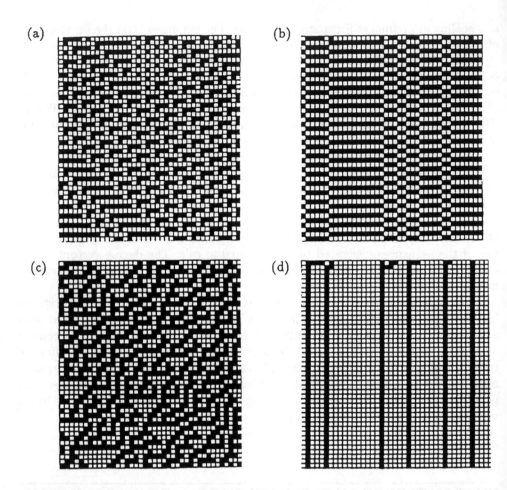

(a)   (b)

(c)   (d)

FIGURE 1 Examples of evolutions of one-dimensional "elementary" (nearest-neighbor, binary site-valued) cellular automata. Each case depicts 45 iterations of a given rule on a finite segment (40 sites) of an infinite lattice. Each row of dots represents the spatial sequence generated at time t, with the top row corresponding to $t = 0$. A black square represents a site value of 1; a white square represents a site value of 0.

(a) Rule 9: $\{001, 010, 100, 101, 110, 111\} \rightarrow 0, \{000, 011\} \rightarrow 1.$
(b) Rule 23: $\{011, 101, 110, 111\} \rightarrow 0, \{000, 001, 010, 100\} \rightarrow 1.$
(c) Rule 30: $\{000, 101, 110, 111\} \rightarrow 0, \{001, 010, 011, 100\} \rightarrow 1.$
(d) Rule 44: $\{000, 001, 100, 110, 111\} \rightarrow 0, \{010, 011, 101\} \rightarrow 1.$
(e) Rule 54: $\{000, 011, 110, 111\} \rightarrow 0, \{001, 010, 100, 101\} \rightarrow 1.$
(f) Rule 60: $\{000, 001, 110, 111\} \rightarrow 0, \{010, 011, 100, 101\} \rightarrow 1.$
(g) Rule 105: $\{001, 010, 100, 111\} \rightarrow 0, \{000, 011, 101, 110\} \rightarrow 1.$
(h) Rule 106: $\{000, 010, 100, 111\} \rightarrow 0, \{001, 011, 101, 110\} \rightarrow 1.$
(i) Rule 126: $\{000, 111\} \rightarrow 0, \{001, 010, 011, 100, 101, 110\} \rightarrow 1.$

(continued)

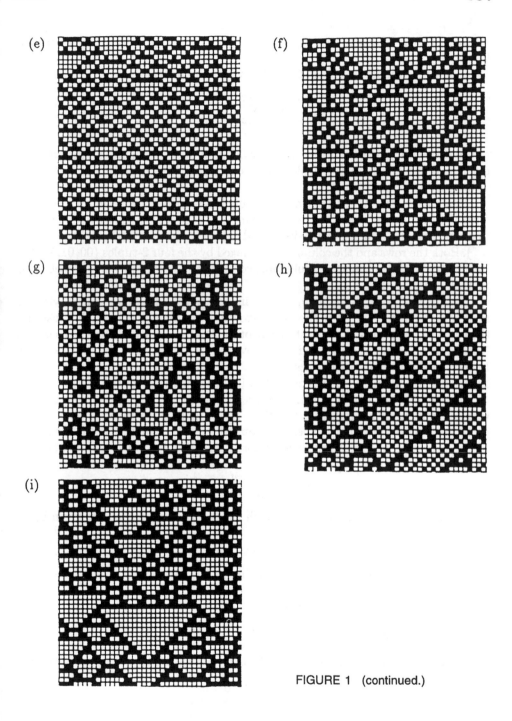

FIGURE 1   (continued.)

where $xyz \to a_i$ indicates that $f(xyz) = a_i$. There is a total of 256 distinct elementary rules.

The conventional labelling scheme[12] assigns the integer

$$R = \sum_{i=0}^{i=7} a_i 2^i$$

to the elementary rule defined by f. The rule number thus assumes an integer value between 0 and 255. Figure 1 depicts some examples of the evolution of elementary automata.

The *rule matrix* M for an elementary rule is designed to represent in convenient form the information contained in the rule table. The matrix is square of dimension 4 (for a general one-dimensional k-site valued rule of radius r, the dimension would be k^{2r}). Both the rows and the columns are indexed by the four 2-tuples $\{00, 01, 10, 11\}$. Each entry in the automaton's rule table corresponds to exactly one element in the matrix.

Specifically, each 3-tuple (y_1, y_2, y_3) in the rule table is associated with the matrix element m_{y_1,y_2,y_2,y_3} in the y_2y_3-th row and the y_2y_3-th column. For instance, the tuple 011 is associated with the element in the 01-th row and the 11-th column. Note that no tuple is associated with, say, the 01-th row and the 00-th column. In fact, of the 16 matrix elements, only half are used to represent information; the other half are "dummy" elements used to induce squareness in the matrix.

The actual element in the rule matrix corresponding to a given a_i is a formal variable similar to an indicator function. Its purpose is to indicate whether the tuple associated with that element is mapped to 0 (in which case the element is set to α) or to 1 (in which case it is set to β). Specifically, define the function $I(x)$ such that

$$I(x) = \begin{cases} \alpha, & \text{if } x = 0 ; \\ \beta, & \text{if } x = 1 . \end{cases}$$

Then the matrix elements associated with the 3-tuples (y_1, y_2, y_3) are given by

$$m_{y_1,y_2,y_2,y_3} = I(a_i),$$

where a_i is the number to which the rule maps the 3-tuple; all other elements are 0's.

For example, for the rule defined by

$$\{000, 111\} \to 0, \qquad \{001, 010, 011, 100, 101, 110\} \to 1 \tag{2.1}$$

(Rule 126 by the labelling convention defined above), the rule matrix is given by

$$M = \begin{pmatrix} \alpha & \beta & 0 & 0 \\ 0 & 0 & \beta & \beta \\ \beta & \beta & 0 & 0 \\ 0 & 0 & \beta & \alpha \end{pmatrix}. \tag{2.2}$$

(See Figure 1 for an example of evolution by the above rule.) The rule matrix M can be viewed as representing the action of the rule on sequences of length 3. The matrix M^n—with multiplication to be defined shortly—then represents the action of the rule on sequences of length $n + 2$ (the resulting sequences are of length n). Thus powers of the matrix are associated, not with temporal iterates of the rule, but with spatial extension of sequences. For any n, the elements M^n enumerate the sequences of length n obtainable from one iteration of the rule.

Matrix multiplication is assumed to be associative and non-commutative; that is, for example,

$$\alpha(\alpha\beta) = \alpha^2\beta \neq \alpha\beta\alpha.$$

Non-commutativity implies that ordering is important: left-multiplication by M corresponds to appending new elements to the left end of given sequences; the opposite is true for right-multiplication. Addition, on the other hand, is taken to be ordinary addition.

For the rule defined by Eq. (2.1),

$$M^2 = \begin{pmatrix} \alpha^2 & \alpha\beta & \beta^2 & \beta^2 \\ \beta^2 & \beta^2 & \beta^2 & \beta\alpha \\ \beta\alpha & \beta^2 & \beta^2 & \beta^2 \\ \beta^2 & \beta^2 & \alpha\beta & \alpha^2 \end{pmatrix}$$

and

$$M^3 = \begin{pmatrix} \alpha^3 + \beta^3 & \alpha^2\beta + \beta^3 & \alpha\beta^2 + \beta^3 & \alpha\beta^2 + \beta^2\alpha \\ \beta^2\alpha + \beta^3 & 2\beta^3 & \beta\alpha\beta + \beta^3 & \beta\alpha^2 + \beta^3 \\ \beta\alpha^2 + \beta^3 & \beta\alpha\beta + \beta^3 & 2\beta^3 & \beta^2\alpha + \beta^3 \\ \beta^2\alpha + \alpha\beta^2 & \alpha\beta^2 + \beta^3 & \alpha^2\beta + \beta^3 & \alpha^3 + \beta^3 \end{pmatrix}.$$

Examination of M^3, for example, indicates that of the sequences of length 5 beginning with 01 and ending with 00, one gives rise to 110 and the other to 111; of those beginning and ending with 01, both give rise to 111.

A variant of the rule matrix defined above has been used to study limit-cycle behavior of one-dimensional automata with periodic boundary conditions.[7] The "invariance" matrix used there has the same structure as the rule matrix here, but its elements serve as "indicators" of the validity of a certain invariance principle for their corresponding tuples. As is true for the rule matrix, higher powers of the invariance matrix provide information on spatial sequences of longer length.

PREIMAGE FORMULAE FOR ELEMENTARY CELLULAR AUTOMATA

Let S be a sequence of length N consisting of all 1's. Consider the problem of computing, for a given elementary automaton rule, the number $N(S)$ of preimages for S. (The general case in which S consists of 0's and 1's is treated in Jen.[8]) In this section the rule matrix of the rule is shown to be a convenient tool for computing $N(S)$.

In the reduced case of sequences of all 1's, the rule matrix defined in the previous section can be simplified by replacing β with 1 and α with 0. (This is possible since the only question is whether a sequence generates another sequence of all 1's; if it does not, then it is omitted from the counting.) For example, the matrix (2.2) can be rewritten in this case as

$$M_1 = \begin{pmatrix} 0 & 1 & 0 & 0 \\ 0 & 0 & 1 & 1 \\ 1 & 1 & 0 & 0 \\ 0 & 0 & 1 & 0 \end{pmatrix}. \tag{3.1}$$

The elements of M_1 "count" the number of sequences of length 3 that are mapped onto 1; similarly, the sum of the elements of M_1^n gives the number of sequences of length $n + 2$ that are mapped onto a sequence of n 1's.

It is clear that the rule matrix provides a technique for enumerating preimages for sequences of all 1's and can easily be revised to do the same for sequences of all 0's. The general case of an arbitrary sequence of 0's and 1's is solved, loosely speaking, by dividing the sequence into contiguous subsequences of 0's and 1's; computing the number of preimages for these simple subsequences; and then combining this information so as to obtain a formula giving the number of preimages for the entire sequence. For example, the preimage formula for any sequence S generated by the rule (2.1) is given by

$$N(S) = \begin{cases} 2F_{b_1+3}, & \text{if } m = 1 ; \\ 2F_{b_m+1}F_{b_1+1}\Pi_{i=2}^{m-1}F_{b_i-1} & \text{otherwise} . \end{cases} \tag{3.2}$$

Here m is the number of distinct blocks of 1's, the b_i's are the number of 1's in the ith block of 1's (the blocks are counted from right to left), and $F_k = F_{k-1} + F_{k-2}$ with $F_0 = 0, F_1 = 1$. (See Jen[8] for details and for a list of preimages formulae for all elementary rules.)

Perhaps the most interesting aspect of a rule matrix is not its use to compute the preimage formula, but what it reveals about qualitative properties of the automaton—including the qualitative form of the preimage formula; the automaton's generation of "gardens-of-Eden" (spatial sequences for which there exist no preimages); and the identification of sequences of maximal probability.

For example, exhaustive treatment of elementary rules indicates that their preimage formulae belong to one of six "types":

A. constant;

B. product of simple terms;

C. product of Fibonacci-like terms;

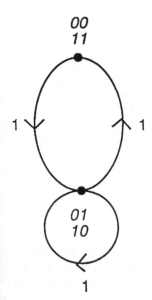

00
11

1   1

01
10

1

FIGURE 2 Graph of rule matrix M of Rule 126 defined by $\{000, 111\} \rightarrow 0, \{001, 010, 011, 100, 101, 110\} \rightarrow 1$.

D. telescoping Fibonacci-like terms;

E. alternating sequences; and

F. complicated recurrence terms.

The easiest way to see the connection between rule matrix and preimage formula type is to look at the graph associated with the matrix. This graph is constructed by assigning a node to each of the four possible 2-tuples 00, 01, 10, and 11; drawing a directed edge between each node x and y such that the (x, y)-th element in the rule matrix is not a "dummy" variable, and then labelling the edge with the (x, y)-th element. Again, the salient features of the problem can be understood by considering only the preimage formula for a sequence of n 1's. The graph corresponding to the matrix M_1 of Eq. (3.1) is depicted in Figure 2.

Suppose (as an exercise) that the problem is to determine the number of preimages $N(S)$ for a sequence of n 1's and that the appropriate rule matrices are associated with the graphs of Figure 3 (one of which is equivalent to the graph for matrix (3.1)). Suppose further that the preimages to be counted are only those starting with the components represented by x and ending with those represented by y, or, equivalently, that the quantity to be computed is the (x, y)-th element of M_1^n. Then the quantity of interest is given in each case by:

a. $N(S) = 1$

b. $N(S) = n$. An admissible path is constructed by traversing the loop $x - x$ for some $n_1 \geq 0$ steps, then following the edge $x - y$, and finally traversing the loop $y - y$ for exactly $n - (n_1 + 1)$ steps. There are n possible positions for the edge $x - y$ to occur in the path, and therefore n possible paths.

c. $N(S) = F_n$, where $F_n = F_{n-1} + F_{n-2}$ with $F_0 = F_1 = 1$. In this case, the number of paths of length n that begin at x and end at y is the sum of the number of paths of length $n - 1$ that end at y and can be lengthened by one traversal of the loop $y - y$, and the number of paths of length $n - 2$ that end at y and can be lengthened by one traversal of the circuit $y - z - y$. Thus the number of paths satisfies a recurrence relation of degree 2 with $F_0 = F_1 = 1$ (in fact, the Fibonacci relation).

d. $N(S) = F_{n-1}$, where $F_k = F_{k-1} + F_{n-3}$ with $F_0 = F_1 = F_2 = 1$. The reasoning is similar to that above.

e. $N(S) = 1$ if n is odd; $N(S) = 0$ if n is even.

The above examples can be taken as representative of the first five "types" of preimage formulae mentioned above. Clearly, the differences in preimage formula type reflect differences in graph-theoretic properties of the appropriate rule matrices.

The preimage formula associated with an automaton can be used to study a number of issues related to preimages and probability distributions. For example, when $N(S) = 0$, the sequence S is by definition a "garden-of-Eden." The preimage formula then can be used to investigate questions such as the existence, number, and detailed structure of gardens-of-Eden. In particular, the *type* of preimage formula provides immediate information in this context. For example, a preimage formula of type (A) or (B)—that is, either constant or a product of non-zero terms related to n, the length of the sequence—clearly corresponds to no garden-of-Eden. By contrast, the expression in Eq. (3.2) for $N(S)$ is easily shown to be 0 if 010 occurs in S.

Similarly, the preimage formula can easily be used to determine the sequences of maximal probability under one iteration of the rule. Again, the type of formula (and therefore the qualitative features of the rule matrix) provides important information. An expression such as Eq. (3.2), for example, is clearly maximized when the sequence consists of all 1's.

In summary, the rule matrix of an automaton can be used to compute a formula for the number of preimages $N(S)$ of an arbitrary sequence S. The form assumed by $N(S)$ is restricted to be one of a small number of possibilities, and the form of $N(S)$ for a given automaton is easily determined by examining properties of its matrix. Moreover the matrix properties determine qualitative features of the automaton's behavior, such as its generation of "gardens-of-Eden" and its sequences of maximal probability.

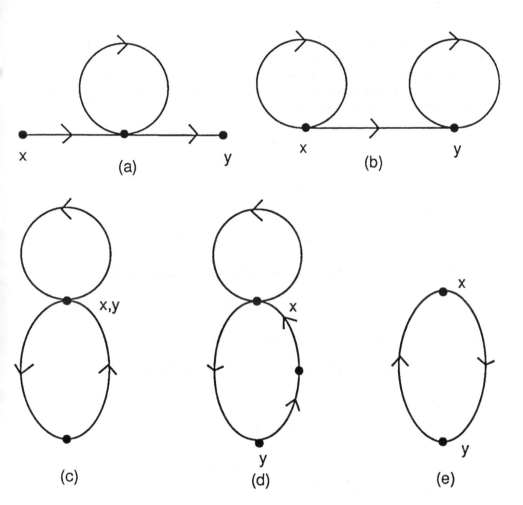

FIGURE 3 De Bruijn graphs for which the number of paths of length n beginning at node x and ending at node y are computed.

DEGREE VECTORS FOR ELEMENTARY RULES

This section uses the rule matrix to construct "degree vectors" for elementary cellular automata; the following section explores the implications of these vectors for a particular "forecasting" problem for spatial sequences generated by the rule.

Like the rule matrix, the degree vectors of a rule are an intrinsic feature of its rule table. The vectors are designed to reflect the extent to which the rule is "one-to-one" versus "many-to-one" on restricted subspaces of the mapping. The degree vectors of a rule are related to many of its fundamental dynamical features, including the size and number of its attractors.[5,9]

Given a rule matrix M, construct the *right degree vector* $\vec{\mu}$ by taking a row-wise sum of the elements, and construct the *left degree vector* $\vec{\nu}$ by taking a column-wise sum. Indexing the components as before,

$$\vec{\mu} \equiv (\mu_{00}, \mu_{01}, \mu_{10}, \mu_{11})$$

and

$$\vec{\nu} \equiv (\nu_{00}, \nu_{01}, \nu_{10}, \nu_{11})$$

are thus defined by

$$\mu_i = \sum_{j} m_{ij} \text{ and } \nu_j = \sum_{i} m_{ij}.$$

For example, the degree vectors for the rule (2.1) are given by

$$\vec{\mu} = \vec{\nu} = (\alpha + \beta, 2\beta, 2\beta, \alpha + \beta).$$

The rule matrix for Rule 44 (see Figure 1), defined by

$$\{000, 001, 100, 110, 111\} \to 0, \qquad \{010, 011, 101\} \to 1, \qquad (4.1)$$

is given by

$$M = \begin{pmatrix} \alpha & \alpha & 0 & 0 \\ 0 & 0 & \beta & \beta \\ \alpha & \beta & 0 & 0 \\ 0 & 0 & \alpha & \alpha \end{pmatrix}, \qquad (4.2)$$

and thus the vectors are found to be

$$\vec{\mu} = (2\alpha, 2\beta, \alpha + \beta, 2\alpha),$$
$$\vec{\nu} = (2\alpha, \alpha + \beta, \alpha + \beta, \alpha + \beta).$$

Note that the right (left) degree vector can be viewed as representing the effect of toggling the right (left) component of any 3-tuple, leaving the other two components fixed. The 00-th component of $\vec{\mu}$ for Rule 44, for example, indicates that no change results from toggling the rightmost component of any 3-tuple 00? (the ? indicates that the rightmost component is unspecified); both tuples 000 and 001 are mapped to 0. Similarly, the 10-th component indicates that toggling in 10? results in a toggled value of the mapping.

At times it will be convenient to use the *0-degree vectors* $\vec{\mu}_0 \vec{\nu}_0$ and the *1-degree vectors* $\vec{\mu}_1 \vec{\nu}_1$. These vectors are constructed from the vectors $\vec{\mu}, \vec{\nu}$ by retaining only the coefficients of the α (in the case of the 0-degree vectors) or the β (for the 1-degree vectors) terms. For example for Rule 44,

$$\bar{\mu}_0 = (2,0,1,2), \qquad \bar{\mu}_1 = (0,2,1,0),$$
$$\vec{\nu}_0 = (2,1,1,1), \qquad \vec{\nu}_1 = (0,1,1,1).$$

Note that the ij-th (ij as usual assuming the values $\{00,01,10,11\}$) component of, say, the left 0-degree vector will be

> 0,   if $\{0ij, 1ij\} \to 1$,
>
> 1,   if either $0ij$ or $1ij$ but not both $\to 1$,
>
> 2,   if $\{0ij, 1ij\} \to 0$;

and the ij-th component of the right 1-degree vector will be

> 0,   if $\{ij0, ij0\} \to 0$,
>
> 1,   if either $ij0$ or $1ij$ but not both $\to 0$,
>
> 2,   if $\{ij0, ij1\} \to 1$.

Similar relations hold for the other two vectors. (Clearly the 16 components of the vectors are not independent; it is easy to show that there are exactly six degrees of freedom.) Thus the degree vectors serve to capture the "1-to-1" versus "2-to-1" nature of the rule.

It follows from the definition of degree vectors that for any "additive" rule[10] (that is, a rule for which the function f is a linear combination of the site values and for which the superposition principle holds), the left and right 0- and 1-degree vectors are given by $(1,1,1,1)$. For a "left-toggle" rule[3,5] (that is, a rule such that a change in the leftmost component of any tuple effects a change in the value to which the rule maps the tuple), the left 0- and 1-degree vectors are given by $\vec{\nu}_0 = \vec{\nu}_1 = (1,1,1,1)$, but the right 0- and 1-degree vectors $\bar{\mu}_0$ and $\bar{\mu}$ will differ.

A FORECASTING PROBLEM FOR ELEMENTARY AUTOMATA

This section uses the degree vectors introduced above to solve a simple forecasting problem for cellular automata. For any string $S = s_1 \cdots s_n$, denote by $N(S)$, as before, the number of preimages of S. Consider the following problem: Append either 0 or 1 to the left and right ends of the sequence so as to minimize (or maximize) the number of preimages of the new extended sequence S^*.

This problem, a much simpler version of the forecasting problem discussed in Zambella and Grassberger,[13] arose in the proof of a particular scaling property for preimages. The constant in the scaling is rule-dependent and provides a measure of the amount of "distortion" induced in a uniform measure by one application of the rule; for details see Jen.[6]

A quick way to try to guess the answer is to examine only a few, say q_l, of the leftmost components of S and a few, say q_r, of the rightmost components of S. The answer will be correct if, for every string S, the leftmost q_l and the rightmost q_r components of S suffice to determine exactly the values $x, y \in \{0, 1\}$ such that

$$N(xSy) = \min_{s^*} N(S^*),$$

where $N(S)$ is, as before, the number of preimages of S, and S^* is any of the four extended sequences associated with S. Note that the minimizing values need not be unique.

The values of q_l, q_r needed for exact solution of the forecasting problem for all 88 distinct nearest-neighbor rules are given in Table 1. The interesting aspect of this table is that the only values assumed by q_l, q_r are $0, 1, 2, 3, \infty$.

As an example of the meaning of Table 1, consider Rule 23 (see Figure 1), defined by

$$\{011, 101, 110, 111\} \rightarrow 0, \{000, 001, 010, 100\} \rightarrow 1. \tag{5.1}$$

For any string S, the minimum number of preimages is attained by appending to each end the "toggled" value of the last component. Here $q_l = q_r = 1$.

The case of Rule 44 (defined by Eq. (4.1)), for which $q_l = 0, q_r = \infty$, typifies the other cases with non-finite values. On the left, preimages are minimized for any string by appending a 1 on the end. On the right, letting S denote an arbitrary string and $\vec{q_r}$ denote the string of length n, n even, consisting of all 1's.

$$N(S00\vec{q_r}0) \leq N(S00\vec{q_r}1),$$

and

$$N(S10\vec{q_r}0) \geq N(S10\vec{q_r}1).$$

That is, in the first case preimages are minimized by appending a 0 on the right, whereas in the second case they are minimized with a 1. Since this "flip-flopping" takes place for all values of n even, it follows that $q_r = \infty$, implying that minimization of preimages for arbitrary strings requires complete knowledge of the string.

A sufficient condition is described in Jen[6] (a slightly stronger form is given there) for the existence of finite q_l, q_r with the minimization properties described above. The weakest form of the condition is presented here in the language of degree vectors:

> Consider an arbitrary nearest-neighbor automaton rule. Suppose that *every* component of the right (left) 0-degree vector is less than or equal to the corresponding component of the right (left) 1-degree vector. Then $q_r = 0(q_l = 0)$ and appending a 0 on the right (left) of the sequence minimizes the number of preimages.

TABLE 1 For each rule, q_l and q_r represent the number of site values on the left and right ends, respectively, of an arbitrary string S required to determine exactly the value that, when appended to S, minimizes the preimages of the extended string.

q_l	q_r	Rule numbers
0	0	0 1 2 3 4 5 6 7 8 9 10 12 15 18 22 24 25 26 30 32 33 34 36 37 40 41 45 50 51 60 62 72 73 74 76 90 94 104 105 106 110 122 126 128 130 132 134 136 138 140 146 150 152 154 160 164 170 204
0	1	28 35 56
0	3	13 14 162
0	∞	11 44
1	1	19 23 29 54 57 77 108 156 178 184 200 232
1	3	78
1	∞	27
2	0	168
2	2	43 142
3	0	42
3	1	58
∞	0	38
∞	1	172
∞	∞	46

The result can obviously be reversed to solve the maximization problem.

The above condition may be progressively strengthened. Suppose that the original condition is not satisfied; that is, not all components of, say, the right 0-degree vector are in the same ordering relationship to their corresponding components in the 1-degree vector, implying that q_r may be greater than 0. If $q_r = 1$, then knowledge of the rightmost component, call it x, of S would suffice to solve the problem. There are two cases: $x = 0, 1$. Consider the rule matrix M^2, representing rule table information for sequences of length 2 generated by the rule. (It is necessary to look at sequences of length 2 since the two cases $x = 0, 1$ must be distinguished in the subsequent analysis.) Construct for M^2 the associated right degree vector $\bar{\mu}$, each component of which is the sum of products of two terms. For $x = 0$, retain only the products beginning with α; this ensures that the sequence generated has a 0 occurring in its rightmost site. Now, as before, compare the 0-degree and the 1-degree vectors. If a consistent ordering can be found, then perform the same steps for $x = 1$. A consistent ordering for $x = 1$ as well implies that $q_r = 1$. The procedure can be continued until the correct value of q_r is found.

The intuition for the sufficient condition is best provided by referring to Rule 23 defined by Eq. (5.1). Consider the problem of appending a value to the right end of a sequence S so as to minimize its number of preimages under this rule. The rule matrix M is given in this case by

$$M = \begin{pmatrix} \beta & \beta & 0 & 0 \\ 0 & 0 & \beta & \alpha \\ \beta & \alpha & 0 & 0 \\ 0 & 0 & \alpha & \alpha \end{pmatrix} \tag{5.2}$$

with right degree vector

$$\vec{\mu} = (2\beta, \alpha + \beta, \alpha + \beta, 2\alpha).$$

The right 0-degree vector is thus found to be $(0, 1, 1, 2)$, implying of course that the right 1-degree vector is given by $(2, 1, 1, 0)$. Since a consistent component-wise ordering does not hold, q_r may be greater than 0 (the condition is only sufficient; it cannot be concluded that q_r is necessarily greater than 0). Consider then the rightmost component of S. If minimization can be achieved given knowledge of this component, then $q_r = 1$.

To determine whether or not $q_r = 1$, use the matrix M^2, given by

$$M^2 = \begin{pmatrix} \beta^2 & \beta^2 & \beta^2 & \beta\alpha \\ \beta^2 & \beta\alpha & \alpha^2 & \alpha^2 \\ \beta^2 & \beta^2 & \alpha\beta & \alpha^2 \\ \alpha\beta & \alpha^2 & \alpha^2 & \alpha^2 \end{pmatrix},$$

and again construct the right degree vector

$$\vec{\mu} = (3\beta^2 + \beta\alpha, \beta\alpha + \beta^2 + 2\alpha^2, \alpha\beta + 2\beta^2 + \alpha^2, \alpha\beta + 3\alpha^2).$$

Consider now two cases:

1. The rightmost component of S is 0. Then the terms of interest in $\vec{\mu}$ are those beginning with α, namely,

$$(0, 2\alpha^2, \alpha\beta + \alpha^2, \alpha\beta + 3\alpha^2).$$

It must be determined whether, for all such sequences, there exists a single value that will perform the required minimization. The 0-degree vector for this case (with components being the coefficients of the terms ending in α) is $(0,2,1,3)$, whereas the 1-degree vector is $(0,1,1,0)$; hence, for any sequence S ending in 0, the number of preimages is minimized by appending a 1 to the right end.

2. The rightmost component of S is 1. The terms to be considered in $\vec{\mu}$ are now those beginning with β, namely,

$$(3\beta^2 + \beta\alpha, \beta\alpha + \beta^2, 2\beta^2, 0).$$

The 0-degree vector for this case (with components being the coefficients of the terms ending in α) is (1,1,0,0), whereas the 1-degree vector is (3,1,2,0); hence, for any sequence S ending in 1, the number of preimages is minimized by appending a 0 to the right end.

Since a consistent ordering has been shown to exist for both cases, it can then be concluded that knowledge of a single component on the right end of an arbitrary sequence S suffices to determine the value that minimizes the number of preimages when appended to S, that is, that $q_r = 1$.

SUMMARY

Two tools—the rule matrix and the degree vectors—have been introduced for the analysis of the behavior generated by elementary cellular automata. Both the rule matrix and the degree vectors are "intrinsic" features of the automaton in that they are determined *a priori* from the function defining the automaton, rather than *a posteriori* from simulations of its evolution.

The rule matrix represents in convenient form the information contained in an automaton's rule table, and its powers capture the action of the automaton applied to spatial sequences of varying lengths. The degree vectors, which are computed from the rule matrix (or its powers), reflect the amount of change induced by "toggling" a component in the spatial sequence upon which the automaton is acting.

The analysis of preimages—including the computation and "classification" of preimage formulae, the generation of gardens-of-Eden, and the identification of sequences of maximal probability—illustrates the use of the rule matrix to study specific problems in this context and to clarify the relationship between detailed features of an automaton's functional definition (via its rule table) and the behavior generated by the automaton. The degree vectors may be used to determine the amount of information needed to solve a simple "forecasting" problem for these systems.

ACKNOWLEDGMENTS

This work was supported in part by the Applied Mathematics Program of the DOE under Contract No. KC-07-01-01. Support was also provided by the Center for Nonlinear Studies at Los Alamos National Laboratory.

REFERENCES

1. Boghosian, B. "Lattice Gases." In *1989 Lectures in Complex Systems*, edited by E. Jen, 371–388. Santa Fe Institute Studies in the Sciences of Complexity, Lect. Vol. II. Reading, MA: Addison-Wesley, 1990.
2. Chen, H., S. Y. Chen, and G. Doolen. "Lattice Gases." In *1989 Lectures in Complex Systems*, edited by E. Jen, 371–388. Santa Fe Institute Studies in the Sciences of Complexity, Lect. Vol. II. Reading, MA: Addison-Wesley, 1990.
3. Guan, P., and Y. He. "Upper Bound on the Number of Cycles in Border-Decisive Cellular Automata." *Complex Systems* **1** (1987): 181.
4. Hurd, L. "Formal Language Characterization of Cellular Automaton Limit Sets." *Complex Systems* **1** (1987): 69.
5. Jen, E. "Global Properties of Cellular Automata." *J. Stat. Phys.* **43** (1986): 219–242.
6. Jen, E. "Scaling of Preimages in Cellular Automata." *Complex Systems* **1** (1987): 1048.
7. Jen, E. "Cylindrical Cellular Automata." *Commun. Math. Phys.* **118** (1988): 569.
8. Jen, E. "Enumeration of Preimages in Cellular Automata." *Complex Systems* **3** (1989): 421–456.
9. Jen, E. "Aperiodicity in One-Dimensional Cellular Automata." *Physica D* **45** (1990): 3–18.
10. Martin, O., A. Odlyzko, and S. Wolfram. "Algebraic Properties of Cellular Automata." *Commun. Math. Phys.* **93** (1984): 219.
11. Wolfram, S. "Statistical Mechanics of Cellular Automata." *Rev. Mod. Phys.* **55** (1983): 601.
12. Wolfram, S. *Theory and Applicators of Cellular Automata*. Singapore: World Scientific Press, 1986.
13. Zambella, D., and P. Grassberger. "Complexity of Forecasting in a Class of Simple Models." *Complex Systems* **2** (1988): 269.

Joel L. Lebowitz,† Christian Maes,‡ and Eugene R. Speer
Department of Mathematics, Rutgers University, New Brunswick, NJ 08903
† also Department of Physics
‡Present Address: Aangesteld Navorser N.F.W.O., Instituut voor Theoretische Fysica, K.U. Leuven, Belgium

Probablistic Cellular Automata: Some Statistical Mechanical Considerations

This chapter originally appeared in *1989 Lectures in Complex Systems*, edited by E. Jen, 401–414. Santa Fe Institute Studies in the Sciences of Complexity, Lect. Vol. II. Reading, MA: Addison-Wesley, 1990. Reprinted by permission.

We sketch some recent developments in the statistical mechanical analysis of Probabilistic Cellular Automata with emphasis on rigorous results.

1. INTRODUCTION

Spin systems evolving in continuous or discrete time under the action of stochastic dynamics are used to model phenomena as diverse as the structure of alloys and the functioning of neural networks. While in some cases the dynamics are secondary, designed to produce a specific stationary measure whose properties one is interested in studying, there are other cases in which the only available information is the dynamical rule. Prime examples of the former are computer simulations, via Glauber dynamics, of equilibrium Gibbs measures with a specified interaction potential. Examples of the latter include various types of majority rule dynamics used as models for pattern recognition and for error-tolerant computations. The present note discusses ways in which techniques found useful in equilibrium statistical mechanics can be applied to a particular class of models of the latter types. These are cellular automata with noise: systems in which the spins are updated stochastically at integer times, simultaneously at all sites of some regular lattice. These models were first investigated in detail in the Soviet literature of the late sixties and early seventies.[14,16] They are now generally referred to as Stochastic or Probabilistic Cellular Automata (PCA),[6,12] and may be considered to include deterministic automata (CA) as special limits.

There are various ways in which one may try to extend the methods of equilibrium statistical mechanics, which have been very successful in elucidating the "interesting" behavior of Gibbs measures, e.g., coexistence of phases and critical phenomena, to more general processes. In particular, it has been recognized for a long time[14,16] that there is an intimate relation between stationary measures of reversible PCA (i.e., those with dynamics satisfying detailed balance) and Gibbs states. We shall use as our starting point here, however, the relation between PCA in d dimensions and Equilibrium Statistical Models (ESM) in $(d + 1)$ dimensions, the extra dimension being the discrete time. This connection has been exploited by Domany[3,4] and others[11,17] to obtain information about the equilibrium properties of some $(d + 1)$-dimensional spin systems from knowledge of specially constructed d-dimensional PCAs. There has also been, more recently, some general study of this connection from a mathematically rigorous point of view.[5,7] The present note is closely related to the work in Lebowitz[9] and Maes[10] and uses the same notation. It consists of three parts: Section 2 presents background material and the basic PCA-ESM connection. Section 3 gives some illustrative examples including some new results. Section 4 considers PCA that satisfy some form of detailed balance including ones which are updated in two or more time steps. These PCA are generalizations of the Domany models where the updating rules satisfy detailed balance conditions with respect to specified Gibbs measures. We refer the reader to Lebowitz[9] and to references given there for additional background and results.

2. PCA FORMALISM

We consider PCA that describe the stochastic discrete time evolution of Ising spin variables on some regular lattice $\mathcal{L}$ which we generally take to be $\mathbf{Z}^d$. We denote the value of the spin at site $i\epsilon\mathbf{Z}^d$ at time $n\epsilon\mathbf{Z}$ by $\sigma_{n,i} = \pm 1$, and write $\underline{\sigma}_n = \{\sigma_{n,i}\}$ for the configuration at time n; we will occasionally let η denote a generic configuration on $\mathbf{Z}^d$. The PCA evolves by simultaneous independent updating of spins. That is, the spin configuration $\underline{\sigma}_{n-1}$ determines the probabilities $p(\sigma_{n,i}|\underline{\sigma}_{n-1})$ of the spin values at each site i at time n, and the conditional probability distribution of the corresponding $\underline{\sigma}_n$, $P(\underline{\sigma}_n|\underline{\sigma}_{n-1})$, is a product measure

$$P(\underline{\sigma}_n|\underline{\sigma}_{n-1}) = \prod_i p(\sigma_{n,i}|\underline{\sigma}_{n-1}). \tag{2.1}$$

Without loss of generality we write

$$P(\sigma_{n,i}|\underline{\sigma}_{n-1}) = \frac{1}{2}(1 + \sigma_{n,i}h_i(\underline{\sigma}_{n-1})), \tag{2.2}$$

with $|h_i(\underline{\sigma}_{n-1})| \leq 1$. We assume that the rules are translation invariant so that $h_i(\underline{\eta})$ is obtained from $h_0(\underline{\eta})$ via a translation by the lattice vector $i\epsilon\mathbf{Z}^d$. We also assume that $h_0(\underline{\eta})$ depends only on the configurations of spins in a finite set U near the origin.

Given $\underline{\sigma}_{n-1}$, the spins at time n are independent with averages given by

$$\langle \prod_{i\epsilon A} \sigma_{n,i}|\underline{\sigma}_{n-1}\rangle = \prod_{i\epsilon A} h_i(\underline{\sigma}_{n-1}). \tag{2.3}$$

When $|h_i(\underline{\eta})| = 1$ for all i and η, we have a deterministic evolution, i.e., a CA. An interesting way of characterizing a general PCA is by specifying a finite collection of deterministic rules M_α together with their probabilities q_α. M_α is a local function of the configuration taking only the values $+1$ and -1, and $\Sigma_\alpha q_\alpha = 1, q_\alpha \geq 0$. The evolution may be pictured as follows: at each site i at time n a choice of deterministic rule is made, independent of other sites or times and of the input $\underline{\sigma}_{n-1}$, so that

$$\sigma_{n,i} = M_{\alpha,i}(\underline{\sigma}_{n-1}) \quad \text{with probability } q_\alpha, \tag{2.4}$$

where $M_{\alpha,i}$ is the translate of M_α with lattice vector i. This procedure gives

$$h_i(\underline{\eta}) \equiv \sum_\alpha q_\alpha M_{\alpha,i}(\underline{\eta}). \tag{2.5}$$

Note that in general many different choices of the deterministic rules M_α may give rise to the same transition probabilities (2.5). The case where there is only one M, i.e., in which $\sigma_{n,i}$ follows $M_i(\underline{\eta})$ with probability $(1 - \epsilon)$ and does the opposite $(-M_i)$ with probability ϵ, is of special interest. We can then write $h_i(\underline{\eta})$ as

$$h_i(\underline{\eta}) = (1 - 2\epsilon)M_i(\underline{\eta}), \quad 0 \leq \epsilon \leq 1, \tag{2.6}$$

which can be interpreted to mean that we follow the deterministic rule M with probability $(1 - 2\epsilon)$ and with probability ϵ choose $\sigma_{n,i} = \pm 1$ independent of $\underline{\eta}$.

One of the main problems for these systems is to characterize their time-invariant probability measure or, more generally, the stationary space-time process generated by the Markov transition rates (2.1). To do this within the framework of statistical physics we might naturally consider $\{\underline{\sigma}_n\}_{n\epsilon\mathbf{Z}}$ as defining a spin configuration $\underline{\sigma}$ on the space-time lattice $\mathbf{Z}^{d+1}$; we will write $x = (n, i)$ for a typical site in the lattice and let $\mathbf{Z}_n^d$ denote the d-dimensional layer corresponding to time n. By the usual convention of cellular automata, we will visualize the time axis in $\mathbf{Z}^{d+1}$ as vertical and oriented so that the past is at the top and the future is at the bottom. If $\rho(\underline{\eta})$ is a measure on the state space of the PCA and we "start" the time evolution with measure $\rho(\underline{\eta})$ on the layer $\mathbf{Z}_{-N}^d$, then the Markov transition rates (2.1) define a measure μ_ρ^{-N} on the set of "future" configurations $\{\underline{\sigma}_n\}_{n \geq -N}$. Suppose now that we choose $\rho = \nu$ which is stationary for the time evolution, i.e.,

$$\sum_{\underline{\eta}'} P(\underline{\eta}|\underline{\eta}')\nu(\underline{\eta}') = \nu(\underline{\eta}), \qquad (2.7)$$

then the measure μ_ν^{-N} will "reproduce" ν in the time direction in the sense that its projection on any $\mathbf{Z}_n^d$, for $n > -N$, will be just $\nu(\underline{\sigma}_n)$. We can then take the limit $N \to \infty$ to produce a measures μ_ν on the set of space-time configurations $\underline{\sigma} = \{\sigma_x\}, x\epsilon\mathbf{Z}^{d+1}$. The measure μ_ν is translation invariant in the time direction and its projection on any $\mathbf{Z}_n^d$ is just ν. Similar conclusions hold when ν is periodic.

The observant reader may have noticed that we have been very cavalier about transition rates $P(\underline{\eta}|\underline{\eta}')$ and measure $\nu(\underline{\eta})$ on the infinite set of spin configurations. It is most convenient to consider this as a limit of PCA defined on a box $\Gamma \subset \mathbf{Z}^d$ with periodic or other specified boundary conditions. This will become clearer in the next section when we consider the corresponding ESM. In case of doubt simply think of the PCA as defined on configuration is a finite periodic box.

FROM PCA TO ESM

It is a simple observation that if the transition probabilities $p_i(\sigma_{n,i}|\underline{\sigma}_{n-1})$ are all strictly positive, i.e., if $|h_o(\underline{\eta})| < 1$ for all $\underline{\eta}$, then μ_ν is a Gibbs measure for the Hamiltonian[3,4]

$$\mathcal{H}(\underline{\sigma}) = \sum_{x\epsilon\mathbf{Z}^{d+1}} H_x(\sigma_x, \underline{\sigma}_{n-1}). \qquad (2.8)$$

Here the single-site energy for $x = (n, i)$ is defined by

$$\exp[-H_x(\sigma_x, \underline{\sigma}_{n-1})] \equiv p_i(\sigma_{n,i}|\underline{\sigma}_{n-1}), \qquad (2.9)$$

and the reciprocal temperature β which usually multiplies the energy in the exponent of Eq. (2.9) has been absorbed into H. The normalization condition of the PCA gives

$$\sum_{\sigma_x=\pm 1} \exp[-H_x(\sigma_x, \underline{\sigma}_{n-1})] = 1. \qquad (2.10)$$

In fact we can always write

$$\exp[-H_0(\sigma_0, \underline{\eta})] = \frac{\exp -\sigma_0 Q_0(\underline{\eta})}{2 \cosh Q_0(\underline{\eta})}, \qquad (2.11)$$

with

$$h_0(\underline{\eta}) = -\tanh Q_0(\underline{\eta}).$$

We can, therefore, think of $h_i(\underline{\eta})$ as the tanh of the "field" acting on the spin at site i when the configuration in the layer "above" it is $\underline{\eta}$. When $|h_0(\underline{\eta})| < 1$ for some configurations $\underline{\eta}$, making the field infinite, we may still be able to consider non-trivial μ_ν—but there will now be certain "hard-core" constraints on the possible configurations $\underline{\sigma}_n$ given $\underline{\sigma}_{n-1}$.

An important consequence of Eq. (2.10) is that the finite-volume free energy for the ESM is identically zero for certain classes of boundary conditions, e.g., for $(d+1)$-dimensional cubes or parallelepipeds with periodic boundary conditions in the space directions, arbitrary configurations (initial conditions) at the top and free boundary conditions at the bottom. More generally, given any region $\Lambda \subset \mathbf{Z}^{d+1}$ and a spin configuration $\underline{\sigma}_{\Lambda^c}$ on the complement of Λ, we can define the finite-volume Gibbs measure $\mu_p(\underline{\sigma}_\Lambda | \underline{\sigma}_{\Lambda^c})$ on the spins $\underline{\sigma}_\Lambda$ in Λ with "PCA boundary conditions" by

$$\mu_p(\underline{\sigma}_\Lambda | \underline{\sigma}_{\Lambda^c}) \equiv \exp\left[-\sum_{x \in \Lambda} H_x(\sigma_x, \underline{\sigma}_{n-1})\right]. \qquad (2.12)$$

N.B. The right side of Eq. (2.12) is properly normalized since the trace over the spins in Λ is by Eq. (2.10) identically one, independent of $\underline{\sigma}_{\Lambda^c}$. Note also that if we weighted the $\underline{\sigma}_{\Lambda^c}$ in Eq. (2.12) according to a probability distribution corresponding to an infinite volume Gibbs measure $\mu_{\mathcal{H}}$ for $\mathcal{H}$ given in Eq. (2.8), the resulting measure on $\underline{\sigma}_\Lambda$ is *not* the same as that obtained by projecting $\mu_{\mathcal{H}}$ on Λ. The difference arises from the fact that many of the interactions *across* the boundary of Λ are omitted from the exponent in Eq. (2.12). Nevertheless, as is well known, in the limit $\Lambda \nearrow \mathbf{Z}^{d+1}$ *all* boundary conditions lead to some Gibbs measure $\mu_{\mathcal{H}}$.

Now if the set of Gibbs measures for $\mathcal{H}$ is unique, i.e., there is only one $\mu_{\mathcal{H}}$, then clearly it is the same as μ_ν. If, however, there is more than one $\mu_{\mathcal{H}}$ as occurs when the ESM has more than one phase, then the question naturally arises as to which boundary conditions give Gibbs measures that coincide with a space-time stationary PCA measure μ_ν for some ν. More generally, what is the correspondence between the set of Gibbs measures $\{\mu_{\mathcal{H}}\}$ and PCA measures $\{\mu_\nu\}$? As remarked above, $\{\mu_\nu\} \subset \{\mu_{\mathcal{H}}\}$. A partial answer to this question was given in Goldstein[7] where it was shown that those measures in each class which are translation invariant or periodic, in *all* $d+1$ directions, *coincide*. What happens for other measures is an open question.

We remark here also that for *any* boundary condition, the finite-volume free energy of the system with interaction $\mathcal{H}$ is at most of the order of the size of the boundary region. Hence the infinite-volume free energy density, which is independent of boundary conditions, is identically zero. In particular, it is analytic in the parameters of the PCA, even when there is a phase transition in the sense of ESM. The same analyticity will hold for the dependence of the free energy on the interaction coefficients entering the Hamiltonian Eq. (2.8). In certain cases this analyticity may be shown to hold separately in the entropy and energy densities, even when there is a phase transition as the parameters change.[9]

3. SOME EXAMPLES

1. An example of a PCA with a particularly simple stationary ν is obtained by letting $\sigma_{i,n} = M_i(\underline{\sigma}_{n-1}) = \Pi_{j\epsilon U(i)}\sigma_{j,n-1}$ with probability $(1 - \epsilon)$ and $\sigma_{n,i} = -M_i$, with probability ϵ for some $\epsilon, 0 < \epsilon \leq 1$, where $U(i)$ is the translate of the neighborhood of the origin U by i. This gives, as per Eqs. (2.6) and (2.11),

$$h_i(\underline{\eta}) = (1 - 2\epsilon) \prod_{j\epsilon U(\imath)} \eta_j \qquad (3.1)$$

and

$$H(\sigma_{n,i},\underline{\sigma}) = -\alpha\sigma_{n,\imath} \prod \sigma_{n-1,j} - \ln[2\cosh\alpha] , \qquad (3.2)$$

with $(1 - 2\epsilon) = \tanh\alpha$. The stationary measure ν for this PCA, corresponding to the projection of the Gibbs measure μ_ν on $\mathbf{Z}^d$, is somewhat surprisingly just the Bernoulli product measure with $\langle\sigma_i\rangle = 0$; c.f. Toom et al.[16] for a special case of this. This can be seen by noting from Eqs. (2.3) and (2.7) that in a periodic box $\Lambda \subset \mathbf{Z}^2$, the product measure ν for which $\langle\eta_j\rangle = 0$ is time invariant. The uniqueness of this stationary measure is easily established, c.f. section 4.[9]

2. The Bennett-Grinstein version of Toom's model[10] is a PCA on the square lattice $\mathbf{Z}^2$ in which every spin becomes (up to errors) equal to the majority, at the previous time, of the spin itself and its northern and eastern nearest neighbor. Errors favoring up spins are made with a probability p, and errors favoring down spins with a probability q. Using Eq. (2.4) we then have for $i = (i_1, i_2)$,

$$h_\imath(\underline{\eta}) = (1 - p - q)\, \mathrm{sgn}\left(\sum_{U(i)} \eta_i\right) + (p - q) , \qquad (3.3)$$

with $U(i) = (i, i + e_1, i + e_2)$, e_1 and e_2 being the unit vectors on $\mathbf{Z}^2$. The resulting ESM on $\mathbf{Z}^3$ has the per-site energy

$$H(\sigma_{n,i},\underline{\sigma}) = -(\beta\sigma_{n,i} + J)\left(\sum_{j\epsilon U(i)} \sigma_{n-1,j} - \prod_{j\epsilon U(i)} \sigma_{n-1,j}\right) - b\sigma_{n,\imath} - \gamma , \qquad (3.4)$$

where

$$\beta = \frac{1}{8} \log \frac{(1-p)(1-q)}{pq} \qquad b = \frac{1}{4} \log \frac{p(1-q)}{q(1-p)},$$

$$J = \frac{1}{8} \log \frac{q(1-q)}{p(1-p)} \qquad \gamma = \frac{1}{4} \log[pq(1-p)(1-q)]. \tag{3.5}$$

Toom[15] has proved that there is more than one invariant measure for this PCA whenever p and q are sufficiently small. A discussion of the phase diagram for this model obtained from computer simulations is contained in Lebowitz[9] and Bennett,[1] from which Figure 1 is taken.

We note here in particular that this system can exist in two phases even when p (or q) is zero. In this case, one of the phases consists of all spins down (up). Such a state has, of course, zero entropy and since its free energy is zero so is its energy (understood as the limit $p \to 0$ for fixed small q). The other state has most spins up and, as q is increased past some critical value $q_c \simeq .07$, it disappears abruptly—corresponding to a first-order-type phase transition. Bennett and Grinstein[1] believe that the transition is in fact first order all along the coexistence curve in Figure 1 except for the symmetric case $p = q$ where it is second order with presumably the same exponents as the two-dimensional Ising model. Nothing, however, is know about this rigorously.

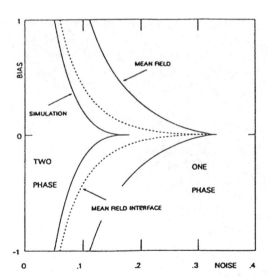

FIGURE 1   In this figure the phase plane is parametrized by the *noise* $p + q$ and the *bias* $(p - q)/p + q$. The transition curves of several mean field approximations to the model are also shown.[1,9]

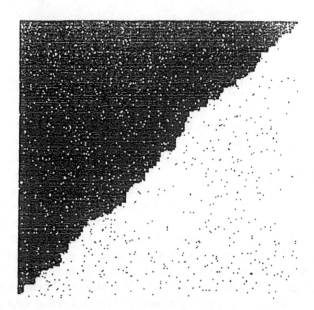

FIGURE 2 Typical configuration of the Toom PCA with + boundary conditions on top and − boundary conditions on the right side at noise level $\epsilon = .02$.

FIGURE 3 Typical configuration of the Toom PCA with + boundary conditions on top and − boundary conditions on the right side at noise level $\epsilon = .09$.

It should be remarked also that the planar region in Figure 1 where the system can exist in (at least) two phases is to be thought of as a two-dimensional region in the four-dimensional parameter space, β, J, b, and γ, of the Hamiltonian Eq. (3.4). Crossing the coexistence line in Figure 1 the free energy remains analytic (since it is zero everywhere in the parameter region specified by Eq. (3.5)). There will, however, be "other" directions of parameter space in which the free energy of the Hamiltonian Eq. (3.4) will have a discontinuous slope—since the ESM has more than one phase.

Finally, we show in Figures 2 and 3 typical configurations of the stationary state of this system in a 256×256 square with $+$ boundary conditions on top and $-$ boundary conditions on the right side à la Eq. (2.12), for $p = q = \epsilon = .02$ and .09 (the boundary conditions on the left and bottom are irrelevant since they do not influence the interior). When there is a bias in favor of the $+$'s ($-$'s), the slope of the interface between the two phases increases (decreases). The behavior of this interface for small noise is discussed in Derrida.[2] It is found that the slope of the interface is asymptotically $\sqrt{q/p}$. Simulations and several approximate models suggest that the width of the interface at distance L from the top right corner grows as L^α, where $\alpha \simeq 1/4$ when $p = q$ and $\alpha \simeq 1/3$ when $p \neq q$. Similar power law behavior of interface width occurs in several related systems. For example, one-dimensional interfaces of length L in two-dimensional systems typically have width $L^{1/2}$, while the width of an initially sharp interface in the infinite plane in many cases grows[14] with time as $t^{1/3}$.

4. STATIONARY MEASURES AND DETAILED BALANCE

Using the normalization condition on the transition probability $P(\underline{\eta}|\underline{\eta}')$, Eq. (2.7) for the stationary ν can be rewritten in the form

$$\sum_{\underline{\eta}'}[P(\underline{\eta}|\underline{\eta}')\nu(\underline{\eta}') - P(\underline{\eta}'|\underline{\eta})\nu(\underline{\eta})] = 0. \tag{4.1}$$

The transition P is said to be reversible or to satisfy detailed balance with respect to ν if the summand in Eq. (4.1) vanishes for all $\underline{\eta}'$, i.e., if

$$P(\underline{\eta}|\underline{\eta}')\nu(\underline{\eta}') = P(\underline{\eta}'|\underline{\eta})\nu(\underline{\eta}) \tag{4.2a}$$

or

$$P(\underline{\eta}|\underline{\eta}') = \frac{S(\underline{\eta}, \underline{\eta}')}{\nu(\underline{\eta}')} \tag{4.2b}$$

with S symmetric,

$$S(\underline{\eta},\underline{\eta}') = S(\underline{\eta}',\underline{\eta}) \,. \tag{4.2c}$$

The condition of detailed balance implies that a motion picture made of the PCA in the stationary state ν will look the same when run forwards or backwards; i.e., the stationary space-time process is reversible.

It was noted by Stavskaja[14] that the detailed balance condition (4.2) implies that

$$\nu(\underline{\eta}) = \nu(-1)\frac{P(\underline{\eta}|-1)}{P(-1|\underline{\eta})}$$

$$= \nu(-1)\prod_i \frac{[1 + \eta_i h_i(-1)]}{[1 - h_i(\underline{\eta})]} \,, \tag{4.3}$$

where we have written -1 for the reference configuration $\eta_i = -1, \forall i$, and have used Eq. (2.2). Stavskaja also introduces a more general detailed balance condition than Eq. (4.2) which is sometimes useful. For any $\underline{\eta}'$ let $S_{\underline{\eta}'}$ be a transformation of the set of $\{\underline{\eta}_i\}$ which can include shifts, inversions, etc. We say that the PCA satisfies "Stavaskaja balance" whenever

$$P(S_{\underline{\eta}'} \cdot \underline{\eta}|\underline{\eta}')\nu(\underline{\eta}') = P(\underline{\eta}'|\underline{\eta})\nu(\underline{\eta}) \,. \tag{4.4}$$

Letting $S_{-1} \cdot \underline{\eta} = \underline{\eta}^*$ we then have in analogy with Eq. (4.3) that Eq. (4.4) implies

$$\nu(\underline{\eta}) = \nu(-1)\prod_i \frac{[1 + \eta_i^* h_i(-1)]}{[1 - h_i(\underline{\eta})]} \,. \tag{4.5}$$

We refer the reader to Toom[16] and Stavskaja[14] for more information on the S-condition.

Returning now to Eq. (4.2) the question naturally arises as to when a given P satisfies Eq. (4.2). It turns out that a very simple criterion can be given using Eqs. (2.9) and (2.11): Eq. (4.2) will be satisfied if $h_i(\underline{\eta})$ has a form of the type frequently used in neural networks,

$$h_i(\underline{\eta}) = \tanh[\lambda_i + \Sigma_j J_{ij}\eta_j] \,, \tag{4.6a}$$

with symmetric inputs,

$$J_{ij} = J_{ji} \,. \tag{4.6b}$$

The corresponding stationary ν is then given by

$$\nu = C \exp\sum_i \{\lambda_i\eta_i + \log 2\cosh[\lambda_i + \Sigma_j J_{ij}\eta_j]\} \,, \tag{4.7}$$

with C a normalization constant. The second term in the exponent in Eq. (4.7) can be expanded into a polynomial of the form

$$\log\{2\cosh[\lambda_0 + \sum_j J_{0j}\eta_j]\} = \Sigma_A J_A^{(d)}\eta_A. \tag{4.8}$$

The sum in Eq. (4.8) goes over all subsets $A \subset U$, with $\eta_A = \prod_{k\in A}\eta_k$, U being the neighborhood of the origin for which $J_{0j} \neq 0$. When the dynamics are translation invariant, $\lambda_i = \lambda$ and $J_{ij} = J(i-j)$, the stationary ν is a Gibbs measure on $\mathbf{Z}^d$ for the Hamiltonian

$$H^{(d)}(\underline{\eta}) = -\sum_{i\in\mathbf{Z}^d}\left(\lambda\eta_i + \sum_A J_A^{(d)}\eta_{A+i}\right) = \sum_i H_i^{(d)}(\underline{\eta}). \tag{4.9}$$

Phase transitions for this d-dimensional equilibrium system will thus correspond to non-ergodicity of the PCA and to phase transitions in the $(d+1)$-dimensional ESM system with Hamiltonian per site $(n,i)\epsilon\mathbf{Z}^{d+1}$ of the form

$$H_{n,i}(\underline{\sigma}) = -2\lambda\sigma_{n,i} - \sum_j \sigma_{n,i}J(i-j)\sigma_{n-1,j} - H_i^{(d)}(\underline{\sigma}_{n-1}). \tag{4.10}$$

It should be noted that the special form of the Hamiltonian $H^{(d)}$ in Eq. (4.9) precludes, in general, the construction of a PCA dynamics which will satisfy detailed balance for an *a priori* given Gibbs measure $\tilde\nu$ corresponding to a Hamiltonian $\tilde H_d$. This is in contrast to continuous-time or sequential updatings where a suitable choice of stochastic dynamics can always be made. This difficulty can be overcome, following Domany,[3,4] when the lattice $\mathcal{L}$ can be divided into two parts, $\mathcal{L} = \mathcal{L}_1 \cup \mathcal{L}_2$, e.g., for $\mathcal{L} = \mathbf{Z}^d$ the even and odd sublattices. Let $\underline{\eta} = (\underline{\eta}^{(1)}, \underline{\eta}^{(2)})$, and assume that the conditional measures on $\mathcal{L}_1$ and $\mathcal{L}_2$, obtained from the measure $\tilde\nu(\underline\eta) = \tilde\nu(\underline\eta^{(1)}, \underline\eta^{(2)})$ in which we are interested, are product measures; i.e.,

$$\tilde\nu\left(\underline\eta^{(\alpha')}|\underline\eta^{(\alpha)}\right) \equiv \frac{\tilde\nu(\underline\eta^{(1)}, \underline\eta^{(2)})}{\tilde\nu_\alpha(\underline\eta^{(\alpha)})}$$
$$= \prod_{i\in\mathcal{L}_{\alpha'}} g_i^{(\alpha)}\left(\eta_i^{(\alpha')}; \underline\eta^{(\alpha)}\right), \text{ with } \alpha' = \alpha + 1 \bmod 2, \tag{4.11}$$

where $\tilde\nu(\underline\eta^{(\alpha)})$ is the trace over $\underline\eta^{(\alpha')}$ of $\tilde\nu(\underline\eta)$. (An example of this is the standard equilibrium Ising model with nearest-neighbor interations.) We can then do the updatings in two steps, first $\mathcal{L}_1$ and then $\mathcal{L}_2$, and choose the transition rates of our PCA so that detailed balance is statisfied for $\tilde\nu$ in the sense of reversibility of the stationary space-time process.

To do this we define the PCA subsystem updatings $P^{(\alpha)}(\underline{\sigma}_{n+1}^{(\alpha)}|\underline{\sigma}_n^{(1)},\underline{\sigma}_n^{(2)})$ as in Eq. (2.1). The full updating, which will in general no longer be of the PCA form (2.1), is then given by applying first $P^{(1)}$, then $P^{(2)}$,

$$\hat{P}(\underline{\sigma}_{n+1}|\underline{\sigma}_n) \equiv P^{(2)}(\underline{\sigma}_{n+1}^{(2)}|\underline{\sigma}_{n+1}^{(1)},\underline{\sigma}_n^{(2)})P^{(1)}(\underline{\sigma}_{n+1}^{(1)}|\underline{\sigma}_n^{(1)},\underline{\sigma}_n^{(2)}). \qquad (4.12)$$

Its "time reversed adjoint" is

$$\hat{P}^+(\underline{\sigma}_{n+1}|\underline{\sigma}_n) \equiv P^{(1)}(\underline{\sigma}_{n+1}^{(1)}|\underline{\sigma}_n^{(1)},\underline{\sigma}_{n+1}^{(2)})P^{(2)}(\underline{\sigma}_{n+1}^{(2)}|\underline{\sigma}_n^{(1)},\underline{\sigma}_n^{(2)}). \qquad (4.13)$$

The detailed balance condition, which ensures that the stationary stochastic process looks statistically the same when run forwards or backwards in time, is then

$$\hat{P}(\underline{\eta}|\underline{\eta}')\nu(\underline{\eta}') = \hat{P}^+(\underline{\eta}'|\underline{\eta})\nu(\underline{\eta}). \qquad (4.14)$$

This condition of detailed balance will clearly be satisfied for measures $\tilde{\nu}$ of the form (4.11) whenever

$$P^{(\alpha)}\left(\underline{\eta}^{(\alpha)}|\underline{\eta}\right)\tilde{\nu}\left(\underline{\eta}^{(\alpha)}|\underline{\eta}^{(\alpha')}\right). \qquad (4.15)$$

The dynamics used by Domany and others[3,4,11,17] for Ising models on different lattices are of this type with

$$P^{(\alpha)}\left(\underline{\eta}^{(\alpha)}|\underline{\eta}\right) = \prod_{i\in\mathcal{L}_\alpha} \frac{\exp -\eta_i\left[\sum_{j\in\mathcal{L}_{\alpha'}} J_\alpha(i-j)\eta_j' + b_\alpha\right]}{2\cosh\left[\sum_{j\in\mathcal{L}_{\alpha'}} J_\alpha(i-j)\eta_j' + b_\alpha\right]}. \qquad (4.16)$$

Here b_α is the external magnetic field on sublattice $\mathcal{L}_\alpha$ and $J_\alpha(k)$ is the pair interaction between a spin on $\mathcal{L}_\alpha$ and a spin on $\mathcal{L}_{\alpha'}$ separated by the vector k.

The right side of Eq. (4.16) can be generalized by replacing the terms in the square brackets there by the more general form, see Eq. (2.11),

$$[\ ] \to \sum_{j\in\mathcal{L}_{\alpha'}} J(|i-j|)\eta_j' + f_i(\underline{\eta}')\eta_i', \qquad (4.17)$$

where $f_i(\underline{\eta}') = 0$ whenever $\sum_{j\in\mathcal{L}_{\alpha'}} J(|i-j|)\eta_j' \neq 0$ and is constant, independent of $\underline{\eta}'$, when $\sum_{j\in\mathcal{L}_{\alpha'}} J(|i-j|)\eta_j' = 0$, e.g., in the case when $J(|i-j|) = J$ for nearest-neighbor sites, half of which are up. The resulting $\hat{P}$ of Eq. (4.12) will then still satisfy detailed balance with respect to the Gibbs measure $\tilde{\nu}$,

$$\tilde{\nu}(\underline{\eta}) = Z^{-1}\exp[-H^{(d)}(\underline{\eta})], \qquad (4.18)$$

and

$$H^{(d)}(\underline{\eta}) = \sum_{i\in\mathcal{L}_1, j\in\mathcal{L}_2} \eta_i J(|i-j|)\eta_j. \qquad (4.19)$$

The generalization (4.17) of the Domany rules permits us to write the stationary measure of a one-dimensional PCA which is updated alternately, on the even and odd sites, according to a majority rule of itself and its two neighbors with some noise, as the Gibbs state of one-dimensional Ising model with nearest-neighbor interactions. It follows then that this PCA will be ergodic and the corresponding two-dimensional ESM will have no phase transitions. The corresponding result for simultaneous updatings for all sites has only been proven recently by Gray.[8]

We note here that $\mathcal{L}_1$ and $\mathcal{L}_2$ can be very different sets, like sites and bonds, with bond variables being on and off according to whether $\eta_i = +1$ or -1. The updating rules can also be quite different and the PCA can then serve as a model of a neural net with two different types of elements. Further generalizations to more than two subsets and to cases where the σ_i can take on more than two values are, of course, also possible.

ACKNOWLEDGMENTS

We thank Jozsef Slawny for useful discussions and Joel Lebowitz thanks Erica Jen for her hospitality at the Santa Fe Institute summer school. Joel Lebowitz and Christian Maes were supported in part by NSF Grant DMR86-12369.

REFERENCES

1. Bennett, C., and G. Grinstein. "Role of Irreversibility in Stabilizing Complex and Nonergodic Behavior in Locally Interacting Discrete Systems." *Phys. Rev. Lett.* **55** (1985): 657.
2. Derrida, B., J. L. Lebowitz, E. R. Speer, and H. Spohn. "Dynamics of an Anchored Toom Interface." *J. Phys. A.* **24** (1991): 4805.
3. Domany, E. "Exact Results for Two- and Three-Dimensional Ising and Potts Models." *Phys. Rev. Lett.* **52** (1984): 871.
4. Domany, E., and W. Kinzel. "Equivalence of Cellular Automata to Ising Models and Directed Percolation." *Phys. Rev. Lett.* **53** (1984): 311.
5. Ferrari, P., J. L. Lebowitz, and C. Maes. "On the Positivity of Correlations in Nonequilibrium Spin Systems." *J. Stat. Phys.* **53** (1988): 295.
6. Georges, A., and P. LeDoussal. "From Equilibrium Spin Models to Probabilistic Cellular Automata." *J. Stat. Phys.* **54** (1989): 1011.
7. Goldstein, S., R. Kuik, J. L. Lebowitz, and C. Maes. "From PCA's to Equilibrium Systems and Back." *Commun. Math. Phys.* (1989).
8. Gray, L. Private communication, Spring, 1989.
9. Lebowitz, J. L., C. Maes, and E. R. Speer. "Statistical Mechanics of Probabilistic Cellular Automata." *J. Stat. Phys.* (1990).
10. Maes, C. In *Mathematical Methods in Statistical Mechanics*, edited by M. Fannes and A. Verdure. Leuven Notes in Mathematical and Theoretical Physics, Vol. 1A. Leuven: Leuven University Press, 1989.
11. Peschel, I., and V. J. Emery. "Calculations of Spin Correlations in Two-Dimensional Ising Systems from One-Dimensional Kinetic Models." *Z. Phys.* **B43** (1981): 241.
12. Rujan, P. "Cellular Automata and Statistical Mechanical Models." *J. Stat. Phys.* **49** (1987): 139.
13. Spohn, H. Private communication, Spring, 1989. See M. Kharder, G. Parisi, and Y.-C. Zhang. "Dynamic Scaling of Growing Interfaces." *Phys. Rev. Lett.* **56** (1986): 889.
14. Stavskaja, O. N. "Gibbs Invariant Measures for Markov Chains on Finite Lattices with Local Interactions." *Math USSR Sobrnik* **21** (1973): 395.
15. Toom, A. L. In *Multicomponent Random Systems*, edited by R. L. Dobrushin and Ya. G. Sinai. New York: Dekker, 1980.
16. Toom, A. L., N. B. Vasilyev, O. N. Stavskaja, L. G. Mitjushin, G. L. Kurdomov, and S. A. Pirogov. "Discrete Local Markov Systems." Preprint, 1989.
17. Verhagen, A. M. V. "An Exactly Soluble Case of the Triangular Ising Model in a Magnetic Field." *J. Stat. Phys.* **15** (1976): 219.

Wentian Li
Santa Fe Institute, 1660 Old Pecos Trail, Suite A, Santa Fe, NM 87501, and
Box 167, Rockefeller University, 1230 York Avenue, New York, NY 10021

Nonlocal Cellular Automata

This chapter originally appeared in *1991 Lectures in Complex Systems*, edited by L. Nadel and D. Stein, 317–327. Santa Fe Institute Studies in the Sciences of Complexity, Lect. Vol. IV. Reading, MA: Addison-Wesley, 1992. Reprinted by permission.

Nonlocal cellular automata are fully discretized and uniform high-dimensional dynamical systems with nonlocal interactions. It is emphasized that although nonlocal interaction is not considered as a correct description of the physical world at its lowest level, at higher levels it, nevertheless, is an important feature for systems in, for example, biology and economics. Many properties of nonlocal cellular automata are investigated in another publication.[9] In this lecture note, I will only highlight a few topics, including the analytic approximation of macroscopic dynamics, systems of coupled selectors, and group meeting problems.

FROM LOCAL TO NONLOCAL DYNAMICAL SYSTEMS

One of the most important aspects of a complex system is its time evolution following a rule which does not change in time. A point of view, though perhaps extreme, is that since all physical laws are fixed (for example, there is no evidence that the gravitational force falls off as $1/r^2$ today—where r is the distance between two mass objects—but falls off as $1/r^3$ tomorrow), whatever has happened on the earth is a realization of a gigantic dynamical system with those fixed physical laws. With this point of view, the evolution of life as well as natural selection can also be modeled by complex dynamical systems with fixed rules, although this modeling will be extremely difficult because the evolution of life is much more complex than practically all model dynamical systems that we have known.

Since physical laws are local (there is no experimental evidence that physical interaction can be accomplished nonlocally), one may argue that we only need locally coupled high-dimensional dynamical systems to model everything. In other words, there is no need to introduce nonlocality in the model dynamical system.

However in a more realistic modeling of the world around us, we do not come down to the very end of the microscopic description. The entities that interact with each other are not quarks, nucleus, atoms, or molecules, but things like neurons in a brain, animals in an ecological system, or agents in a stock market. As the level of description increases, two notions have been changed. First, the dynamics rule may not be fixed in time (they are not the golden, universal, time-invariant physical laws any more). Second, the interaction between entities may not be local.

If the dynamical rule is not time-invariant, it can be very hard to describe and to study the resulting dynamical system, unless the *dynamics of the rule* is describable. In other words, we need two sets of dynamical systems: at the higher level, there is a dynamics of the rules, and at the lower level, there is a dynamics of the entities. Many evolutionary models are of such nature. The complexity of the system results from the interplay between higher-level and lower-level dynamics. One can even imagine three or more levels of dynamics, in which the entity of the higher level is the rule of the level one step lower.

These multi-level dynamical systems are fascinating systems to study. But they are outside the realm of this lecture note. For the time being, to start from the simplest scenario, I will assume that the lower-level rules are not changed. An explanation for this assumption is that the higher-level rules function on a much longer time scale, so that during this time scale, the lower-level rules can be considered as unchanged.

The issue I want to address here is the following: what happens when nonlocality is introduced to a dynamical system? One should know that the locality of interaction is a terrible assumption for many systems with a high-level description. For example, the transmission of signal from one neuron to another in the brain is through axons and dendrites. The distance between two neurons is an irrelevant piece of information concerning whether or not the two are connected to each other.

Note that the nonlocality at this level of the description (interaction between neurons) does not contradict the locality at the microscopic level: the traveling chemical signals do obey local physical and chemical laws. This fact, however, does not prevent us from explicitly incorporating the nonlocality into the modeling process when describing the interaction between neurons.

Similarly, two agents or brokers in a stock market can communicate via telephone line regardless of how far or close the two are to each other. Again, there is no contradiction with the locality of the physical laws. Admittedly, the electrical signal does take a longer time to travel for a longer telephone line than a short one, but the difference is so small compared with the time scale of the stock market activities, that this fact is irrelevant. The more important information is who makes phone calls to whom (whether the connectivity is one or zero) than the actual distance between them.

There are many, many other examples. What we have learned from this discussion is that when the level of description of a system is increased, one sometimes needs to explicitly introduce nonlocality to the modeling. This nonlocality does not violate the locality at the lowest level description—the physical description.

CELLULAR AUTOMATA AS A FULLY DISCRETIZED AND UNIFORM HIGH-DIMENSIONAL DYNAMICAL SYSTEM

What is a cellular automaton? With so many introductory articles and books existing on this topic, I will refer the reader to the original publications (e.g., Toffoli and Margolus,[14] and Wolfram[17,18]). To put it into simple terms, one can say that cellular automata are high-dimensional, fully discretized, synchronous, uniform, and locally coupled dynamical systems. There are high-dimensional dynamical systems that are not fully discretized, such as partial differential equations, coupled differential equations, and lattice maps (e.g., Crutchfield and Kaneko[4]). There are also high-dimensional, fully discretized dynamical systems that are not synchronous or uniform. The model systems I will introduce are high-dimensional, fully discretized, synchronous, and uniform dynamical systems with *nonlocal* connections.[9]

There exist other names that can describe nonlocally coupled, high-dimensional dynamical systems; for example, *automata networks*, or simply, *networks*. I will use the name "nonlocal cellular automata" to have a closer reference to the locally coupled cellular automata, in order to emphasize the uniformity and synchronousness of the system.

Suppose the state value of the component i at time t is x_i^t, and the total number of components in the system is N, then the state configuration of the system consists of state value for each component: $(x_1^t, x_2^t, \ldots, x_N^t)$. An n-input nonlocal cellular automaton is defined by the rule $f(.)$:

$$x_i^{t+1} = f(x_{j_1(i)}^t, x_{j_2(i)}^t, \ldots, x_{j_n(i)}^t),$$ (1)

which says that each component i updates its state value by checking the state values of n other components, which have indexes $j_1(i), j_2(i), \ldots, j_n(i)$. Knowing the state values of these components, and knowing the rule $f(.)$ which is written as a *rule table* (a list of all possible n-component configurations as well as which state value they lead to), we are able to determine what the state value of the component i is at the next time step (x_i^{t+1}).

There are other types of networks previously studied. One of them, studied by Walker and Ashby, that might be called "Ashby nets,"[15,16] is also uniform and synchronous, but not all inputs are randomly chosen—one input is always the component itself. More about Ashby nets will be discussed in the next section.

Another type of nets, that might be called the "Kauffman nets," is studied in Kauffman.[11,12] These nets are synchronous, but not uniform: the rule acting on one component may differ from that on another component. It is well known that different rules can lead to different dynamical behaviors, so mixing all of them into one system leads to rather poor statistics. If the number of inputs (n) and the number of components (N) are fixed, and we ask the question of what the "typical" transient and cycle times are, there will be no "good" answer. The median value (see, for example, Press et al.[13] for a definition of the median as well as the mean value) of a wide-spreading distribution of these statistical quantities, as used in Kauffman,[11] may not give a true "typical" value. Numerical results show that median cycle lengths for these nets are quite different from the mean cycle lengths, though I will not include these results nor discuss this type of net further here. (See update in the 1996 edition.)

WIRING DIAGRAM

Besides the dynamical rule, the wiring diagram of a network also plays an important role in determining the dynamics. It could happen that with the same rule, some wiring diagrams lead to one type of dynamics, while other wiring diagrams lead to another. When we talk about dynamics of a nonlocal cellular automaton rule, there is an implication that almost all wiring diagrams ("typical" or randomly chosen) lead to the same dynamics.

It is in an analogous situation to local cellular automata. For local cellular automata, we also talk freely about the dynamics of a rule, without specifically mentioning the initial configuration. It is again implied that almost all typical or randomly chosen initial configurations lead to the same dynamics. This idea is essential to the concept of "attractor"; that is, whatever the initial conditions, they are all attracted to the same limiting behavior.

The wiring diagram dictates where to take inputs for each component. In some sense, it determines the direction of information flows. Obviously wiring diagrams with different topological structures will transmit information in different ways, and dynamical behaviors can also be different.

For example, if one assumes that for each component i, one of its inputs is always itself:

$$\text{for all } i\text{'s } j_1(i) = i, \text{ but other } j_k(i)\text{'s are random } (k = 2, 3, \ldots, n), \qquad (2)$$

then the wiring diagram will not be completely random. I called this type of wiring diagram *partially local* or *partially nonlocal*.[9]

Some nonlocal cellular automata with partially local wiring diagrams were studied in Walker and Ashby.[15,16] These are 3-input, 2-state, nonlocal cellular automata with the second input being the component itself

$$\text{for all } i\text{'s } j_2(i) = i, \text{ but } j_1(i) \text{ and } j_3(i) \text{ are random}. \qquad (3)$$

It has been shown that for many 3-input, 2-state rules, fully nonlocal wiring diagrams and partially local wiring diagrams lead to different dynamics.[9]

Another issue related to the wiring diagram is the discussion on how dynamics are affected by changing the number of inputs. It is numerically shown that the number of inputs is important in determining the dynamical behavior.[8] If one randomly picks a rule, the more inputs one has, the more likely the dynamics are chaotic. For local cellular automata, the increase of the number of inputs will increase the percentage of rules that are chaotic. More detailed discussions are in Li, Packard, and Langton.[8]

Now back to the discussion of nonlocal cellular automata. Even though each component is supposed to receive n inputs, a particular realization of the random number generator may actually assign two inputs to be the same. If this happens, the rule as applied to that particular component will have one less number of input than it should have. And if many other components also have this *degeneracy* of inputs, it is more likely that the resulting dynamics will act as if the number of inputs is smaller. Some specific examples of the difference between the degeneracy-permitted and distinct-input diagrams are presented in Li.[9]

ANALYTIC APPROXIMATION OF MACROSCOPIC DYNAMICS

The ultimate method to study a dynamical system is to run the time evolution following the rule that updates the state value for each component. The simulation for 3-input, 2-state, nonlocal cellular automata has been carried out and the results are summarized in Li.[9]

If one is only interested in dynamics of *macroscopic* quantities, for example, the density of state 1, some alternative dynamical equations for that macroscopic quantity can be derived. These dynamical equations for macroscopic quantities are *not* equivalent to the original dynamics rules, but they will, in many cases, provide valuable information to the original dynamics.

The dynamical equation for the density of state 1 can be called *return map*:

$$d^{t+1} = F(d^t) \tag{4}$$

where d^t is the density of state 1 at time t, and the $F(.)$ is determined either by actually running the rule $f(.)$ or by some approximation schemes. Note that different $f(.)$ ruless can give the same macroscopic $F(.)$ dynamic.

One approximation scheme called *mean-field theory* assumes that all inputs are independent of each other, and the probability for having state 1 when the n inputs contain m state 1 and $n - m$ state 0 is estimated by counting the percentage of input configurations (containing m state 1 and $n - m$ state 0) that are mapped to state 1. For a general introduction, see Gutowitz.[6]

To illustrate this approximation scheme, let me use the following rule as an example (the triplet is the value of the three inputs, and the number below the triplet is the value to be updated to):

$$\begin{array}{cccccccc} 111 & 110 & 101 & 100 & 011 & 010 & 001 & 000 \\ 1 & 0 & 1 & 1 & 1 & 0 & 0 & 0 \end{array} \tag{5}$$

When all three inputs are 1, the state value will be 1; when two inputs are 1 and one input is 0, two out of three configurations will be mapped to 1; when one input is 1 and two inputs are 0, one out of three configurations will be mapped to 1; and when all three inputs are 0, the state value will never be 1. It is easy to show that one can approximate the return map by

$$d^{t+1} = (d^t)^3 + 2(d^t)^2(1 - d^t) + d^t(1 - d^t)^2. \tag{6}$$

Simple manipulation shows that it leads to

$$d^{t+1} = d^t; \tag{7}$$

that is, the density of state 1 does not change with time.

In fact, some important information can be extracted from this approximation of return maps. If the return map has a stable fixed-point solution equal to zero, the limiting density of state 1 should be zero or very low. That is the case when the original system has a fixed-point dynamics with zero-density or low-density spatial configurations.

If the return map has a non-zero, stable, fixed-point solution, there are two possibilities for the original dynamics: (1) the original system has a fixed-point dynamics with a spatial configuration about half filled with 0s and half with 1s, and (2) the original system is chaotic, with some kind of "thermal equilibrium states" being reached. Even though the state value for each component changes constantly, the density of 1s is nevertheless a constant.

I have yet to discover a return map with chaotic solutions. Generally speaking, it is very difficult for *macroscopic quantities to fluctuate chaotically*. Occasionally, numerical simulation shows that macroscopic quantities such as the density of state 1 do fluctuate irregularly. Nevertheless, it is always because these simulations are carried out for systems with finite sizes. The magnitude of these irregular fluctuation decreases as the system becomes larger. And, in principle, they will disappear in the infinite size limit. See, however, the discussions in Bohr et al.[3] and Kaneko.[10]

Although the return map is not equivalent to the original dynamical rule, it can provide valuable information. Because of the low dimensionality of the return maps, it is easier to study its own "bifurcation" phenomena (how dynamics of the return maps change with the parameter). From these studies, one can then understand some aspects of the bifurcation phenomena in the original system. A study of nonlocal cellular automata rule space, following this strategy, is carried out in Li.[9] In particular, it is partially understood why some bits in the rule table, which are called "hot bits" in Li and Packard[7] are more important than others. It is because the hot bits change the form of the return map more drastically than other bits.[9]

SYSTEMS OF COUPLED SELECTORS

The cellular automaton rule defined in Eq. (5) can be written in another form:

$$x_i^{t+1} = \begin{cases} x_{j_1(i)}^t & \text{if } x_{j_2(i)}^t = 0 \,, \\ x_{j_3(i)}^t & \text{if } x_{j_2(i)}^t = 1 \,. \end{cases} \tag{8}$$

In other words, if the second input is in state 0, the rule transmits the state value from the first input; if the second input is in state 1, the rule transmits the value from the third input. This rule can be called a *selector*, or a *multiplexer*, with the second input called a *control input*; that is, it decides which input to select.

This rule turns out to be the most intriguing 3-input 2-state nonlocal cellular automata. A typical spatial-temporal pattern for this rule is shown in Figure 1. Although the limiting dynamics is periodic, the transient dynamics looks chaotic. This combination of long chaotic transients and simple limiting dynamics is typical for systems on the "edge of chaos."

From Figure 1, we can see some dark and light horizontal stripes. In order for the dark stripe to form, if one component has state 1, other components also tend to have state 1 so that the total number of components with state 1 is increased. This seemingly simple fact implies the existence of certain cooperation among components. Indeed, other components do not have reasons to follow suit when one component switches from 0 to 1, unless they are dragged into doing so.

The emergence of higher level structures is also a hallmark of the edge-of-chaos systems.

The transient time for systems of coupled selectors is observed to increase with the size of the system. More careful simulation shows that the increase is almost linear:

$$T_{av}(N) \sim N \tag{9}$$

where $T_{av}(N)$ is the mean transient time for systems with size N. If we exclude all degeneracies in choosing inputs, it has been observed that the increase of transient time is more than linear[9]:

$$T_{av}(N) \sim N^{1.2}. \tag{10}$$

There are many open issues concerning the transient behavior and I will not discuss them in length here. Briefly, there are questions regarding how large the system size should be in order to trust the scaling; what the distribution of transient times is for a fixed size; whether this distribution with respect to wiring sampling is different from that of initial configuration sampling; whether the mean transient time is a better quantity than the median transient time, and whether one should take the logarithm first, then do the average; how the permission of degenerate inputs change the result; and so on.

time $T + 1 \rightarrow 2T$

time $0 \rightarrow T$

FIGURE 1   A spatial-temporal pattern of the coupled selectors. The configuration of the system (horizontal string) consists of either 1 (black) or 0 (white). And the updating of the configuration is represented by showing the configuration at each time step (time is increased going down).

In Figure 1, the limiting configuration has a very low density of state 1. If we change the wiring and initial condition, and run the simulation again, it is possible that the limiting configuration can have a very high instead of a very low density of state 1. These two types of configurations are called *consensus states*. It is not clear before finishing the simulation which consensus state will be reached. In fact, it has been observed that the density could wander up and down in such a way that the system almost hits the high-density consensus state before turning the trend to eventually reach the low-density one!

GROUP MEETING PROBLEMS

Imagine a group of people having a meeting. The goal of the meeting is to find a consensus opinion: either most of the people vote yes, or most of them vote no. Requiring a 100 percent yes-vote or no-vote may not be realistic, so some compromise is made: we allow the meeting to finish whenever the density of yes or no is higher or lower than a certain threshold value.

The system of coupled selectors discussed in the last section can be recast into a group meeting problem. At the beginning of the meeting, each person votes yes or no randomly. Then each person chooses three other persons (he can also choose himself) as his or her "inputs." Each one of the three inputs is labeled as either the first, the second, or the third input. The second person is most important: whenever he or she votes no, the first person's vote will be followed; and whenever he or she votes yes, the third person's vote is followed.

This somehow bizarre way for a group meeting to proceed is, nevertheless, not as trivial as one might have thought. First of all, will the group meeting ever reach a consensus? By the result presented in the last section, the answer is yes. But this is true only when the three inputs are chosen randomly. There are examples where a consensus is never reached. For example, if the wiring is partially local, i.e., $j_2(i) = i$, it is almost always the case that the dynamics is chaotic and the density of state 1 is around 0.5.[9]

Second, even if a consensus is reached, do we know *which one*?

For the system of coupled selectors, we do not know beforehand whether it is an all-yes or all-no state that is reached. Both low- and high-density configurations are "traps" or "attractors" of the dynamics. If we consider the fluctuation of the density as a random walk (though it is a *deterministic* random walk because initial configuration, wiring diagram, and dynamical rule are fixed during the updating), it has a 50–50 chance to reach either the low-density or the high-density configuration.

The system of coupled selectors is not the only system to have two consensus states. Actually, there exists a large class of "unbiased" rules that behave similarly (by "unbiased," I mean that the rule does not have any reason to prefer either one of the consensus states). Interestingly, one such rule, a 7-input 2-state local cellular automaton, called Gacs-Kurdyumov-Levin rule, was proposed more than ten years ago,[5] defined as the following:

$$x_i^{t+1} = \begin{cases} \text{majority among } x_i^t, x_{i-1}^t, \text{ and } x_{i-3}^t & \text{if } x_i^t = 0; \\ \text{majority among } x_i^t, x_{i+1}^t, \text{ and } x_{i+3}^t & \text{if } x_i^t = 1. \end{cases} \tag{11}$$

It has been shown that Gacs-Kurdyumov-Levin rule (11) has two attractors: all-zero and all-one configurations. The all-zero consensus state will be reached if the initial density of state 1 is smaller than 0.5; and the all-one consensus state will be reached if the initial density is larger than 0.5 (this result is proved for Gacs-Kurdyumov-Levin rule (11) in the infinite size limit).

Similar to the systems of coupled selectors, consensus states may not be reached for Gacs-Kurdyumov-Levin rule if the wiring diagram is modified. For example if the majority is chosen among x_i^t, x_{i-1}^t, and x_{i-2}^t when $x_i^t = 0$, and among x_i^t, x_{i+1}^t, and x_{i+2}^t when $x_i^t = 1$, then the limiting density will be more or less the same with the initial density. So, if the initial configuration is random, the limiting density will be around 0.5 instead of 0 or 1. We can see easily that the Gacs-Kurdyumov-Levin rule can also be translated to a group meeting problem. If the periodic boundary condition is used, we are going to have a "roundtable group meeting"!

With limited space and time, I can only introduce a few topics on nonlocal cellular automata. There are other major topics that are not covered here, for example, viewing nonlocal cellular automata as computers, the structure of nonlocal cellular automata rule space. For interested readers, see my paper for more details.[9] I hope I have conveyed to readers the richness of dynamical behaviors for dynamical systems with nonlocal interaction, and I hope more people will share my excitement in studying these systems.

ACKNOWLEDGMENTS

The work at Santa Fe Institute was funded by MacArthur Foundation, National Science Foundation (NSF grant PHY-87-14918), and Department of Energy (DOE grant DE-FG05-88ER25054). This article was written at Rockefeller University where I was supported by DOE grant DE-FG02-88-ER13847.

UPDATE IN THE 1996 EDITION

Recent results show that the "wide-spreading distribution" of transient and cycle lengths in Kauffman nets mentioned in the 1992 edition of this lecture note is actually inverse power-law distributions.[1,2] The exponents of these two inverse power-law distributions decrease with the number of components in the nets (N), effectively increasing the median transient and cycle length. Since the mean of a power-law distribution may diverge from the tail when the exponent of the power-law function is too small, we now understand why the mean value of the transient or the cycle length in Kauffman nets were not measured in the original study of this system.

REFERENCES

1. Bhattacharija, and S. Liang. "Median Attractor and Transients in Random Boolean Nets." *Physica D* (1996): to appear.
2. Bhattacharija, and S. Liang. "Power-Law Distributions in Kauffman Net." (1996): submitted.
3. Bohr, T., G. Grinstein, Y. He, and C. Jayaprakash. "Coherence, Chaos, and Broken Symmetry in Classical, Many-Body Dynamical Systems." *Phys. Rev. Lett.* **58(21)** (1987): 2155–2158.
4. Crutchfield, J., and K. Kaneko. "Phenomenology of Spatial-Temporal Chaos." In *Directions in Chaos*, edited by B.-L. Hao. Singapore: World Scientific, 1987.
5. Gacs, P., G. L. Kurdyumov, and L. A. Levin. "One-Dimensional Uniform Arrays that Wash Out Finite Islands." *Probl. Peredachi. Info.* **14**: (1978): 92–98.
6. Gutowitz, H. "Hierarchical Classification of Cellular Automata." *Physica D* **45(1-3)** (1990): 136–156.
7. Li, W., and N. Packard. "Structure of the Elementary Cellular Automata Rule Space." *Complex Systems* **4(3)** (1990): 281–297.
8. Li, W., N. Packard, and C. Langton. "Transition Phenomena in Cellular Automata Rule Space." *Physica D* **45** (1990): 77–94.
9. Li, W. "Phenomenology of Non-local Cellular Automata." *J. Stat. Phys.* **68(5/6)** (1992): 829–882.
10. Kaneko, K. "Globally Coupled Chaos Violate the Law of Large Numbers but Not the Central-Limit Theorem." *Phys. Rev. Lett.* **65** (1990): 1391–1394.
11. Kauffman, S. A. "Metabolic Stability and Epigenesis in Randomly Constructed Genetic Nets." *J. Theor. Biol.* **22** (1969): 437–467.

12. Kauffman, S. A. "Emergent Properties in Randomly Complex Automata." *Physica D* **10** (1984): 145–156.
13. Press, W. H., B. P. Flannery, S. A. Teukolsky, and W. T. Vetterling. *Numerical Recipes in C.* Cambridge, MA: Cambridge University Press, 1988.
14. Toffoli, T., and N. Margolus. *Cellular Automata Machines—A New Environment for Modeling.* Cambridge, MA: MIT Press, 1987.
15. Walker, C. C., and W. R. Ashby. "On Temporal Characteristics of Behavior in a Class of Complex Systems." *Kybernetik* **3** (1966): 100–108.
16. Walker, C. C. "Behavior of a Class of Complex Systems: The Effect of System Size on Properties of Terminal Cycles." *J. Cybernetics* **1(4)** (1971): 57–67.
17. Wolfram, S. "Statistical Mechanics of Cellular Automata." *Rev. Mod. Phys.* **55** (1983): 601–644.
18. Wolfram, S., ed. *Theory and Applications of Cellular Automata.* Singapore: World Scientific, 1986.

Alan C. Newell
University of Arizona, Department of Mathematics, Tucson, AZ 85721

The Dynamics and Analysis of Patterns

This chapter originally appeared in *Lectures in the Sciences of Complexity*, edited by L. Nadel and D. Stein, 107–173. Santa Fe Institute Studies in the Sciences of Complexity, Lect. Vol. I. Reading, MA: Addison-Wesley, 1989. Reprinted by permission.

1. INTRODUCTION

A. GENERAL

Patterns of an approximately periodic nature appear everywhere. One sees them in cloud formations, on the surface of the saguaro cactus, in the ripples of sand in deserts and near the edge of shallow water, in ocean gravity waves. These lectures are primarily directed toward patterns which arise when a system is driven away from some equilibrium state and undergoes sudden transitions which, with increasing external stress, break the symmetry of the simple, least stressed state, in more and more complicated ways. From such a series of bifurcations, one obtains organized structures which in some sense represent optimal configurations for the system at given values of its intrinsic parameters and the external stress parameter.

For most of these lectures, we will use convection in fluids, either driven by temperature differences of by electric forces, as our working model. This model, together with the analysis and language which emerges from its investigation, serves as a paradigm for analyzing similar behavior in a broad range of situations. I will also use simple phenomenological models which, although they are not direct reductions of any set of equations which describe accurately the microscopic behavior of a real system, are nevertheless useful because one can work out all the mathematical details simply and explicitly and thereby avoid clouding the main ideas with cumbersome calculations.

The level of difficulty increases as one progresses through the set of lectures and as more realistic situations are addressed. However, most readers should easily be able to handle the ideas in the introduction and the elementary mathematical models of the first two chapters. Once the reader gets to Chapters 4 and 5, he should also have started to read some of the recommended papers in the literature. As the references are listed in alphabetical order, the recommended papers are identified by numbers. It includes the papers numbered 14, 15, 16, 21, 34, 42, 52, 56, 58, 59, 65, 66, 71, and 72.

B. WHAT ONE WOULD LIKE TO ACCOMPLISH

Look at the patterns in Figure 1, taken from the beautiful paper of Heutmaker and Gollub.[34] It contains a series of pictures, taken from above, of water convecting in a circular container of approximately 3 mm depth and 42 mm radius. The pattern reflects the fact that the motion of the fluid consists of a series of cellular rolls, whose width is of the order of the depth, rolls which are bent in order to fit into the circular box. The rolls are least damped at the boundary when their axes are perpendicular to its tangent. The bright lines result from the light being focused along contours of phase where the fluid is coldest and the motion is downward. Notice that one cannot simply fill the circle with straight parallel rolls nor circular patches. There must be singularities in the pattern such as the dislocation shown in Figure 2, taken from the paper of Cross and Newell.[21] Our goal is to understand such a pattern by answering questions like: Does it reach an equilibrium? If so, is it unique? If not unique, what is the set of possible final states and do they share any common (e.g., topological) features? If it does not reach equilibrium, what keeps the pot stirring? What is the nature of these instabilities? Can the pattern settle down to a time-dependent state which is periodic or at least statistically steady (e.g., a strange attractor)?

Look at Figure 3. This is taken from a series of experiments by Ribotta[57] of convection in liquid crystals. Figure 3 is an (x,t) plot of the formation of a dislocation. The wavelength is simply too long and the pattern adjusts by lowering the amplitude where the rolls are widest and then inserting a new roll. This is very like the dislocation one sees in (x,y) space; for example, Figure 11. The pattern in Figure 3 consists of very slow left-moving waves (the crests move very slowly leftward as t increases) and the wavelike and elastic character of the system is seen in the relaxation oscillations of the crests. There are two qualitative, fundamental differences between these experiments and those of Gollub. First, in nematic liquid crystals, there is a preferred horizontal direction so that roll patterns in all directions are not equally likely, and the system does not possess the rotational symmetry of the water convection experiment. Second, the instability can and often does set in as a set of traveling waves, rolls which change their rotation sense as a periodic

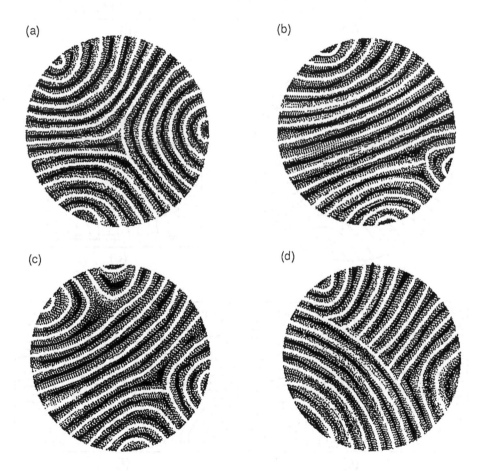

(a)   (b)

(c)   (d)

FIGURE 1 A reproduction of four convection patterns taken from Heutmaker and Gollub[34] (with permission from Academic Press LTD., London England) exhibiting defect structures. (a) $\epsilon^2 = 1.6, t = 40\tau_H$ (pattern stationary); disclination separating three circular patches. (b) $\epsilon^2 = 1.56, t = 1.91\tau_H$ (pattern still changing); disclination separating two upper foci from circular patch about lower focus. (c) $\epsilon^2 = 1.56, t = 49.7\tau_H$ (pattern stationary); four sidewall foci and two disclinations. (d) $\epsilon^2 = 0.10, t = 45.6\tau_H$ Grain Boundary. Reprinted by permission of the American Physical Society.

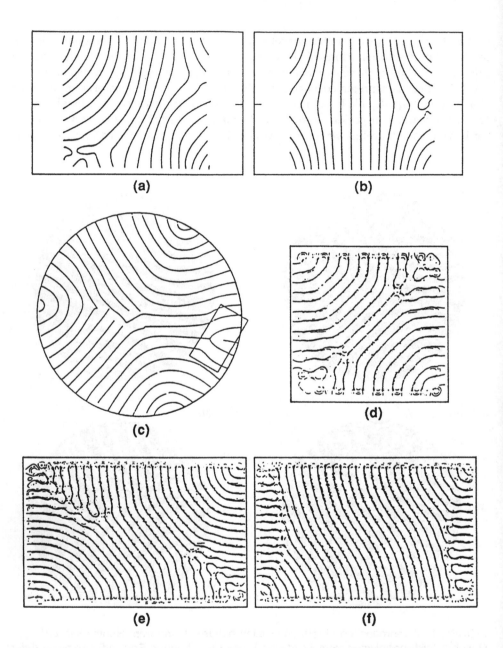

FIGURE 2   Configuration of convective rolls in large aspect ration cells. (a) and (b) are experimental results of Gollub et al.[30] at $R/R_c = 4, \sigma = 2.5$; (c) of Croquette et al.[24] at $R/R_c = 1.4, \sigma = 380$; (d), (e) and (f) are from numerical simulations of a model equation by Greenside et al.[32] The boxed region in (c) displays a dislocation. Reprinted by permission of Elsevier Science Publishers, Physical Sciences and Engineering Division.

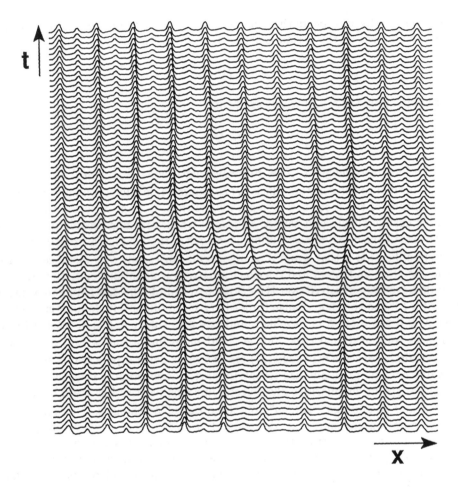

FIGURE 3 A dislocation showing the insertion of a roll pair. Reprinted by permission of *Journal de Physique.*

function of time. This means that the resulting patterns are not simply diffusion dominated but can have wavelike (dispersion, nonlinear focusing) behavior. One of the great advantages of the liquid crystal experiment is that it provides a situation in which the aspect ratio is huge (the wavelength of the pattern is of the order of 1/10 mm, whereas the box size is of the order of centimeters) and therefore one can avoid boundary-dominated effects. A disadvantage is that, to date, a generally accepted set of equations for the microscopic behavior has not been found.

How can one hope to describe these patterns analytically? The key is to notice that when the aspect ratio $\Gamma = L/\lambda$ (ratio of width box L to wavelength of pattern

λ) is large, locally it appears that the field w (velocity, temperature, electric) is a 2π-periodic function (the period is arbitrarily chosen to be 2π)

$$w = f(\theta, A) \tag{1.1}$$

of a phase variable $\theta(x, y, t)$ whose gradient

$$\vec{k} = \nabla\theta \tag{1.2}$$

gives the local wavelength (λ, the roll wavelength is $2\pi/k, k = |\vec{k}|$) and direction $\hat{k} = \vec{k}/k$ normal to the constant phase (θ) contours. A, locally a constant, represents the amplitude of the motion. The main idea is that, in large aspect ratio systems, A and $\vec{k}$ change slowly and satisfy equations which are macroscopic in the sense that the small-scale (λ), almost periodic features of the field are averaged over. The great simplification is that $\vec{k}$ and A change significantly only over distances of order L whereas, if we had integrated the exact microscopic equations of motion, the field w would have to be resolved over the distance λ. Couple the depth and area savings (order $(\lambda/L)^3$) to the time change (order $(\lambda/L)^2$) in diffusively dominated systems, order λ/L if the motion is wavelike), and one sees that one has at least a $(\lambda/L)^4$ savings in computational cost. Further, and more important, $\vec{k}$ and A (together with certain large-scale drift fields $\vec{U}$) are the most natural parameters to describe the patterns. The price one has to pay in such a description is that there are certain places (point singularities like defects, line singularities like grain boundaries) where $\vec{k}$ and A undergo "sudden" (on the order of λ) changes. The hope is that the configuration and motion of defects and grain boundaries, which are perfectly well behaved when seen from the point of view of the microscopic description, can be captured by allowing for singular and weak solutions of the macroscopic equations. In other words, near the defect in Figure 2, does the microscopic field w, as $d/\lambda = \sqrt{(x - x_0)^2 + (y - y_0)^2}/\lambda \to \infty$, approach a weak solution of the macroscopic equation with the property that, on a closed loop surrounding x_0, y_0 at a distance that is large with respect to λ but small with respect to L, $\int_c \vec{k} \cdot d\vec{s} = 2\pi$? (Notice $\int_c \vec{k} \cdot d\vec{s} = [\theta]$, the jump in phase as we go around the singularity and is equal to 2π; simply count the number of maximum phase contours one crosses if one passes the singularity on two different sides.) In other words, from a set of microscopic field equations with no singularities, we have produced, by averaging over the pattern, macroscopic fields $\vec{k}$ which have both smooth and particle-like properties.

To achieve such a description is the hope. We are far from that point at which we can say that we have a rigorous, self-consistent theory. This is one of the many great challenges facing the theoretician.

C. OUTLINE OF LECTURES

The lectures follow closely the survey article "The Dynamics of Patterns"[47] and much of that material is contained here. However, more time will be spent on discussing the information one gets from linear and weakly nonlinear stability theory, and several elementary examples which show the reader how to handle the mathematics in simple cases are worked out in detail. We break the discussions into two main categories.

I. CONVECTION NEAR ONSET which looks at the behavior of the solution near that value R_c of the external stress parameter R (e.g., the Raleigh number) at which the simple (conduction) solution first loses its stability. In this case, we seek to describe the behavior of the system by writing down ordinary differential equations or simple partial differential equations for the amplitudes (the order parameters) of modes belonging to a special set A which play important roles in the dynamics near R_c. The determination of the set A is discussed in detail. In this analysis, the small parameter which makes the problem tractable is

$$\epsilon^2 = \frac{|R - R_c|}{R_C} \qquad \text{or} \qquad \epsilon^2 = R \qquad \text{if} \qquad R_c = 0 \qquad (1.3)$$

the fractional amount by which R exceeds or is less than the critical value R_c. There are two subclasses I_a and I_b. In I_a, the aspect ratio Γ is order one and, in this case, the set A has only a finite number of elements. I_b, Γ is large, more definitely $\Gamma\epsilon$ is large, so that the dimension of A is very large. Whereas there are theorems which assure us that in I_a the dynamics of the original equation is well approximated by the dynamics of the elements of A, there are no theorems for case I_b, although we shall argue that the description we use correctly captures the behavior of the full system. The proof that it does is still open.

II. CONVECTION FAR FROM ONSET in which ϵ^2 is order one but Γ, the aspect ratio, is large. Here, as we have already described, the basic idea is to use the fact (if it is true) that there exists stable, periodic solutions (1.1) in the infinite geometry case and to look for solutions in which the wavevector $\vec{k} = \nabla\theta$ (θ changes over distances λ but its gradient changes over distances $L = \lambda\Lambda$) changes slowly, that is, over distances that are long with respect to the wavelength. The analysis used can be described as the Whitham theory for patterns as Whitham[70,71] used it first to describe the modulation of neutrally stable fully nonlinear wavetrains. The major difference with Category I is that the amplitude parameter A tends to be slaved to the modulus of the phase gradient except near defects when it assumes a life of its own. Therefore, once the wavevector is known, the whole field w can be reconstructed using Eq. (1.1) providing there are no large-scale mean drift fields $\vec{U}$. When there are, the field becomes a "sum" of $f(\theta, A)$ and $\vec{U}$, and the equations for $\vec{k}$ and $\vec{U}$ are coupled. On the other hand, as $R \to R_c$, the amplitude A becomes less and less slaved and eventually satisfies a partial differential equation of its own. In this limit, mean drift effects tend to be less important.

Basically what we shall find is that $\vec{k}$ satisfies a nonlinear diffusion equation with coefficients which depend on its modulus k. When these coefficients change sign, the solution f loses its stability. The range of wavenumbers $(k_L(R), k_R(R))$ for which the solution is stable is called the stability band. It depends on R (and also on other parameters in the problem). The region of R, k (see Figures 10, 13, and 14) for which the solutions (1.1) exist and are stable, is called the Busse balloon.[14,15,21] Interesting behavior occurs when, through various influences, the wavenumber is forced (locally) to leave the balloon, in which case the field responds by triggering the relevant instability which attempts to bring the wavenumber back to the balloon. (I should warn the reader that this presupposes that the instabilities are long wavelength only. I will return to this point later.) If the wavenumber everywhere in the box can remain within the Busse balloon, then the motion tends to relax to an equilibrium state. If not, then it tends to relax to a chaotic attractor.

D. CONTENTS OF CHAPTER

2. LINEAR STABILITY THEORY

A. GENERAL DISCUSSION

Suppose we have a system governed by the nonlinear equation

$$F\left(\frac{\partial}{\partial t}, \frac{\partial}{\partial z}, \vec{\nabla}_H, R, u\right) = 0. \tag{2.1}$$

In Eq. (2.1), t is time, z represents directions (by analogy with the convective problem we call these vertical) which are finite in extent, $\vec{\nabla}_H$ represents the gradient along $\vec{x} = (x, y)$ which are directions infinite in extent (we call these horizontal), R is a (set of) stress parameter(s), and u is the (vector) field. We begin with a simple exact solution u_0 of Eq. (2.1) which, again by analogy with convection, we call the conduction solution. It is natural to ask when u_0 loses stability to linear perturbations. Set $u = u_0 + w$ and ignore terms quadratic in w and its derivatives.

We obtain (for simplicity I will assume u_0 is zero and L does not depend on x, y, z, t explicitly)

$$L\left(\frac{\partial}{\partial t}, \frac{\partial}{\partial z}, \vec{\nabla}_H, R\right) w = 0. \tag{2.2}$$

We look for solutions for Eq. (2.2)

$$w = e^{\sigma t} e^{i\vec{k}\cdot\vec{x}} \phi_n(z) \tag{2.3}$$

where $\phi_n(z)$ is one of a set of shapes compatible with boundary conditions on w, and obtain a complex dispersion relations

$$L(\sigma = \nu - i\omega, \vec{k}, n, R) = 0 \tag{2.4}$$

which gives the growth rate ν

$$\nu = \nu(\vec{k}, n, R) \tag{2.5}$$

and frequency ω

$$\omega = \omega(\vec{k}, n, R) \tag{2.6}$$

as functions of $\vec{k}$ and R. We ask: for which modes $\vec{k}$ and at which value R_c does the growth rate ν first become positive? The boundary of this domain $\nu(\vec{k}, n, R) = 0$ is called the set of neutral stability surfaces $R_n = R(\vec{k}, n), n = 1, 2, \ldots$ and we call the point or set of points on these surfaces for which R is minimal over n and $\vec{k}$ the critical mode values $\{n_c \vec{k}_c\}$. Sometimes the growth rate will first become positive when $\omega \equiv 0$, in which case one makes a transition from a damped to excited state with no oscillations. This is called exchange of stabilities by Chandrasekhar.[17] If $\omega \not\equiv 0$, then this situation is called overstable. Neither is a very good name. If there is more than one $\vec{k}$ value for which R is minimal, we say we have a degeneracy. Often, degeneracies reflect symmetries which the original system enjoys. The complex dispersion relation (2.4) and the neutral stability curve

$$\nu(\vec{k}, n, R) = 0 \tag{2.7}$$

are very important.

B. SWIFT-HOHENBERG MODEL ON A FINITE LINE

Consider the exact solution $w = 0$ of

$$\frac{\partial}{\partial t} w + \left(\frac{\partial^2}{\partial z^2} + 1\right)^2 w - Rw + w^3 = 0 \tag{2.8}$$

with boundary conditions

$$w = \frac{\partial^2 w}{\partial z^2} = 0, \qquad z = 0, L. \tag{2.9}$$

I use the notion z to denote spatial directions of finite extent. Linearizing about $w = 0$ (set $w = 0 + w'$ in Eq. (2.8) and ignore all terms not linear in w'; drop $'$ from w') and obtain

$$\frac{\partial w}{\partial t} + \left(\frac{\partial^2}{\partial z^2} + 1\right)^2 w - Rw = 0, \tag{2.10}$$

$$w = \frac{\partial^2 w}{\partial z^2} = 0, \quad z = 0, L. \tag{2.11}$$

Use separation of variables

$$w(z,t) = T(t)Z(z) \tag{2.12}$$

to find solutions and obtain (substitute Eq. (2.12) into Eq. (2.10), divide across by TZ)

$$+\frac{1}{T}\frac{dT}{dt} = -\frac{1}{Z}\left[\left(\frac{d^2}{dz^2} + 1\right)^2 Z - RZ\right] = \sigma$$

where σ must be constant. The nontrivial solutions (eigenfunctions) to

$$\left(\frac{d^2}{dz^2 + 1}\right)^2 Z + (\sigma - R)z = 0 \tag{2.13}$$

with boundary conditions $Z = d^2 Z/dz^2 = 0, z = 0, L$ are (Eq. (2.13) is a fourth-order, constant coefficient equation; look for a solution as a linear combination of exponentials and apply boundary conditions)

$$Z_n = \sin\frac{n\pi z}{L}, \tag{2.14}$$

and only occur for a very special set of σ's (called the spectrum or eigenvalues of the boundary value problem (2.13))

$$\sigma_n = R - \left(\frac{n^2\pi^2}{L^2} - 1\right)^2. \tag{2.15}$$

The corresponding time behavior is $T = A_n e^{\sigma_n t}$. Therefore the general solution to Eq. (2.16) can be written

$$w(z,t) = \sum_{n=1}^{\infty} A_n e^{\sigma_n t} \sin\frac{n\pi z}{L} \tag{2.16}$$

where the σ_n are given by Eq. (2.15). I now discuss the neutral stability curves $\sigma_n = 0$ for various n in the R, L plane (see Figure 4). For $L = L_B$, if we raise R, then the first mode to destabilize is $n = 1$, which occurs at $R_c = ((\pi^2/L_B^2) - 1)^2$. For $R < R_c, R = R_c, R > R_c$, the position of the eigenvalues (growth rates) are as shown in Figure 4. We will shortly be asking what happens when $R = R_c(1 + \epsilon^2 \alpha)$. Note that linear theory tells us that mode $n = 1$ will grow at a rate $\sigma_1 = \epsilon^2 R_c \alpha t$, but all other modes decay at the much faster rates $e^{\sigma_n t}$ where $\sigma_n = ((\pi^2/L_B^2) - 1)^2(1 + \epsilon^2 \alpha) - ((n^2 \pi^2/L_B^2) - 1)^2$ is a negative number of order one. The idea is then that even when we include the nonlinear term, we can approximate the solution to the nonlinear equation (2.8) by $A_1(t) \sin(\pi x/L_B)$.

If we choose L to be L_D defined by $(\pi^2/L_D^2) - 1 = -(4\pi^2/L_D^2) + 1$, then as R is raised to $R_c = ((\pi^2/L_D^2) - 1)^2$, both eigenvalues σ_1 and σ_2 become positive at the same time. If we now were to ask what happens to the nonlinear solution in the vicinity of $L = L_D, R_c = ((\pi^2/L_D^2) - 1)^2$ (i.e., suppose $R = R_c(1 + \epsilon^2 \alpha), L = L_D(1 + \epsilon^2 \beta)$, then we have to include both modes $A_1 \sin(\pi z/L)$ and $A_2(t) \sin(2\pi z/L)$ in the leading order approximation to the field w.

The choice of A, namely the set of modes which determine the behavior of the fully nonlinear solution to Eq. (2.8) near a bifurcation value R_c of the stress para-

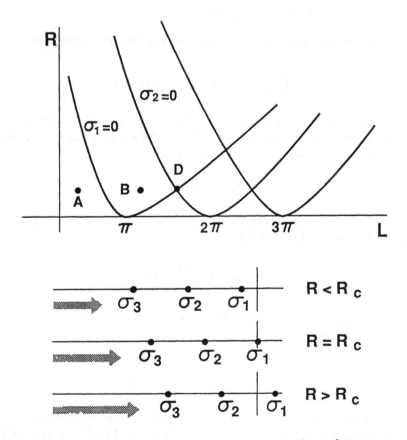

FIGURE 4 Neutral stability curves and growth rates as functions of stress parameter R.

meter, is made by looking at all those modes whose growth or decay rates ν (namely, the real part of σ) are less than ϵ^2, i.e., $|\sigma| \leq O(\epsilon^2)$ when R is within ϵ^2 of R_c. If the initial instability has an oscillatory behavior, then ω, the imaginary part of σ, will be nonzero. It is also important to include in A neutral modes, namely those modes whose real part is zero for a range of R in the neighborhood of R_c. These modes may not grow as R is raised, but they are not damped either and so can play important catalytic roles in the behavior of the nonlinear system.

C. THE LORENZ EQUATIONS

Let us now consider a finite-dimensional system, a set of three nonlinear o.d.e.'s,

$$\frac{dX}{dt} = -pX + pY,$$
$$\frac{dY}{dt} = rX - Y - XZ, \tag{2.17}$$
$$\frac{dZ}{dt} = -bZ + XY.$$

Begin with the simple solution $X = Y = Z = 0$ and examine its linear stability properties (set $X = 0+X'$, etc., substitute into Eq. (2.17), ignore nonlinear products and, for convenience, drop primes)

$$\frac{dw}{dt} = Aw, \quad w = \begin{pmatrix} X \\ Y \\ Z \end{pmatrix}, \quad A = \begin{pmatrix} -p & p & 0 \\ r & -1 & 0 \\ 0 & 0 & -b \end{pmatrix}. \tag{2.18}$$

Let $w = \hat{w}e^{\sigma t}$, and nontrivial solutions of Eq. (2.18) mean that σ must be one of the three eigenvalues of A

$$\sigma_{1,2} = -(p+1) \pm \sqrt{(p+1)^2 + 4p(r-1)}, \tag{2.19}$$
$$\sigma_3 = -b.$$

Note that as r is increased to one, the largest eigenvalue σ_a increases to zero. For the subsequent analysis, it will be important to calculate the right and left normalized eigenvectors of A when $r = 1$. They are

$$\hat{u}_1 = \frac{1}{\sqrt{2}}\begin{pmatrix} 1 \\ 1 \\ 0 \end{pmatrix}, \quad \hat{u}_2 = \begin{pmatrix} -p \\ 1 \\ 0 \end{pmatrix}\frac{1}{\sqrt{1+p^2}}, \quad \hat{u}_3 = \begin{pmatrix} 0 \\ 0 \\ 1 \end{pmatrix}. \tag{2.20}$$

The corresponding left eigenvectors are

$$\hat{v}_1^T = \frac{1}{\sqrt{1+p^2}}(1,p,0), \quad \hat{v}_2^T = \frac{1}{\sqrt{2}}(-1,1,0), \quad \hat{v}_3^T = (0,0,1), \tag{2.21}$$

and Eqs. (2.20) and (2.21) are biorthogonal bases; i.e., $\hat{v}_i^T \hat{u}_j = c_i \delta_{ij}; c_1 = c_2 = (1+p)/(2(1+p^2)^{1/2}), c_3 = 1$.

Note that at $r = 1 + \epsilon^2 \alpha$, the general solutions to Eq. (2.18) has the form

$$A_1 e^{\sigma_1 t} \hat{u}_1 + A_2 e^{\sigma_2 t} \hat{u}_2 + A_3 e^{\sigma_3 t} \hat{u}_3$$

which, at time t such that $\epsilon^2 t = O(1)$, asymptotically approaches

$$A_1 e^{\sigma_1 t} \hat{u}_1$$

because $\sigma_2 \simeq -2(p+1)$ and $\sigma_3 = -b$. So all initial vectors $w = A_1 \hat{u}_1 + A_2 \hat{u}_2 + A_3 \hat{u}_3$ very quickly move so that they lie along the direction of $\hat{u}_1$. This observation is the key to examining the nonlinear behavior of the system (2.17) near $r = 1$. Think of an arbitrary initial vector

$$w = A_1 \hat{u}_1 + A_2 \hat{u}_2 + A_3 \hat{u}_3.$$

First it aligns itself along $\hat{u}_1$, but then we will see that due to nonlinear effects it will also get a $\hat{u}_2$ and $\hat{u}_3$ component. However, and this is the important point, the coefficients of $\hat{u}_2$ and $\hat{u}_3$, namely A_2, A_3 will depend algebraically on A_1, the coefficient in front of $\hat{u}_1$. Therefore, the solution to the nonlinear problem will lie on a manifold (in fact, in this case a curve) $A_2 = A_2(A_1), A_3 = A_3(A_1)$, which is called the center manifold and which, for A_1 very small, starts out from the origin along $\hat{u}_1$ (in the negative or positive directions).

D. SWIFT-HOHENBERG MODEL ON AN INFINITE LINE

$$\frac{\partial w}{\partial t} + \left(\frac{\partial^2}{\partial x^2} + 1 \right)^2 w - Rw + w^3 = 0, \qquad -\infty < x < \infty. \qquad (2.22)$$

Consider that the domain is infinite in one horizontal direction and that the allowed solutions of Eq. (2.22) are smooth and bounded as $x \to \pm\infty$. The stability of the conduction solution $w = 0$ is investigated by setting $w = 0 + w'$, inserting in Eq. (2.22), and ignoring quadratic terms. Dropping the prime,

$$\frac{\partial w}{\partial t} + \left(\frac{\partial^2}{\partial x^2} + 1 \right)^2 w - Rw = 0 \qquad (2.23)$$

solutions exist in the form

$$w(x,t) = e^{\sigma t + ikx} \qquad (2.24)$$

for any real k as long as

$$\sigma(k, R) = R - (k^2 - 1)^2. \qquad (2.25)$$

The minimum value R for which $\sigma(k, R)$ first becomes positive if $R = 0$ which occurs when $k^2 = 1$. We draw the neutral stability curve in Figure 5. Note that the spectrum σ is continuous. Thus far $R > R_c$, say, $R = \epsilon^2\alpha$, there is a finite bandwidth (infinite number) of modes of width

$$1 - \frac{1}{2}\epsilon\sqrt{\alpha} < k < 1 + \frac{1}{2}\epsilon\sqrt{\alpha} \tag{2.26}$$

and whose growth rates σ are all of order ϵ^2. Even though the most critical mode $k = 1$ grows the fastest initially, all the modes in the band (2.26) must be included in the nonlinear analysis. I will show you shortly how to incorporate them all in a clever way. This finite bandwidth effect is one kind of degeneracy we have to deal with in infinite geometries. Even if the geometry is finite, say, of length L, and the modes are quantified $k = n\pi/L$, there would still be a large number of modes (or order the bandwidth $\epsilon\sqrt{\alpha}$ divided by the distance $k_c L/2\pi$ between the modes)

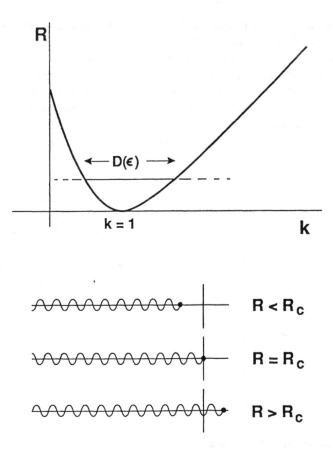

FIGURE 5   Neutral stability curve and growth rates as function of stress parameter R.

equal to $\sqrt{\alpha}\Gamma\epsilon$ where Γ is the aspect ratio. Thus in the limit $\Gamma \gg 1/\epsilon$, the modes are so close together that we might as well think of them as a continuum.

E. THE TWO-DIMENSIONAL SWIFT-HOHENBERG MODEL

$$\frac{\partial w}{\partial t} + (\nabla^2 + 1)^2 w - Rw + w^2 = 0. \tag{2.27}$$

Consider those fields $w(x, y, t)$ which are smooth and bounded at infinity. The stability of the conduction solution is found by linearizing Eq. (2.27) about $w = 0$ setting

$$w(x, y, t) = e^{\sigma t + i\vec{k} \cdot \vec{x}}, \tag{2.28}$$

and obtaining

$$\sigma(\vec{k}, R) = R - (k^2 - 1)^2. \tag{2.29}$$

Note that, because of the rotational degeneracy (x and y derivatives in Eq. (2.27) occur only in the Laplacian), the growth rate only depends on the modulus of k. Therefore the minimum value of R for which σ is zero occurs on a circle of $\vec{k}$ modes, namely, all those which lie on the circle $|\vec{k}| = k = 1$. This means that we now have a second kind of degeneracy (see Figure 6(b)) and in the nonlinear analysis we must include in principle all the modes which lie on the circle. It will turn out that, because of nonlinear interactions, these modes will interact with each other. Some combinations will be suppressed, others will be favored. *It is the combination*

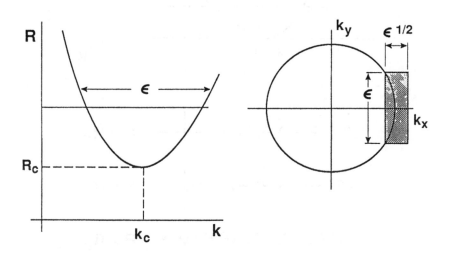

FIGURE 6   Neutral stability curve in $R, \vec{k}$ plane.

of $\vec{k}$ modes that determines the physical configuration of the convection pattern. One mode $\vec{k}_1 = (1,0)$ is a roll; two of equal amplitude give rise to rectangular cells and three of equal amplitude 120° apart give rise to hexagonal structures (see Chandrasekhar's[17] book).

F. THE OBERBECK-BOUSSINESQ EQUATIONS

The equations which describe the motion of the velocity field $\vec{u}(\vec{x}, t)$, the temperature field $\Theta(\vec{x}, t)$, the pressure field $p(\vec{x}, t)$ and density field $\rho(\vec{x}, t)$ in a fluid are

$$\frac{\partial \rho}{\partial t} + \vec{u} \cdot \nabla \rho + \rho \nabla \cdot \vec{u} = 0, \tag{2.30}$$

$$\rho \left(\frac{\partial \vec{u}}{\partial t} + \vec{u} \cdot \nabla \vec{u} \right) = -\nabla P - \rho g \hat{z} + \rho \nu \nabla^2 \vec{u}, \tag{2.31}$$

$$\frac{\partial \Theta}{\partial t} + \vec{u} \cdot \nabla \Theta = \kappa \nabla^2 \Theta, \tag{2.32}$$

$$\rho - \bar{\rho} = -\alpha(\Theta - \bar{\Theta}). \tag{2.33}$$

In Eq. (2.31), ν is the viscosity, $\hat{z}$ the unit vector in the vertical direction; in Eq. (2.32) κ is thermometric conductivity; and in Eq. (2.33) α is the coefficient of cubic expansion and $\bar{\rho}$ and $\bar{\Theta}$ are reference densities and temperatures. The geometry is that of an infinite, horizontal layer of fluid of depth d, and the boundary conditions on the bottom and top plates are

$$\Theta = \bar{\Theta} + \Delta, \qquad z = 0 \qquad \Theta = \bar{\Theta}, \qquad z = d \tag{2.34}$$

and

$$u = v = w = 0 \text{ at } z = 0, d. \tag{2.35}$$

Often, Eq. (2.35) is replaced by what turns out to lead to a simpler mathematical problem

$$\frac{\partial u}{\partial z} = \frac{\partial v}{\partial z} = w = 0 \text{ at } z = 0, d \tag{2.36}$$

which corresponds to the upper and lower boundaries being simply stress free rather than rigid. We will work with the latter conditions, although I expect the reader to carry through the linear and nonlinear analysis for the former.

The conductive state is given by

$$\vec{u} = 0, \qquad \Theta_c - \bar{\Theta} = -\frac{\Delta}{d}(z - d),$$

$$\rho_c - \bar{\rho} = \frac{\alpha \Delta}{d}(z - d), \qquad \frac{dP_c}{dz} = -\rho g. \tag{2.37}$$

The heat flow across the layer is proportional to $-d\Theta_c/dz$. To test its stability, we set

$$\vec{u} = \vec{u}, \qquad \theta = \Theta_c + \theta, \qquad P = P_c + p, \qquad \rho = \rho_c + \bar{\rho}\delta\rho. \tag{2.38}$$

We will also nondimensionalize the problem by setting

$$\vec{x} \to d\vec{x}, \quad \vec{u} \to \frac{\kappa}{d}\vec{u}, \quad t \to \frac{d^2}{\kappa}d,$$

$$\theta \to \frac{\Delta}{R}\theta, \quad p \to \rho_c \frac{\nu\kappa}{d^2}p \tag{2.39}$$

where

$$R = \frac{\alpha g d^3 \Delta}{\nu\kappa}$$

is the Rayleigh number (nondimensional temperature difference). We find

$$\nabla \cdot \vec{u} = 0,$$

$$\frac{1}{P_0}\left(\frac{\partial u}{\partial t} + u \cdot \nabla u\right) = -\nabla p + \theta\hat{z} + \nabla^2\vec{u},$$

$$\frac{\partial\theta}{\partial t} + u \cdot \nabla\theta - Rw = \nabla^2\theta, \tag{2.40}$$

$$\delta p = -\alpha\bar{\rho}\theta,$$

where $p_0 = \nu/\kappa$ is the Prandtl number and $\vec{u} = (u, v, w)$. The conductive solution is now $\vec{u} = \theta = p = 0$.

It is convenient to write an equation for the vorticity $\vec{\omega} = (\xi, \eta, \zeta)$

$$\frac{1}{P_0}\left(\frac{\partial\vec{\omega}}{\partial t} + \nabla \times (u \cdot \nabla)u\right) = \nabla\theta \times \hat{z} + \nabla^2\vec{\omega} \tag{2.41}$$

which eliminates the pressure and shows that horizontal gradients of temperature produce torques in the fluid. Again taking the curl of Eq. (2.41) and then its $\hat{z}$ component, we obtain (∇_1^2 is $(\partial^2/\partial x^2) + (\partial^2/\partial y^2)$)

$$\left(\frac{\partial}{\partial t} - \nabla^2\right)\nabla^2 w - \nabla_1^2\theta = \frac{1}{P_0}\hat{z}(\nabla \times \nabla \times (u \cdot \nabla)u). \tag{2.42}$$

Now apply $(\partial/\partial t) - \nabla^2$ and use the equation for temperature to obtain

$$\left(\frac{\partial}{\partial t} - \nabla^2\right)\left(\frac{1}{P_0}\frac{\partial}{\partial t} - \nabla^2\right)\nabla^2 w - R\nabla_1^2 w =$$

$$- \nabla_1^2(u \cdot \nabla)\theta + \frac{1}{P_0}\left(\frac{\partial}{\partial t} - \nabla^2\right)(\nabla \times \nabla \times (u \cdot \nabla)u) \cdot \hat{z}. \tag{2.43}$$

The linear stability analysis on the conduction solution can be carried out directly on Eq. (2.43) by ignoring the RHS. If we take stress-free boundary conditions (2.36), then the boundary conditions on w are greatly simplified to give

$$w = \frac{\partial^2 w}{\partial z^2} = \frac{\partial^4 w}{\partial z^4} = 0 \text{ at } z = 0, 1.\tag{2.44}$$

Thus we can look for solutions

$$w(x, y, z, t) = e^{\sigma t} e^{i\vec{k}\cdot\vec{x}} \sin n\pi z$$

and obtain a quadratic equation for σ as function of $k^2 = |\vec{k}|^2, n^2$ and R. It can be shown that the real part of σ will first become positive when σ itself is purely real and the neutral stability curve $\sigma(k, n, R) = 0$ is given by

$$R = \frac{(k^2 + n^2\pi^2)^3}{k^2}\tag{2.45}$$

which is minimized when $n = 1, k = k_c = \pi/\sqrt{2}$. Therefore, when $R = R_c(1 + \epsilon^2 \alpha)$, all modes in an annulus of width $\epsilon\sqrt{\alpha}$ about $\kappa = \pi/\sqrt{2}$ have order ϵ or less growth or decay rates. See Figure 6.

The reader should consult Chandrasekhar[17] for a more complete discussion of the linear stability problem.

3. CONVECTION NEAR ONSET; WEAKLY NONLINEAR THEORIES

A. GENERAL STRATEGY

We now ask in each of these five cases what happens when

$$R = R_c(1 + \epsilon^2 \alpha) \text{ or } R = \epsilon^2 \alpha \text{ or } R_C = 0\tag{3.1}$$

and the growth of the linearly unstable modes is affected by nonlinear interactions of an unstable mode both with itself and with others. The first thing we must do is decide which set A of modes should be included in the leading approximation to the solution of the nonlinear problem. In examples 3A and 3B, the choice is not difficult because it is clear that only a finite number of directions in the function space in 3A or one direction in the vector space in 3B can be important. For example in 3A, if $L = L_B$ and $R = R_B$, a value just above the neutral stability curve for the $\sin(\pi x)/L$ mode, i.e., $R = R_c(1 + \epsilon^2 \chi)$, $R_c = ((\pi^2/L_B^2) - 1)^2$, then the directions $\sin(n\pi x/L_B), n \geq 2$ are so heavily damped at the linear approximation that they never have a chance to play active roles in the dynamics. Because of nonlinearity, they will be excited by multiple products of $\sin(\pi x/L_B)$, but their amplitudes are slaved to (i.e., determined algebraically by) the amplitude of $\sin(\pi x / L_B)$. So even though solutions of Eq. (2.8) are potentially infinite dimensional, the dynamics in the neighborhood of L_B, R_B are eventually one dimensional. Similarly the dynamics near $r = 1$ in example 3B are also one dimensional. In these cases we take the leading order approximations to the nonlinear solutions to be

$$A_1(T = \epsilon^2 t) \sin \frac{\pi x}{L_B} \tag{3.2}$$

and

$$B(T = \epsilon^2 t)\hat{u}_1 \tag{3.3}$$

respectively. The amplitudes $A_1(T = \epsilon^2 t)$ and $B(T = \epsilon^2 t)$ are slowly varying functions of time, i.e., A_t and B_t are $O(\epsilon^2)$. For small A_1 and B, they begin by growing at an exponential rate as suggested by linear theory. However, once they grow to finite amplitudes, their time behaviors are modified by nonlinear feedback. If the feedback is negative, namely, tends to saturate the growth, then the time behavior of A_1 and B will continue to be slow.

In example 3C, things are a little more complicated because the spectrum is continuous and there is a continuum of modes, e^{ikx}, in an ϵ neighborhood of $k = 1$ which can play a role in the dynamics. How do we include this finite bandwidth? One way would simply be to take a sum over a discrete number of modes in the band; that is, A would consist of the infinite sum

$$\sum_j A_j(T = \epsilon^2 t)e^{i(1+\epsilon K_j)x} + (*). \tag{3.4}$$

[(*) refers to complex conjugate and is included if the solution is to be real.] However, we note that Eq. (3.4) an be written more simply as

$$A(X = \epsilon x, T = \epsilon^2 t)e^{ix} + (*) \tag{3.5}$$

which not only includes Eq. (3.4) as a special case, but also contains the full band of modes. The price we pay is that the dynamics of $A(X, T)$ is described by a partial differential equation. The big plus is that not only have we correctly captured the behavior near R_c, but the p.d.e. that $A(X, T)$ satisfies is universal and holds for a broad class of problems.

In example 2D, we have an additional degeneracy to the finite bandwidth one, and even the latter becomes slightly more complicated. Consider the mode $\vec{k} = (1, 0)$ lying on the critical circle. If $R = \epsilon^2 \chi$, there is an order ϵ bandwidth of modes in the k_x direction which can grow, and an order $\sqrt{\epsilon}$ bandwidth in the k_y direction. This can be seen from Figures 6(a) and 6(b). Therefore, the amplitude A of e^{ix} must be taken to be a slowly varying function of x through $X = \epsilon x$ and a slowly varying function of y throughout $Y = \sqrt{\epsilon} y$. It is interesting that the isotropy produced by the rotational symmetry of the problem leads to an anisotropy in the envelope equation for $A(X = \epsilon x, Y = \sqrt{\epsilon} y, T = \epsilon^2 t)$.

On the other hand, if there had been no such symmetry in the original problem and the minimum value R_c of R had occurred at a unique value $\vec{k}_c$ of $\vec{k}$ so that the R vs. $\vec{k}$ neutral stability surface $R l\sigma = \nu(\vec{k}, R) = 0$ had non-zero curvature in both the x and y directions, then the x and y directions would both scale the same, i.e., $A(X = \epsilon x, Y = \epsilon y, T = \epsilon^2 t)$, although the coefficients of A_{XX} and A_{YY}, reflecting the different curvatures of the neutral stability surface at its minimum might be different.

Whereas there is a satisfactory way in which to deal with the sideband degeneracy, there is, as yet, no really satisfactory way to deal with the rotational degeneracy question. What one would like is to have a natural envelope-type description for functions, the support of whose Fourier transform lies in an ϵ-annulus about the circle $k = k_c$. One can formally write things down in a Fourier-Bessel series, but this does not seem convenient. What we do instead is to look at finite sums of modes lying on the critical circle. At first, we ignore the sideband degeneracy and simply write the leading order approximation to $w(x, y, t)$ as

$$\sum_{|\vec{k}_j|=1} A_j(T) e^{i\vec{k}_j \cdot \vec{x}} + (*), \tag{3.6}$$

write down a set of o.d.e.'s for $A_j(T)$ [I will indicate how shortly], and determine which particular configurations (sets of solutions to these equations which will correspond to rolls, squares, rectangles, hexagons, higher-order multiple rolls) are most stable. Having determined the likely form of the solution, we then add the envelope structure which allows for modulations of amplitudes over distances of ϵ^{-1} times the roll wavelengths.

However, a caveat. Once modulations are introduced, it is important to test their stability against other platforms. For example, rolls which have the structure $e^{ik_c(1+\epsilon)x}$ may be unstable to cross rolls $e^{ik_c y}$ even though the center mode $e^{ik_c x}$ is not. The envelope equation which describes the behavior of the wavepacket $A(X = \epsilon x, Y = \sqrt{\epsilon} y, T = \epsilon^2 t) e^{ik_c x}$ will only describe correctly the behavior of the system when the modes which enter into the dynamics come from the neighborhood of $(k_c, 0)$.

Having established the structure of the active set A, we now address the question of how to find the equations for the mode amplitudes. To do this, we take advantage of the fact that the parameter ϵ is small and make more rigorous what we mean when we say "leading order approximation to the solution of the nonlinear problem" by constructing an asymptotic expansion for $w(x, t)$. We give now three definitions. (The reader should also consult the books by Olver,[50] Bleistein,[9] Erdelyi,[28] and Copson[18] on asymptotic expansion.) We say the sequence $\{\mu_n(\epsilon)\}$ is asymptotic as $\epsilon \to 0$ if $\mu_{n+1}(\epsilon) = o(\mu_n(\epsilon))$ for each n as $\epsilon \to 0$. The little o means that the limit of the ratio μ_{n+1}/μ_n goes to zero as ϵ tends to zero. For example, the sequences $\{\epsilon^n\}$ and $\{1, \epsilon, \epsilon^2 \ln \epsilon, \epsilon^2, \ldots\}$ are asymptotic. Next we say that the formal series

$$\sum_{n=0}^{\infty} \mu_n(\epsilon) w_n \qquad (3.7)$$

is an symptotic expansion of w if

$$w - \sum_{0}^{k} \mu_n(\epsilon) w_n = o(\mu_k(\epsilon)) \text{ for each } k \text{ as } \epsilon \to 0. \qquad (3.8)$$

Usually w depends on other variables x, t, etc. If Eq. (3.8) holds uniformly in x, t over their domains, then we say that the series (3.8) is a uniform asymptotic expansion for $w(x, t; \epsilon)$ and write

$$w(x, t; \epsilon) \sim \sum_{0}^{\infty} \mu_n(\epsilon) w_n(x, t; \epsilon). \qquad (3.9)$$

The symbol $\sim$ does not mean equals. In many cases the series (3.9) actually diverges so that $\sim$ really means that, near $\epsilon = 0, w(x, t; \epsilon)$ looks like $\sum_0^k \mu_n(\epsilon) w_n$ plus a remainder which is much smaller than $\mu_k(\epsilon)$ as $\epsilon \to 0$.

We will develop solutions to the nonlinear problems when R is close to critical by looking for expansions (3.9). The choice of elements in the sequence $\{\mu_n(\epsilon)\}$ will be determined by various balances which occur. For example, if the nonlinear cubic term in Eq. (2.8) which is of leading order $\mu_0^3 w_0^3$ has to balance the linear growth $\epsilon^2 \alpha \mu_0 w_0$ coming from the term Rw, then μ_0 must be ϵ and it follows that $\mu_n(\epsilon)$ will be $\epsilon^{n+1}, n = 0, 1, \ldots$. The first iterate w_0 in the series consists of a linear combination of the active modes belonging to the set A determined by the linear stability problem. [In general, the solution will be $w = w_c$ (conduction) $+ \sum \mu_n(\epsilon) w_n$, but we always rewrite the original nonlinear equation for the difference $w - w_c$ (conduction).] In this case, w_0 is indeed the solution of the linear stability problem for w_c (conduction) or, as we have written the equations $w - w_c$ (conduction), the solution of the linearized problems (2.8), (2.17), (2.22), (2.27), and (2.30–33). The latter iterates $\{w_n\}, n \geq 1$ are determined by substituting Eq. (3.9) into the nonlinear equation and equating the various powers of ϵ. [Usually the sequence $\mu_n(\epsilon)$ consists of powers of ϵ; if not, one can still separate out the equations for w_1, w_2, etc. by using the definition (3.8).] The equation for $w_n, n \geq 1$ has the form

$$L w_n = g_n \qquad (3.10)$$

where the g_n contain nonlinear interactions (products of mode amplitudes) and the slow modulations of the mode amplitudes $(\partial A/\partial T, \partial A/\partial X, \text{etc.})$ and L is some linear operator. In addition, there will be boundary conditions on w_n. The existence of solutions w_n will require the satisfaction of a solvability condition known as the Fredholm alternative. [For example, note that $y'' + \pi^2 y = \sin \pi x, y(0) = y(1) = 0$

has no solution. As an exercise, determine the value of A for which $y'' + \pi^2 y = \sin \pi x + Ax, y(0) = y(1) = 0$ has a solution; hint, write x in a Fourier sine series.] The Fredholm Alternative Theorem tells us the necessary and sufficient condition under which solutions to the boundary value problem exist. The reader should consult books on this; for example, I suggest Stakgold's book *Green's Functions and Boundary Value Problems* (Wiley Interscience, 1979). Briefly stated, the theorem says that if the homogeneous boundary value problem $Lw_n = 0$ has nontrivial solutions, then the adjoint boundary value problem $L * w = 0$ with appropriate adjoint boundary conditions will also have the same number of nontrivial solutions, and solutions to Eq. (3.10) only exist when the RHS of Eq. (3.10), namely, g_n, is orthogonal to the solutions of the adjoint problem. Necessity is usually easy to prove. It follows from the definition of adjoint. Sufficiency requires one to prove that g_n is in the range of the operator L, not always an easy task for infinite-dimensional problems. If L is a real matrix, L^* is its transpose and, if v is a vector such that $v^T L = 0$ then $L^* v = 0$, so the adjoint solutions are the left eigenvectors of the matrix L with zero eigenvalue.

Application of the solvability conditions will tell us how the amplitudes of the active modes evolve slowly in time. On solving the iterates $\{w_n\}, n \geq 1$, one obtains an asymptotic series for the solution

$$w(x, t; \epsilon) \sim \sum_0 \mu_n(\epsilon) w_n(x, t; \epsilon). \qquad (3.11)$$

Let H be the Hilbert space to which the solutions $w(x, t; \epsilon)$ belong. Then, as we have said, H is divided into two subsets, the active A and passive P modes. The leading order approximation w_0 to w contains a linear combination of active modes with amplitudes $A_1, \ldots, A_N$ whose slow evolution is determined by the solvability conditions. The amplitudes of all the modes in P will be determined by solving for $w_n, n \geq 1$ as functions of the amplitudes $A_1, \ldots, A_N$. This manifold, which says how the passive mode amplitudes are related to the active ones, is called the Center Manifold. The key element in the success of the method is that, in the Hilbert space of solutions, any point not originally on the Center Manifold will move quickly to it in a time scale short with respect to ϵ^{-2}, the time scale on which points move along the manifold.

We have said so far only that Eq. (3.11) is asymptotic. If finite amplitude solutions exist, we will expect that it should converge for a finite range of ϵ. Indeed, we will often find this to be the case. In example 3C, the weakly nonlinear analysis of the Lorenz system, it will actually terminate when the active mode amplitude reaches its saturation value.

As we have said, the solvability condition will give rise to a set of o.d.e.'s and in some cases a set of p.d.e.'s for the mode amplitudes. We say, then, that evolution equations for the mode amplitudes, called the order parameters of the system, arise as asymptotic solvability conditions. Many of the canonical equations of mathematical physics (e.g., Korteweg-deVries, nonlinear Shrödinger) arise in precisely the same way. They all have universal features in the sense that their structure is only determined by symmetries of the original problem from which they were derived and only the values of the coefficients and not the form of the equation depend on individual details.

We now apply this strategy to the examples 2(a) through 2(e).

B. THE SWIFT-HOHENBERG MODEL ON A FINITE LINE

Let $R = R_c(1 + \epsilon^2\alpha)$, $R_c = ((\pi^2/L_B^2) - 1)^2$ and

$$w = \mu_0 w_0 + \mu_1 w_1 + \mu_2 w_2 + \ldots \qquad (3.12)$$

with

$$w_0 = A(T = \epsilon^2 t) \sin\frac{\pi z}{L_B}. \qquad (3.13)$$

Insert in Eq. (2.8). At $O(\mu_0)$, we have

$$\left(\frac{\partial^2}{\partial z^2} + 1\right)^2 w_0 - R_c w_0 = 0, \qquad w_0 = \frac{d^2 w_0}{dz^2} = 0, \qquad z = 0, L_B, \qquad (3.14)$$

automatically satisfied by the choice (3.13). Assuming $\mu_0\epsilon \ll \mu_1$, we get the same equation for w_1 and we can take $w_1 = 0$ without loss of generality because adding another slowly varying function of T times $\sin(\pi z/L_B)$ can be accommodated in A in Eq. (3.13). At the next order, the balance between the growth $\epsilon^2\alpha\mu_0 w_0$ term and the nonlinear term $\mu_0^3 w_0^3$ forces the choice $\mu_0 = \epsilon$. It is then clear that $\mu_n = \epsilon^{n+1}, n \geq 0$. At $O(\epsilon^3)$, we have

$$\left(\frac{\partial^2}{\partial z^2 + 1}\right)^2 w_2 - R_c w_2 = -\frac{\partial w_0}{\partial T} + R_c\alpha w_0 - w_0^3 \qquad (3.15)$$

with boundary conditions $w_2 = d^2 w_2/dx^2 = 0, z = 0, L_B$. The first term on the RHS comes from the time derivative; $\partial w/\partial t \sim \epsilon(\partial w_0/\partial t) = \epsilon^3(\partial w_0/\partial T)$. The RHS has the form

$$-\left(\frac{\partial A}{\partial T} + \alpha A - \frac{3}{4}A^3\right)\sin\frac{\pi z}{L_B} + \frac{1}{4}A^3\sin\frac{3\pi z}{L_B}. \qquad (3.16)$$

The Fredholm alternative condition says that the existence of a solution w_2 demands the coefficient of $\sin(\pi x/L_B)$ is zero. [Simply look for solutions $w_2 = B_1\sin(\pi z/L_B) + B_3\sin(3\pi z/L_B)$. One finds $0 \cdot B_1 = -(\partial A/\partial T) + \alpha A - (3/4)A^3$, $(((9\pi^2/L_B^2) - 1)^2 - ((\pi^2/L_B^2) - 1)^2)B_3 = (1/4)A^3$.] Therefore we must choose

$$\frac{dA}{dT} = R_c\alpha A - \frac{3}{4}A^3. \qquad (3.17)$$

Eq. (3.17) is called the Stuart-Watson[65] equation. It is clear that for large T,

$$A \rightarrow \pm\sqrt{\frac{4}{3}R_c\alpha}. \qquad (3.18)$$

Now what happens if we continue? $AO(\epsilon^5)$, there will again be terms proportional to A^5 which multiply $\sin(\pi z/L_B)$. To remove them, we realize that the RHS of Eq. (3.17) is simply the first term and asymptotic expansion for dA/dT. In general we write

$$\frac{dA}{dT} = F_3(A) + \epsilon^2 F_5(A) + \dots. \qquad (3.19)$$

We find $F_3 = R_c\alpha A - (3/4)A^3$, F_5 is proportional to A^5, and so on. However, the saturation value of A is mainly determined (up to $O(\epsilon^2)$) by the zeros of $F_3(A)$.

The interates w_{2n} will all be linear combinations of $\sin((2r + 1)z/L_B), r = 1,\dots,n$ with coefficients proportional to a series in A containing powers $2r + 1$ to $2n+1$. One can see that it converges because the amplitudes will be proportional to A^{2n+1}/n^4. Thus the final asymptotic series (3.12) will have a finite radius of convergence. In this case, it is possible to go through and make the explicit computations. In general, it has to be done numerically.

However, the important information is really obtained in the asymptotic solvability condition (3.17). From it, we know that the solution $A = 0$ becomes unstable at $\alpha = 0$ and that it bifurcates to one of two finite amplitude states. I will come back to a graphical picture of Eq. (3.17) in a minute, but first I want to address a small point which may have concerned you. You might say: why did you look for solutions to w_2 which were time independent? If you had included time, then the LHS of Eq. (3.15) would have been a partial differential equation and, when we substitute $w_2 = B_1 \sin(\pi z/L_B) + B_3 \sin(3\pi z/L_B)$, we do not get $0 \cdot B_1$ on the left-hand side as coefficient of $\sin(\pi z/L_B)$; rather we get dB_1/dt. All right, let us do this. We get

$$\frac{dB_1}{dt} = \left(-\frac{\partial A}{\partial T} + R_c\alpha A - \frac{3}{4}A^3\right). \qquad (3.20)$$

Now since the RHS is a function of $T = \epsilon^2 t$, we integrate and obtain

$$B_1 = \frac{1}{\epsilon^2} \int^T \left(-\frac{\partial A}{\partial T} + R_c\alpha A - \frac{3}{4}A^3\right) dT.$$

Now we find that w_2 is order $1/\epsilon^2$ and so in the asymptotic expansion

$$w = \epsilon w_0 + \epsilon w_1 + \epsilon^3 w_2 + \dots$$

the third term is as big as the first. This violates the definition of a *uniform* asymptotic expansion and the only recourse is to insist on Eq. (3.17). You might also say:

suppose we did not want the slow time behavior of A and took A to be a constant. Then

$$B_1 = t(R_c\alpha A - \frac{3}{4}A^3)$$

and the asymptotic expansion (3.12) loses its uniformity after a time $\epsilon^2 t = O(1)$. Again we would be forced to conclude that $A^2 = (4/3)R_c\alpha$ is the only allowable saturation amplitude. This result was first found by Malkus and Veronis.

EXERCISES.

1. Let $L = \pi, R = \epsilon^2\alpha$. Calculate the finite amplitude state exactly.
2. Let R_D, L_D be such that $(n^2\pi^2/L_D^2) - 1 = -((n+1)^2\pi^2/L_D^2) + 1$, i.e., at the intersection of the neutral stability curves $\sigma_n = \sigma_{n+1} = 0$. Let $R = R_D(1+\epsilon^2\alpha)$, $L = L_D(1 + \epsilon^2\beta)$ and discuss the finite amplitude behavior of the system.
3. Let L become large, say equal to $(n+\beta)\pi, n \gg 1, 0 < \beta < 1$. Notice the difference between successive wavenumbers is of order L^{-1} and small and, moreover, the difference in growth rates of the modes $\sin((n+r)\pi z/L)$ is of order r^2/n^2. All wavenumbers in the band $r\pi = \varepsilon L$ will have growth (or decay) rates of order ε^2 and must be included in the active set A. In the limit $\varepsilon L \to \infty$, the number of active modes becomes infinite and the problem must be handled in a manner similar to that introduced in Section F. See also remark 3 in 4A.

C. SUPERCRITICAL, SUBCRITICAL, AND TRANSCRITICAL BIFURCATIONS AND THE ROLE OF IMPERFECTIONS

Consider a mechanical system in which R is near its lowest critical value R_c, and at this value linear stability analysis suggests that only one of the normal modes of the system is about to make a transition from a damped (and perhaps oscillatory) or purely oscillating state to one which grows exponentially. In a manner parallel to the approach followed in 3B, we can describe the behavior of the system near $R = R_c$ by an equation for the amplitude A of the mode in transition,

$$\delta A = I + \alpha A + \gamma A^2 - \beta A^3 = F(\chi, A). \tag{3.21}$$

In Eq. (3.21), δ stands for d/dt if the transition is from a damped to excited state, and d^2/dt^2 if the transition is from a neutral (oscillatory) to excited state. The parameter α measures $R - R_c$ and for our discussion we will assume it to be real; γ and β, both taken positive, measure the nonlinear reaction of the system. In the example in 3B, there was no significant quadratic interaction and γ was zero. In the context of elastic shells, the quadratic term results from the influence of a nonlinear elastic foundation; the cubic term usually arises from a self-modal interaction. We call parameter I the geometric imperfection after Koiter, who introduced the term to account for the imperfections which may be present in the shell before loading.

As we point out later in our narrative, this constant term can result from many factors. *It plays two very important roles.*

Consider Figures 7 and 8. Figure 7 is the curve $F(\alpha = R - R_c, A) = 0$ with $\gamma = 0$ and represents what is called *supercritical bifurcation*. If $I = 0$, there is only one root $A = 0$ for $\alpha = R - R_c < 0$ and three, $A = 0, A = \pm\sqrt{\alpha/\beta}$ for $\alpha > 0$. The curves CD and CE represent stable solutions of Eq. (3.21); CF is unstable. The transition at the bifurcation point is non-analytic; $\alpha < 0, A = 0; \alpha > 0, A = \pm\sqrt{\alpha/\beta}$. If $I > 0$, then the curve $F(\alpha, A)$ is the dotted curve OGD (stable) and EHF (EH stable, HF unstable). Note that now the transition is smooth and analytic. This is the first important role of the geometric imperfection. For example, its presence due to the effect of cylinders of finite length affects the onset of Taylor vortices in the supercritical bifurcation of a flow between rotating cylinders. The Taylor vortices can be seen as ghostly apparitions at subcritical values of R, the Taylor number; however, they are amplified rather rapidly, albeit smoothly, when R is close to R_c.

Figure 8 shows the curve of $\gamma \neq 0$. This kind of bifurcation is called *transcritical*. Again, if $I = 0$, the "parabola" ECD in Figure 7 is simply displaced so that its vertex C is at $\chi = -\gamma^2/4\beta, A = \gamma/2\beta$. The portion of the curve R_cC is unstable; CD and R_cE are stable; R_cF is unstable. The stability properties of the various branches can be simply understood. Let A_0 be an equilibrium solution (3.21). Then if $A = A_0 + \rho$, $d\rho/dt = (\partial F/\partial A)_0\rho$ to first order. But, $(\partial F/\partial \alpha) + (\partial F/\partial A)\cdot(dA/d\alpha) = 0$ and, since $(\partial F/\partial \alpha) = A$, we have $(d\rho/dt) = -A_0(dA/d\alpha)_0^{-1}\rho$. Hence, for $A_0 > 0(< 0)$, the

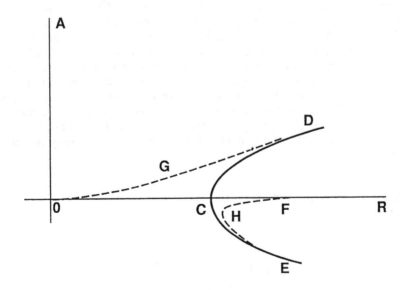

FIGURE 7   Bifurcation diagram near a supercritical bifurcation.

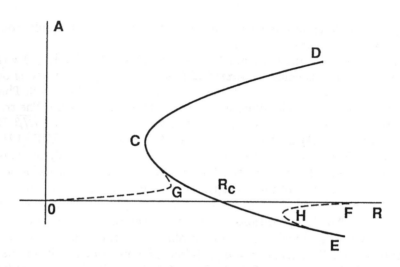

FIGURE 8   Bifurcation diagram near a transcritical bifurcation.

branch is unstable (stable) if A decreases with α, stable(unstable) if A increases with α. In the situation depicted in Figure 8, one can have a *subcritical* bifurcation; that is, for values of $\alpha < 0$, there is a possibility that, if perturbations are large enough, the system can transfer from the stable solution OR_c to the stable solution CD. The phenomenon of hysteresis is also present. In the case of transcritical bifurcation, the geometric imperfection I plays a second important role. When $I > 0$, the curve $F(\alpha, A) = 0$ changes as shown by the dotted line in Figure 8. The curve has two separate branches OGCD (OG stable, GC unstable, CD stable) and EHF (EH stable, HF unstable). Moreover, we also note that if I is sufficiently large, the curve OGCD describes a single-valued relation between the amplitude A and the Reynolds number R. However, the important point is that the imperfection provides a way for the system to reach the neighborhood of the unstable saddle points without the benefit of large disturbances. If α or R is increased beyond the value corresponding to G, the system will be attracted to the only possible stable configuration on CD.

D. THE LORENZ EQUATIONS

Let $r = 1 + \epsilon^2 \alpha$; we find $w = (X, Y, Z)^T$ satisfies

$$(A_c + \epsilon^2 \delta A)w = \frac{\partial w}{\partial t} + \begin{pmatrix} 0 \\ XZ \\ -XY \end{pmatrix} \qquad (3.22)$$

where

$$A_c = \begin{pmatrix} -p & p & 0 \\ 1 & -1 & 0 \\ 0 & 0 & -0 \end{pmatrix}$$

has right eigenvectors $\hat{u}_1, \hat{u}_2,$ and $\hat{u}_3$ and corresponding eigenvalues

$$0, \qquad -2(p+1), \qquad -b, \qquad \text{and} \qquad \delta A = \alpha \begin{pmatrix} 0 & 0 & 0 \\ 1 & 0 & 0 \\ 0 & 0 & 0 \end{pmatrix}.$$

Let

$$w = \epsilon w_0 + \epsilon^2 w_1 + \epsilon^3 w_2 + \dots$$

(the same reasoning as in 3B gives $\mu_n = \epsilon^{n+1}$) with w_0 consisting of the set of active modes (just one),

$$w_0 = B(T = \epsilon^2 t)\hat{u}_1. \tag{3.23}$$

Inserting Eq. (3.23) into Eq. (3.22), we find the order ϵ balance is automatically satisfied at $O(\epsilon^2)$,

$$A_c w_1 = \begin{pmatrix} 0 \\ X_0 Z_0 \\ -Z_0 Y_0 \end{pmatrix} = \frac{B^2}{2} \begin{pmatrix} 0 \\ 0 \\ -1 \end{pmatrix}$$

from which we find

$$w_1 = \frac{B^2}{2b}\hat{u}_3. \tag{3.24}$$

At $O(\epsilon^3)$,

$$A_c w_2 = -\delta A w_0 + \begin{pmatrix} 0 \\ X_0 Z_1 + X_1 Z_0 \\ -X_0 Y_1 - X_1 Y_0 \end{pmatrix} + \frac{\partial w_0}{\partial T}$$

$$= \frac{dB}{dT} \frac{1}{\sqrt{2}} \begin{pmatrix} 1 \\ 1 \\ 0 \end{pmatrix} - \alpha \frac{B}{\sqrt{2}} \begin{pmatrix} 0 \\ 1 \\ 0 \end{pmatrix} + \frac{B^3}{2\sqrt{2b}} \begin{pmatrix} 0 \\ 1 \\ 0 \end{pmatrix}. \tag{3.25}$$

The solvability condition for w_2 is that the RHS is orthogonal to the left eigenvector v_1^T of A_c with eigenvalue zero, which when normalized is $\hat{v}_1^T = (1/\sqrt{1+p^2})(1, p, 0)$. Applying this condition we find the Stuart-Watson equation

$$\left(1 + \frac{1}{p}\right)\frac{dB}{dT} = \alpha B - \frac{1}{2b}B^3. \tag{3.26}$$

After a long time T,

$$B \to \pm\sqrt{2\alpha b},$$

and then w_2 and all later iterates are zero. Therefore the asymptotic series (3.23) terminates and we find

$$\begin{pmatrix} X \\ Y \\ Z \end{pmatrix} \rightarrow \begin{pmatrix} \pm\sqrt{b(r-1)} \\ \pm\sqrt{b(r-1)} \\ (r-1) \end{pmatrix},$$ (3.27)

an exact solution of the Lorenz equations which exists for all, but is only stable for a finite range of r.

E. THE SWIFT-HOHENBERG MODEL ON AN INFINITE LINE

We now consider

$$\frac{\partial w}{\partial t} + \left(\frac{\partial^2}{\partial x^2}+1\right)^2 w - Rw + w^3 = 0$$ (3.28)

on $-\infty < x, \infty$ and simply ask that all solutions w are smooth and bounded at infinity. Linear stability theory indicates that, when $R = \epsilon^2\alpha$, there is an ϵ band of unstable wavenumbers k in the vicinity of the most unstable wavenumber $k = 1$. Let

$$w = \epsilon w_0 + \epsilon^2 w_1 + \epsilon^3 w_2 + \dots$$ (3.29)

where

$$w_0 = A(X = \epsilon x, T = \epsilon^2 t)e^{ix} + (*).$$ (3.30)

((*) is complex conjugate.) We find Eq. (3.28) is exactly satisfied at $O(\epsilon)$, $w_1 = 0$ from the $O(\epsilon^2)$ balance and at $O(\epsilon^3)$,

$$\left(\frac{\partial^2}{\partial x^2}+1\right)^2 w_2 = \alpha w_0 - w_0^3 - \frac{\partial w_0}{\partial T} + 4\frac{\partial^2 A}{\partial X^2}e^{ix} + 4\frac{\partial^2 A^*}{\partial X^2}e^{-ix}.$$ (3.31)

Solvability demands that the coefficients of both e^{ix} and e^{-ix} on the RHS of Eq. (3.31) are zero. Therefore

$$\frac{\partial A}{\partial T} - 4\frac{\partial^2 A}{\partial X^2} = \alpha A - 3A^2 A^*.$$ (3.32)

This is a special case of the Newell-Whitehead-Segel (NWS) equation.[42,43]

EXERCISE.

1. Consider the one-dimensional Swift-Hohenberg equation on a long interval $(-L/2, L/2)$ on which $w = \partial^2 w/\partial x^2 = 0$ at each boundary and $L \gg \epsilon^{-1}$. Show that the same Eq. (3.32) holds. What boundary conditions would $A(X)$ have? The distance from the boundary in which $A(X)$ makes its transition from its boundary value to its value in the bulk is called the healing length. See 4A, remark 3, and reference 20.

F. THE TWO-DIMENSIONAL SWIFT-HOHENBERG MODEL

Consider

$$\frac{\partial w}{\partial t} + (\nabla^2 + 1)^2 w - Rw + w^3 = 0 \tag{3.33}$$

where $\nabla^2 = (\partial^2/\partial x^2) + (\partial^2/\partial y^2)$, and w is bounded as $x^2 + y^2 \to \infty$. We wish to examine the behavior of this system when $R = \epsilon^2 \alpha > 0$. Let

$$w = \epsilon w_0 + \epsilon^2 w_1 + \epsilon^3 w_2 + \cdots . \tag{3.34}$$

As we have noted, there are now two degerneracies; one is orientational, the other is finite bandwidth. Let us treat the first one by simply asking that w_0 is a finite sum of modes, all of whose $\vec{k}$ vectors lie on the unit circle

$$w_0 = \sum_{|\vec{k}_j|=1}^{n} A_j(T) e^{i \vec{k}_j \cdot \vec{x}} + (*) . \tag{3.35}$$

In order to bring in some new features, I am also going to include a term $-\gamma \epsilon w^2$ in Eq. (3.28). Substitute Eq. (3.34) and we find at $O(\epsilon)$ things are automatically satisfied; at $O(\epsilon^2)$, $w_1 = 0$ and at $O(\epsilon^2)$ the solvability condition demands that all coefficients of $e^{i\vec{k}_j \cdot \vec{x}}, j = 1, -N$ in the equation for w_2 are zero.

$$\frac{dA_j}{dT} = \alpha A_j + \gamma A_l^* A_n^* - 3A_j \left(A_j A_j^* + 2\sum_{n \neq j} A_n A_n^* \right) . \tag{3.36}$$

The terms in the amplitude equation arise from several kinds of nonlinear interactions.

Triad : $e^{-i\vec{k}_l \cdot \vec{x}} \cdot e^{i\vec{k}_m \cdot \vec{x}} \to e^{-i\vec{k}_j \cdot \vec{x}}$, when $\vec{k}_l + \vec{k}_m = -\vec{k}_j$.

In order that $\vec{k}_j, \vec{k}_l, \vec{k}_m$ all lie on the unit circle, they must lie 120° apart. The configuration in which w_0 is the sum of the three such modes with equal amplitudes gives equiphase contours which look hexagonal (see Chandrasekhar). The existence of a term such as $\epsilon \gamma w^2$ in the equation reflects the breaking of the $w \to -w$ symmetry and this is also true of hexagonal patterns. The flow can be up (w positive) or down (w negative) in the center. It is the rotational symmetry of this problem that leads to the quartic terms giving rise to hexagonal patterns. If the underlying system had a different symmetry, quadratic terms could give rise to other patterns such as the diamond shapes which occur when one compresses a cylindrical shell in a direction parallel to its axis.

The next type of nonlinear interactions are quartic and arise from cubic terms w_0^3 or quadratic products $w_0 w_1$ if w_1 were nonzero and contained terms like

$$e^{i(\vec{k}_l + \vec{k}_m) \cdot x}$$

where $|\vec{k_l} + \vec{k_m}| \neq 1$.

$$\text{Modal self interaction} \quad e^{i\vec{k_j}\cdot\vec{x}} \cdot e^{i\vec{k_j}\cdot\vec{x}} \cdot e^{-i\vec{k_j}\cdot\vec{x}} \rightarrow e^{i\vec{k_j}\cdot\vec{x}}$$

$$\text{Modal interaction} \quad e^{i\vec{k_j}\cdot\vec{x}} \cdot e^{i\vec{k_n}\cdot\vec{x}} e^{-i\vec{k_n}\cdot\vec{x}} \rightarrow e^{i\vec{k_j}\cdot\vec{x}} .$$

The first of these lead to the terms $A_j^2 A_j^*$ in the amplitude equation, the second to the terms $A_j A_n A_n^*$.

I am going to let the reader analyze Eq. (3.36) by looking for equilibrium solutions and then testing their stability. If $\gamma = 0$, you will find that a solution which shares the energy between any subset $r > 1$ of modes is unstable. Only the single roll $A_1 = \sqrt{\alpha/3}, A_j = 0, j = 2, \ldots, N$ is stable, it can have any direction. The reader should also consult the paper of Busse.[14,15]

If one looks for the equation for the modulation of a single roll whose wavevector lies along the x direction $\vec{k} = (1, 0)$, then we take

$$w_0 = A(X = \epsilon x, Y = \sqrt{\epsilon} y, T = \epsilon^2 t)e^{ix} + (*) \tag{3.37}$$

and obtain

$$\frac{\partial A}{\partial T} - 4\left(\frac{\partial}{\partial X} - \frac{i}{2}\frac{\partial^2}{\partial Y^2}\right)^2 A = \alpha A - 3A^2 A^* \tag{3.38}$$

known as the Newell, Whitehead, Segel equation.[42,43] I will refer the reader to the references for a more detailed derivation of Eq. (3.38) from the full Oberbeck-Boussinesq equations, but will discuss some of its properties in the next section. I will finish this chapter with three comments.

The first concerns the stability of solutions to Eq. (3.38) such as

$$A = \sqrt{\frac{\alpha - 4\kappa^2}{3}}e^{i\kappa X} .$$

One must be careful to test stability not only in Eq. (3.38) but also against other modes $e^{i\vec{k_j}\cdot\vec{x}}$ where $\vec{k_j}$ lies in the annulus about $k = k_c$.

The second concerns the generation of mean flows by slow gradients of the intensities of the periodic fields. We find in certain circumstances terms like $(\partial/\partial Y)$. $(\partial/\partial X)AA^*$ can drive velocity fields which are not periodic but vary over distances ϵ^{-1}.

Finally, I want to stress again that the Eqs. (3.32) and (3.38) have a universal character. In other words, the form and structure of the equation only depends on very general features of the problem (e.g., symmetries, whether the onset was through exchange of stabilities or through growing waves). The coefficients depend on things like the curvature of the neutral stability surface, the dispersion relation of the most unstable wave, etc., and in many cases they simply define appropriate time and length scales.

You will find therefore that these equations of the Ginzburg-Landau type turn up all over the place: convection in ordinary fluids, in binary fluids, in liquid crystals, in shell buckling, in superconductivity, in describing atmospheric patterns, in reaction-diffusion systems, and in biological patterns.

4. DISCUSSION AND PROPERTIES OF THE ENVELOPE EQUATIONS

A. NEWELL-WHITEHEAD-SEGEL (NWS) EQUATION

Here R_c occurs for a zero value of the wave frequency ω. There is no group velocity translation of the envelope and consequently it changes over times of the order of the growth rate ϵ^2 of the convective instability. If $R = R_c(1 + \epsilon^2 \hat{a})$, and mean drift effects which we will discuss shortly are absent, the amplitude $A(X = \epsilon x, Y = \sqrt{\epsilon y}, T_2 = \epsilon^2 t)$ of the mode $\exp i k_c x$ satisfies the NWS equation[42,43]

$$\frac{\partial A}{\partial T_2} = -\gamma_r \left(\frac{\partial}{\partial X} - \frac{i}{2k_c} \frac{\partial^2}{\partial Y^2} \right)^2 A = \alpha A - \beta A^2 A^* \qquad (4.1)$$

where $\gamma_r = (1/2)(\partial \nu/\partial R) \cdot (\partial^2 R/\partial k^2)$ and $\alpha = \hat{a}R(\partial \nu/\partial R)$ are positive, and β is the Landau constant, which here is real and positive meaning that nonlinear terms act to saturate the linear growth. All coefficients are evaluated at critical. $(\partial \nu/\partial R)$ is the rate at which the growth rate changes as R changes and $(d^2 R/dk^2)$ is the curvature of the neutral stability curve (Figure 6a) at critical. The spatial derivate terms can be understood by taking $A = A_O(T_2) \exp(iK_x X + iK_y Y)$, which means that microscopic field w has wavevector $k_c + \epsilon K_x, \sqrt{\epsilon}K_y$ whose growth rate $\nu(\sqrt{(k_c + \epsilon K_x)^2}) + \epsilon K_y^2, R_c(1 + \epsilon^2 \hat{a})$ is $(1/2)(\partial^2 \nu/\partial k^2)\epsilon^2(K_x + (1/2k_c)K_y^2)^2 + \epsilon^2 \hat{a}(\partial \nu/\partial R)R$ because ν and $(\partial \nu/\partial k)$ are zero at critical. Also, at critical $(\partial^2 \nu/\partial k^2) = -(\partial \nu/\partial R)(d^2 R/dk^2)$. This is exactly the linear growth rate predicted by Eq. (4.1). These equations are relaxational and can be derived from the Lyapunov functional

$$F[A, A^*] = \int \alpha A A^* - \frac{1}{2}\beta A^2 A^{*2} - \gamma_r \left| \left(\frac{\partial}{\partial X} - \frac{i}{2k_c} \frac{\partial^2}{\partial Y^2} \right) A \right|^2 dX\, dY, \qquad (4.2)$$

the global minimum of which corresponds to one of the two saturated states $A = \pm\sqrt{\alpha\beta^{-1}}$. Nontrivial solutions of these equations, therefore, must involve intermediate time behavior or nongradient effects introduced by boundary conditions or other forcing effects. In many of these situations the equations have been remarkably successful in predicting what is observed. Some of these situations are:

1. The velocity of the front connecting the stable solution $A = \sqrt{\alpha\beta^{-1}}$ and the unstable solution $A = 0$ is $2\sqrt{\alpha\gamma_r}$. This value, derived by an argument of Kolmogoroff (see Dee and Langer[26]), is independent of the nature of the stable state and depends only on the behavior of the equation near its unstable saddle point $A = 0$ and, of course, the existence of a heteroclinic orbit including the unstable manifold at $A = 0$, joining $A = 0$ to whether the stable state may be.
2. The shape and stability of domain walls joining two stable states $A = \sqrt{\alpha\beta^{-1}}$ (see Newell land Whitehead[42,43]).

3. The effects of distant sidewalls on the range of allowable wavenumbers in the bulk of the fluid.[20] The idea here is that near boundaries the amplitude of the convection state will be small compared to its value of $\sqrt{\alpha\beta^{-1}}$ in the bulk and that this will restrict the amount of phase winding which the pattern can experience. Consider the one spatial dimension case. Construct the two first integrals of the equations for the stationary solution of the complex envelope $A = r \exp i\phi$. One obtains

$$\gamma_r \left(\frac{1}{2}r_X^2 + \frac{1}{2}\frac{h^2}{r^2} \right) + \frac{1}{2}\alpha r^2 - \frac{1}{4}\beta r^4 = E,$$

$$r^2\phi_x = h.$$

From its value in the bulk where r^2 is approximately $\alpha\beta^{-1}$ and $r_x = 0$, E can be estimated and it is clearly bounded. To keep it bounded near the wall, the amount of phase winding, governed by the value of the constant h, can be no more so as to keep h^2/r^2 of order one. Hence, h, must be no more than the value which $|A|$ takes at the boundary. In realistic situations, Cross et al.[20] have estimated this to be of order ϵ. Therefore, the change $[k]$ in wavenumber across the box can be at most $[k] = [\phi_x] = [\epsilon\phi_X]$ which is of order $\epsilon^2 = (R - R_c)/R_c$. This is much narrower than the width $\sqrt{(R - R_c)/R_c}$ allowed when sidewall effects are not taken into account. We have recently verified this prediction by a numerical integration of the Oberbeck-Boussinesq equations.[5] It should also be emphasized that these arguments are only valid when the box width L is long with respect to the healing length $d\epsilon^{-1}$; i.e., $\Gamma\epsilon \gg 1$.

4. The effect of tapered boundaries, which effectively means that the stress parameter R is slowly changing in X from values of above to below critical, can also be successfully captured using the modulation equations. These predictions[55,56] also hold up for values of R which significantly exceed critical although in that case the CN equations (5.4) must be used. In 3.E.ii, I will give an example.

B. MEAN DRIFT

The envelope equation (4.1) was derived as a solvability condition which preserved the uniformity of the asymptotic expansion for w's; specifically, it ensures that the part of w_2 with spatial periodicity $(2\pi/k_c^2)\vec{k}_c$ exists and is bounded. These solutions have the same horizontal periodicity as the most unstable mode of the linearized stability problem and the solvability condition (4.1) removes that part of their vertical (z) structure which would resonate with the solutions of the linear homogeneous equation which governs the linear stability of the conductive solution. But there is also another solution of this linear operator, which is neutral in its time dependence and which arises because of the horizontal translational symmetry of the infinite box problem. In the case of stress-free rather than rigid boundary conditions, it corresponds to a constant horizontal velocity field $\bar{u}$. Now, although it is not possible to add a net global momentum to the fluid, the slowly varying property of the envelope $A(X, Y, T_2)$ means that it is possible for slow gradients of the intensity AA^* and other densities, like the current $i(A\nabla A^* - A^*\nabla A)$ to produce mean drift fields which locally (on scales of the box size) look constant but which are in fact slowly varying over the whole box in such a way as to respect conservation of total momentum. This mean drift field is most easily calculated by considering the equation for the mean vertical vorticity $\zeta = (\nabla \times \vec{u}) \cdot \hat{z}$ which reads,[59]

$$\left(-\frac{1}{\sigma}\frac{\partial}{\partial t} + \nabla^2\right)\zeta = -\frac{4}{\sigma}\gamma\epsilon^{7/2}\frac{\partial}{\partial Y}\left(\frac{\partial}{\partial X}AA^* - \frac{i}{2k_c}\frac{\partial}{\partial Y}\left(A^*\frac{\partial A}{\partial Y} - A\frac{\partial A^*}{\partial Y}\right)\right).$$
(4.3)

In Eq. (4.3), σ is the Prandtl number and γ is a positive constant computed from quadratic products of the vertical structure $\phi(z)$ and its derivatives whose vertical average $\bar{\gamma} = \int \gamma(z)dz$ is nonzero. In turn, the divergence-free mean drift $\vec{u} = (\psi_y, -\psi_x)$, calculated from Eq. (4.3) and $-\zeta = \psi_{xx} + \psi_{yy}$, will affect the envelope A through the addition of the term

$$i\epsilon^{-2}\vec{k} \cdot \vec{u}A$$
(4.4)

on the left-hand side of Eq. (4.1). A good way to think of this is to remember that the negative time derivative of the phase of A (which corresponds to a frequency) is Doppler-shifted through an amount $\epsilon^{-2}\vec{k} \cdot \vec{u}$ by the mean drift.

Observe that the mean current is scaled by $(d^2/\kappa) \cdot (1/\sigma)$ or (d^2/ν), the viscous diffusion velocity, and is zero in the infinite Prandtl number limit. However, for values of ϵ of order one, it has significant effects up to $\sigma = 10$. For ϵ small, on the other hand, it will affect the envelope equation to leading order only in the case when the boundary conditions on the horizontal boundaries at $z = 0, d$ are stress free. In that case, integration of Eq. (4.3) over $(0, d)$ gives

$$\left(-\frac{1}{\sigma}\frac{\partial}{\partial t} + \nabla^2\right)\zeta = -\frac{4}{\sigma}\epsilon^{7/2}\frac{\partial}{\partial Y}\left(\frac{\partial}{\partial Y}AA^* - \frac{i}{2k_c}\frac{\partial}{\partial Y}\left(A^*\frac{\partial A}{\partial Y} - A\frac{\partial A^*}{\partial Y}\right)\right) \quad (4.5)$$

where ∇^2 is the horizontal Laplacian ($\int_0^d (\partial^2\zeta/\partial z^2)$ integrates to zero because of the stress-free boundaries) and $\bar{\gamma}$ is exactly one. Inserting the appropriate scaling $\vec{u} = \epsilon^2(\bar{u}, \sqrt{\epsilon}\bar{v})\zeta = \epsilon^{5/2}\Omega = -\epsilon^{5/2}(\partial\bar{u}/\partial Y)$,

$$\left(-\frac{\epsilon}{\sigma}\frac{\partial}{\partial T_2} + \epsilon\frac{\partial^2}{\partial X^2} + \frac{\partial^2}{\partial Y^2}\right)\Omega = -\frac{4}{\sigma}\frac{\partial}{\partial Y}\left(A^*\frac{\partial A}{\partial Y} - A\frac{\partial A^*}{\partial Y}\right).$$
(4.6)

On the other hand, when the horizontal boundaries at $z = 0, d$ are rigid, ζ has a nontrivial z dependence $(1/2)z(z-d)$ found by integrating Eq. (4.5) and using only the $\partial^2\zeta/\partial z^2$ term on the left-hand side. In this case, the mean vertical vorticity $\bar{\zeta}$ and the mean drift $\vec{u}$ (the vertically averaged field calculated from Eq. (4.5)) are scaled down by a further factor of ϵ, i.e., $\bar{\zeta} = O(\epsilon^{7/2})$, $\vec{u} = \epsilon^3(\bar{u}, \sqrt{\epsilon}\bar{v})$ and produce a commensurably smaller effect. We find

$$\bar{\zeta} = -\epsilon^{7/2}\frac{\partial \bar{u}}{\partial Y} = \frac{4}{\sigma}\bar{\gamma}\epsilon^{7/2}\frac{\partial}{\partial Y}\left(\frac{\partial}{\partial X}AA^* - \frac{i}{2k_c}\frac{\partial}{\partial Y}\left(A^*\frac{\partial A}{\partial Y} - A\frac{\partial A^*}{\partial Y}\right)\right). \quad (4.7)$$

In solving Eq. (4.7), one must be careful only to include the solenoidal part (divergence-free part) of the vector field $\vec{u}$. In the case of stress-free boundaries, the additional term in the envelope equation is the same order as the existing terms, namely, $ik_c\bar{u}A$ added to the left-hand side of Eq. (2.2); for rigid boundaries, the term is $i\epsilon k_c\bar{u}A$. Bernoff[7] has shown that it is necessary to add $O(\epsilon)$ correction terms to the Siggia-Zippelius[59] expression (4.5) if one wants to bring the results into agreement with the stability calculations of Busse, Bolton, and Clever.[16] The necessity for adding mean drift in the full Oberbeck-Boussinesq equations is seen when one averages the horizontal momentum equations in Eq. (2.40) first over the horizontal period of the pattern and then over the depth. This leaves terms which are slow gradients

$$\sigma^{-1}\left(\frac{\partial}{\partial X}\langle u_0^2 \rangle + \frac{\partial}{\partial Y}\langle u_0 v_0 \rangle\right) \supset \sigma^{-1}\left(\frac{\partial}{\partial X}\langle u_0 v_0 \rangle + \frac{\partial}{\partial Y}\langle v_0^2 \rangle\right)$$

which can only be balanced by the addition of a Poiseuille-like mean flow together with a large-scale, depth-independent gradient. These effects disappear at infinite Prandtl numbers and are also small when $R - R_c$ is small. However, for Prandtl numbers less than ten and when $R - R_c$ is order one, they are very important and give rise to a real, qualitative difference in the pattern behavior (e.g., the skew-varicose instability; see Section 5).

C. OVERSTABLE WAVES AND THE COMPLEX GINZBURG LANDAU (CGL) EQUATION

Here the situation can be dramatically different because of the wavelike properties of the envelope equation. This new feature, first noted by Newell and Whitehead[42] in their 1969 paper and quantified in a sequence of later papers,[43,44] involves a modulational instability[39,43] which can overcome the gradient properties of the NWS equation (4.1). The finite amplitude saturated states are no longer constants independent of the spatial variables, but can be very complicated depending on the size of the box in which the patterns occur. Indeed, for very large boxes, the asymptotic state of the system can be turbulent even for values of R close to critical. The physical model we use in these discussions is that of convection in binary mixtures (e.g., He$_3$ and He$_4$ or water-alcohol) where the Soret effect inhibits buoyancy and the different diffusion rates of heat and concentration cause convection cells to oscillate rather than overturn indefinitely. The local oscillations are propagated as dispersive traveling waves.

In this situation, we have shown that the equation for the envelope of a traveling wave satisfies[11,12]

$$
\frac{\partial A}{\partial T} = w_c' \left(\frac{\partial}{\partial X} - \frac{i}{2k_c} \frac{\partial^2}{\partial Y^2} \right) A - \frac{iw_c'\epsilon}{2k_c} \frac{\partial^2 A}{\partial X^2}
$$
$$
- \frac{\epsilon}{2} \left\{ \frac{\partial \nu}{\partial R} R_c'' + i \left(w_c'' - \frac{w_c'}{k_c} \right) \right\} \left(\frac{\partial}{\partial X} - \frac{i}{2k_c} \frac{\partial^2}{\partial Y^2} \right)^2 A \qquad (4.8)
$$
$$
= \epsilon \alpha A - \epsilon(\beta_r + i\beta_i)A^2 A^* - i\epsilon k_c \bar{u} A
$$

with $\bar{u}$ given by

$$
\left(-\frac{1}{\sigma} \frac{\partial}{\partial T} + \frac{\partial^2}{\partial Y^2} \right) \bar{u} = \frac{4}{\sigma} \left(\frac{\partial}{\partial X}(AA^*) - \frac{i}{2k_c} \frac{\partial}{\partial Y} \left(A^* \frac{\partial A}{\partial Y} - A \frac{\partial A^*}{\partial Y} \right) \right) \qquad (4.9)
$$

if the horizontal boundaries at $z = 0, d$ are stress free and

$$
\bar{u} = -\frac{4}{\sigma} \epsilon \bar{\gamma} \left(\frac{\partial}{\partial X} AA^* - \frac{i}{2k_c} \frac{\partial}{\partial Y} \left(A^* \frac{\partial A}{\partial Y} - A \frac{\partial A^*}{\partial Y} \right) \right) \qquad (4.10)
$$

if the boundaries are rigid. The corresponding Y velocity $\bar{v}(\vec{u} = \epsilon^\alpha(\bar{u}, \sqrt{\epsilon}\bar{u}). \ \alpha = 2$ is stress free, $\alpha = 3$, rigid) must be chosen so that $\vec{u}$ is solenoidal. Notice that in the stress-free case, the right-hand side of Eq. (4.10) is $-(4/\sigma w_c')(\partial/\partial T)AA^*$ to leading order.

If the original system does not have rotational symmetry and a unique $\vec{k}_c$ is chosen, then X and Y each scales with ϵ, $X = \epsilon(x - (\partial w/\partial k_x)t)$, $Y = \epsilon(y - (\partial w/\partial k_y)t)$, $T = \epsilon^2 t$ and the appropriate equation is (without mean drift)[45]

$$
\frac{\partial A}{\partial T} - \frac{1}{2} \left(\frac{\partial \nu}{\partial R} \frac{\partial^2 R}{\partial k_x^2} + i \frac{\partial^2 A}{\partial k_x^2} \right) \frac{\partial^2 A}{\partial X^2} - \left(\frac{\partial \nu}{\partial R} \frac{\partial^2 R}{\partial k_x \partial k_y} + i \frac{\partial^2 w}{\partial k_x \partial k_y} \right) \frac{\partial^2 A}{\partial X \partial Y} -
$$
$$
\frac{1}{2} \left(\frac{\partial \nu}{\partial R} \frac{\partial^2}{\partial k_y^2} + i \frac{\partial^2 w}{\partial k_y^2} \right) \frac{\partial^2 A}{\partial Y^2} = \alpha A - (\beta_r + i\beta_i)A^2 A^* . \qquad (4.11)
$$

It is best to begin by discussing the one-dimensional case[43,62] in which the Y dependence is suppressed by the box being narrower in that direction. Since mean vorticity production depends on variation in the Y direction (see Eq. (4.3)), $\bar{u} = 0$ and we can write Eq. (4.8) as

$$
\frac{\partial A}{\partial T} + w_c' \frac{\partial A}{\partial X} - \frac{\epsilon}{2} \left(\frac{\partial \nu}{\partial R} R_c'' + i w_c'' \right) \frac{\partial^2 A}{\partial X^2} = \epsilon \alpha A - \epsilon(\beta_r + i\beta_i)A^2 A^* . \qquad (4.12)
$$

In Eqs. (4.8) and (4.12), w_c' and w_c'' are the group velocity and dispersion, respectively, estimated at critical. There is no *a priori* reason that a dual equation for B, the envelope of the left-moving traveling wave, should not be adjoined to Eq. (4.12). In that case B would move with group velocity $-w_c'$, all other terms would be the same except for an additional cross-coupling term $(\delta_r + i\delta_i)BB^*A$ added to Eq. (4.12) and a similar term $(\delta_r + i\delta_i)AA^*B$ added to the equation for B. In many circumstances, however, the complex coefficients β and δ are such that the presence of a finite amplitude, right-moving field A will suppress the left-moving field and vice versa. In these cases, standing wave solutions are unstable in the bulk of the fluid and it is only near the boundaries where wave reflection[22] is important or in situations where one field is modulationally unstable (see comments in following paragraph) and leads to the creation of source and sink-like defects of both fields A and B,[18,19] that both fields exist together.

For the moment, therefore, we will neglect the effect of boundaries and concentrate on the properties of a single traveling wavepacket. In particular, I want to introduce the reader to a somewhat surprising result, the impact of which has not been generally appreciated although the result itself has been known for 15 years. It concerns the asymptotic state of Eq. (4.12). In the absence of dispersion and the imaginary part of the nonlinear term, the equation is a gradient flow which minimizes the Lyapunov function (2.3) when $|A| = \pm\sqrt{\alpha\beta_r^{-1}}$. Domain walls which link these two states will slowly move and cancel each other out, leaving the system in one of the two constant states. Only near the boundary where A, though "half" of a domain wall solution, relaxes to an almost zero state consistent with boundary conditions, is A dependent on the space coordinate. The existence of the gradient flow and relaxation to a constant state does not survive the presence of the imaginary terms. Observe that in the absence of all terms with real coefficients, Eq. (4.12) is the nonlinear Schrödinger equation well known for its focusing properties. Using light propagation as an analogy, the perturbation in the refractive index is $-\beta_i|A|^2$ and, if β_i is negative, this means that light will intensify in regions of larger amplitude. Indeed, it is easy to show that, if $\beta_i\omega_c'' < 0$, the X-independent envelope which, in terms of the original field variable w, corresponds to a monochromatic wavetrain with a nonlinear frequency modulation, is unstable and, in the one-dimensional case, a series of nonlinear Schrödinger solitons emerge from the instability. When both real function and imaginary terms are present, this modulational or Benjamin-Feir instability can still prevail. If[39,44,45]

$$\beta_i\omega_c'' + \beta_r\frac{\partial\nu}{\partial R}R_c'' < 0, \tag{4.13}$$

then the X-dependent, monochromatic wave solution of Eq. (4.12), namely,

$$A(X,T) = \sqrt{\alpha\beta_r^{-1}}\exp(-i\beta_i\alpha\beta_r^{-1}\epsilon T), \tag{4.14}$$

is unstable to a sideband disturbance $e^{iKX+\epsilon\sigma T}$ with growth rate σ given by

$$\sigma^2 = \alpha\beta_r^{-1}\left(\beta_i\omega_c'' + \beta_r\frac{\partial\nu}{\partial R}R_c''\right)K^2 - \frac{1}{4}\left(\omega_c''^2 + \left(\frac{\partial\nu}{\partial R}R_c''\right)^2\right)K^4$$

$$= \frac{1}{4}\left(\omega_c''^2 + \left(\frac{\partial\nu}{\partial R}R_c''\right)^2\right)K^2(K_M^2 - K^2). \tag{4.15}$$

What are the consequences of the instability? The main one is that the asymptotic state of the system for long times is a lot more complicated. Just how complicated depends on the box size L and how many unstable modes N, with wavenumbers $n\pi/L, 1 \leq n \leq N$, fit under the unstable band $0 < K^2 < K_m^2$. Although the shapes can differ appreciably from the almost constant state, the dynamics is governed by N active complex-valued amplitudes with N proportional to L and each mode has the shape of solitary wavelike pulse of approximate width $2\pi\sqrt{2K_M^{-1}}$, the fastest growing wavelength of the modulational instability, and an amplitude governed by the balance among linear growth, diffusion and nonlinear saturation. Several authors have looked at the increasing complication of the flow states as a function of L, but I should warn the reader that most have incorrectly assumed that the dynamics will be symmetric about the midpoint of the box $X = L/2$. What is interesting is that even for low values of L in which only two linearly unstable modes lie under the Benjamin-Feir unstable band, Powell (Ph.D. thesis) has found a nontrivial behavior in which the system alternately visits a turbulent state close to the saturation amplitude of Eq. (4.14) and then intermittently returns to the neighborhood of the zero state in a manner reminiscent of a S'ilnikof loop. Such behavior is indeed observed near the threshold of the oscillatory (traveling wave) branch of binary mixture convection.[1]

It is also interesting to mention what happens when the opposite inequality in Eq. (4.13) obtains the modulationally stable case. In that situation, even though the X-independent states are linearly stable, there are nearby stable solutions which have the form of finite amplitude depressions which travel on a background of the $|A| = \sqrt{\alpha\beta_r^{-1}}$ state. These are called dark solitons in the optics literature as the intensity decreases where they are largest. Binary mixture convection can exhibit both behaviors near threshold; in this wave case, one must allow for the possibility of waves in both directions because, even though a constant wavetrain in one direction suppresses the traveling wave in the other, the presence of defects caused by modulational instabilities requires both waves.

Another consequence of the modulational instability is that it is not necessary for β_r to be positive in order to obtain a bounded state. The reason for this is most easily seen by considering the power balance,

$$\frac{\partial AA^*}{\partial T} = 2\epsilon\alpha AA^* - 2\epsilon\beta_a A^2 A^{*2} - \epsilon\nu_R R_c'' \frac{\partial A}{\partial X} \cdot \frac{\partial A^*}{\partial X} + \frac{\partial F}{\partial X} \qquad (4.16)$$

where the flux terms are

$$F = -\omega_c AA^* + \frac{1}{2}\epsilon\nu_R R_c'' \frac{\partial}{\partial X} AA^* + \frac{i\omega_c''\epsilon}{2}\left(A^*\frac{\partial A}{\partial X} - A\frac{\partial A^*}{\partial X}\right).$$

Observe the role of the modulational instability. Because X-independent solutions are unstable, the linear growth can be balanced by diffusion terms even in the absence of nonlinear saturation. Formally $\int |\partial A/\partial X|^2$ can be bounded by $q^2 \int |A|^2$

where $q \propto L^{-1}$ is related to the lowest eigenvalue of the operator d^2/dX^2 in a box of length L but, in fact, because of the nature of the modulational instability $q \propto K_M^{-1}$ which is much larger. This observation is important because it is often the case in doubly diffusive mixtures that the real part, β_r, of the Landau constant is very small. Indeed, the asymptotic state of Eq. (4.12) when $\beta_r = 0$ and $\beta_i \omega_c'' < 0$ will be a sequence of solitary wave pulses as observed in the numerical experiments of Bretherton and Spiegel.[13]

Finally, we mention that Nozaki and Bekki[48] have found an interesting class of solutions of the CGL equation in the forms of solitary waves, holes, and shocks. It would be very interesting to show that, when Eq. (4.13) obtains, the asymptotic state of the system can somehow be viewed as a (nonlinear) superposition of these solitary waves and that the chaotic behavior is simply a loss of phase information on where the pulses are exactly located. It would be of further interest to ask whether the shock solutions, which join the unstable state $A = 0$ to the stable attracting state, are still valid when the modulational instability is operative. Imagine the following situation. The flow is initiated from the state $A = 0$ with a pulse at $X = 0$. Near $X = 0$, the system will grow to $A = \sqrt{\alpha \beta_r^{-1}} \exp(-i\beta_i \alpha \beta_r^{-1} \epsilon T)$ and this solution will propagate into the unstable region with speed $2\epsilon \sqrt{\alpha \gamma_r}(1 + (\gamma_i^2/\gamma_r^2))^{1/2}$ with $\gamma_r = (1/2)(\partial \nu/\partial R)R_c''$ and $\gamma_i = (1/2)\omega_c''$. Until the length of the state $|A|^2 = \alpha \beta_r^{-1}$ exceeds the value π/K_M, this state is stable. However, as the length behind the shock increases, the modulational instability sets in and the solution behind the shock becomes more and more complicated. The question is: do the usual arguments about the speed of heteroclinic orbits joining stable to unstable states still obtain or can the turbulence behind the shock initiate instabilities ahead of the shock before the shock gets there. If disturbances can propagate ahead of the shock and prematurely trigger the instability in the unstable $A = 0$ state, then what is the new shock speed and how is its shape modified?

D. COMMENTS ON THE TWO-DIMENSIONAL CGL EQUATION (4.8) AND APPLICATIONS TO CONVECTION IN THE BINARY MIXTURES

Let us now turn to the two-dimensional equation (4.8). At this point, I want to draw an analogy with the propagation of light in a medium of refractive index $n = n_0 + \delta n(\vec{x}, |E|^2)$. In that context, the envelope of $E(\vec{x}, t)$ of a wavetrain $\exp(i\omega t - i(\omega n_0/c)z)$ propagating in the z direction satisfies

$$i\frac{\partial E}{\partial t} + \nabla^2 E + \delta n E = 0 \qquad (4.17)$$

where ∇^2, representing wave deffraction, is the Laplacian in the plane $\vec{x}(x, y)$ of the wave. I want to make two points about the solutions of this equation. The first is that the rays which are perpendicular to the plane of propagation when δn is zero or constant will bend toward the regions of higher refractive index when δn varies across the plane of propagation. In Eq. (4.8) the effective refractive index correction δn is $-\beta_i|A|^2$. For a right-traveling wave with $\omega_c' > 0$, this means that the largest refractive index, when $\beta_i < 0$, occurs along the centerline $Y = M/2$ where $|A|$ is largest. (The domain of the Y coordinate is taken to be $[0, M]$.) Therefore, we expect that when $\beta_i < 0$, the wave crests should focus toward the center line. This is exactly what is observed in the binary mixtures convection. The states which appear just above threshold have a slightly asymmetrical football shape[58] which extends from a fixed value X_c of X to a focus X_f which is approximately the distance of a healing length $d\epsilon^{-1}$ from the right-hand boundary. (See Figure 9, taken from Ahlers et al.[4]) The wave crests move through the football-shaped envelope. We have not yet written down solutions to the full equation (4.8) which correspond to this shape, consisting of a point source on the right to an almost Y-independent shape on the left. Presumably the left boundary is kept in place as the balance of two velocities; ω_c', the group velocity which sweeps the packet to the right and

$$2\epsilon\sqrt{\alpha\gamma_r}\left(1 + \frac{\gamma_i^2}{\gamma_r^2}\right)^{1/2}, \qquad \gamma_r = \frac{1}{2}\frac{\partial\nu}{\partial R}R_c'', \qquad \gamma_i = \frac{1}{2}\omega_c'',$$

the velocity at which a front moves into the unstable $A = 0$ region in the left half of the box.

The second point in connection with Eq. (4.17) is that if the medium is nonlinear with a Kerr nonlinearity $\delta n = \beta|E|^2, \beta > 0$, then for almost all initial conditions with sufficient power $\int EE^* d\vec{x}$, the envelope $E(\vec{x}, T)$ will collapse in the form of a self-similar pulse $f(t)^{-1}R(\sqrt{(x^2 + y^2)}(f(t))^{-1})$ with $f(t) \to 0$ in a finite time. This behavior contrasts sharply with the one-dimensional situation where $\nabla^2 = \partial^2/(\partial x^2)$ in which case the asymptotic state of the system consists of solitons and radiation.

What analogous behavior can be expected from Eq. (4.8)? First, we point out that the effect of diffraction and the modulational instability is much stronger in the direction Y perpendicular to the direction X of wave propagation. Moreover, one can expect the modulational stability to set in with a wavenumber in the Y direction of approximately $4\alpha\beta_i(\partial\nu/\partial R)R_ck_c(\omega_c')^{-1}$. In the binary mixture convection experiments of Steinberg and Moses,[62] this would correspond to six roll widths

FIGURE 9   The football-shaped wavepackets observed in binary fluid convection by Steinberg and Moses, and Ahlers, Cannel, and Heinrichs.[4] Reprinted by permission of the America Physical Society.

which is about the box width M at which one sees significant three-dimensional effects.

The modulational instability in the transverse direction can also lead to other interesting possibilities such as a right-moving wave in the top of the box $M/2 < y < M$ and a left-moving one in the lower half $0 < y < M/2$. This solution may be related to the so-called zipper state seen in early experiments. Second, we emphasize that as the box becomes wider and larger, the two-dimensional and more violent modulational instability should be seen. No one has yet worked out what the stabilizing effects of diffusion and nonlinear saturation will be, but we can expect a much more disordered behavior.

Finally, we point out that the modulational instability can play an important role in triggering *local*, subcritical, finite amplitude instabilities.[45] The idea is that even though amplitudes are initially well below some critical threshold, the focusing effect of the modulational instability can give rise to local amplitudes which exceed the threshold. We would predict on this basis that in the parameter ranges where both β_i and β_r are less than zero, that the solution can access a higher amplitude branch (involving higher-order terms in Eq. (4.8)) at subcritical values of R.

5. CONVECTION FAR FROM ONSET
A. THE BASIC IDEAS

We now turn to a description of patterns when the stress parameter is well above critical but still in a range where the field is dominated *locally* (see Figures 1 and 2) by what would appear to be straight parallel rolls. We know, for example, that, in a convecting fluid with a horizontal geometry, there exist linearly stable, spatially periodic solutions in certain ranges of the Rayleigh number (R), Prandtl number (σ), wavenumber (k) parameter space. This region of a parameter space is known as the Busse balloon[15,16] because of its cross-section shape in (R, k) with σ fixed. A more appropriate name, connoting the three-dimensionality of R, σ, k space, would be the Busse windsock. The existence of stable, periodic solutions of the governing Oberbeck-Boussinesq equations when the inverse aspect ration $\Gamma^{-1} = \mu^2$ is zero is the starting point of the analysis for the case when μ^2 is small but finite. The basic idea stems from the work of Whitham,[70,71] and it has been applied to reaction-diffusion situations by Howard and Kopell[35] and to the convection problem by Cross and Newell.[21,46] The notion is simple. The stationary, spatially periodic solution for the vertical velocity, say, in an infinite geometry has the form

$$w(\vec{x} = (x,y), z, t) = f(\theta = \vec{k} \cdot \vec{x} + \theta_0, z, A) \tag{5.1}$$

where A and θ_0 are constant phase and amplitude parameters. For example, because of translation invariance in the infinite horizontal geometry, θ_0 is arbitrary. The amplitude A, on the other hand, is determined by the balance of buoyancy and dissipation and reflects the saturated amplitude of the roll solution. The vector $\vec{k}$ is the roll wavevector; its direction is perpendicular to the roll axes, and its amplitude divided by 2π is the inverse of the roll wavelength. In practice, the solution $f(\theta, z, A)$ is developed as a Galerkin approximation

$$f(\theta, z, A) = \sum_{m,n} A_{mn} e^{in\theta} \phi_m(z) \tag{5.2}$$

where n runs through integer 1 through N, the truncation level, and m runs between $-N + n$ and $N - n$. All the A_{mn} can be related to one parameter, A.

What happens when μ^2 is not zero? A glance at Figures 1 and 2 will convince the reader that the pattern is much more complicated, involving curved rolls, defects in the pattern such as roll dislocations, and other singular solutions such as disclinations and grain boundaries in which, usually near a boundary, rolls of one alignment abut a sequence of rolls with a perpendicular alignment. However, one also observes that in the bulk of the pattern (away from defects and other singularities), the roll pattern locally looks like a set of parallel straight rolls. As one moves around the convecting box, the pattern wavevector $\vec{k}$ changes by an order-one amount, but, except at singularities, it changes slowly, its gradient being proportional to μ^2, the inverse aspect ratio (Figures 1 and 2). We take advantage of this observation analytically by returning to the solution (5.1) of the Oberbeck-Boussinesq equationss and developing a neighboring solution of Eq. (5.1) in which the wavevector $\vec{k} = \nabla\theta$ and A, constant in Eq. (5.1), are allowed to be slowly varying quantities which depend on the variables $X = \mu^2 x$, $Y = \mu^2 y$ and $T = \mu^4 t$, the horizontal diffusion time scale. The vertical velocity field is now written

$$w(\vec{x} = (x, y), z, t) = f\left(\theta = \frac{\Theta(X, Y, T)}{\mu^2}, A(X, Y, T), Z\right) + \mu^2 w^{(1)} + \mu^4 w^{(2)} + \cdots \tag{5.3}$$

with $\nabla_{\vec{X}}\Theta = \nabla_{\vec{x}}\theta = \vec{k}(X, Y, T)$. Equations for $\vec{k}$ and A are developed by demanding that f is 2π periodic in θ and that the asymptotic expansion (5.3) for the macroscopic field variable remains well ordered in space and time. In addition, we must append to these equations an equation for the mean drift velocity $\vec{u}(X, Y, T)$ which plays a nontrivial role in the pattern dynamics. I will not derive the equations here but will refer to the reference.[21]

It is encouraging that the equations for the macroscopic variables $(\vec{k}, A, \vec{u})$ which describe the pattern dynamics, have a canonical structure that does not depend in any crucial way on the details of the underlying microscopic dynamics, but rather on the symmetries associated with the dominant microscopic fields. In this sense, the equations are universal and describe a broad variety of situations with similar symmetries. There is clearly a tremendous advantage, from the points of view of both understanding and computation, when one can describe the complicated dynamics of microscopic fields that change over a roll wavelength by much more slowly varying macroscopic or averaged fields which obey universal equations. However, there is a price to be paid for simplification. Because one cannot smoothly tile a plane with patches consisting of rolls (or whatever the dominant pattern structure may be), one has to deal with singularities, which arise as focus singularities in the center of circular patterns, dislocations joining different roll patches or disclinations separating patches of nearly circular rolls (see Figure 1). These singularities play a central role in the evolution of the total pattern and it is necessary to model their dynamics and interaction with the smooth part of the wavevector field. For example, the most common defect is the dislocation at the center of which the phase is undefined but, on any curve C which surrounds it, the phase changes by 2π, i.e., $\int_c \vec{k} \cdot d\vec{x} = 2\pi$. These vortex-like singularities play an important role in helping the pattern adjust its local wavenumber. For example, it is generally felt that if, locally, the pattern wavenumber is too large, a readjustment takes place by the nucleation of two dislocations which climb in opposite directions to the boundary, or around a circle where they meet and annihilate, thereby eliminating a roll pair and relaxing the stress on the pattern.

An important question is: Does one have to return to the microscopic dynamics in order to handle the singularities? If this were the case, the advantages of using slowly varying order parameters in the bulk of the medium is clearly somewhat nullified by having to reintroduce microscopic variables at singularities. However, this may not be necessary. It is this author's conjecture that, away from the singularities, the smooth (on the short scales) microscopic solutions have a slowly varying asymptotic structure which can be described by *singular solutions of the macroscopic equations*. If this turns out to be true, the entire pattern dynamics can be dealt with using the smooth and singular solutions of the macroscopic field equations. However, to date, we have not yet succeeded in developing a unified theory that incorporates both fields (the smooth part of the wavevector field) and particles (the singular part, e.g., vortices and other defects). Nevertheless, there have been some promising developments in this direction and, in 5E and 5F, I mention two results which support the conjecture.

B. EXPERIMENTAL OBSERVATIONS

Over the past decade, a series[2,8,24,29,30,33,34] of very sophisticated and careful experiments have given us some reliable details about pattern evolution in both rectangular and circular boxes with reasonably large aspect ratios (between 10 and 50). In particular, the control of the parameters has been such that each experiment can be run for times up to 50 horizontal diffusion time scales. Here I will give a brief overview of the observations, relying principally on the recent article of

Heutmaker and Gollub.[34] (See also lectures by Ahlers, this volume[1]). They have a cylindrical cell of aspect ratio 14. The convecting fluid is water at room temperature with a Prandtl number of 2.5. They find three regimes. In the first, $R < 1.2R_c$, they observe somewhat surprisingly, that the pattern remains time dependent and aperiodic. This feature was noticed earlier by Ahlers and Behringer.[2] After several horizontal diffusion time scales, the patterns simplify, and appear to contain weakly circular roll patches about sidewall foci and the different patches appear to be mediated by grain boundaries. Figure 1(d) is a typical state. An analysis of the wavenumber band in the pattern shows that a significant portion lies to the left of the zig-zag and cross-roll instabilities. The shape of the cell appears to have some influence. In a square cell of the same aspect ratio, some of the runs appear to stabilize after 100 horizontal diffusion times.

In the second regime, $1.2R_c < R < 4.5R_c$, the pattern will also stabilize, but the final structure may not be unique. Again, both in the rectangular and circular geometries, the textures are dominated by circular patches which (usually) surround sidewall foci. The rolls are more bent and most of the roll axes are perpendicular to the boundaries. After transients, there would appear to be a minimal number of defects and disclinations which mediate and separate the different circular patches. The pattern takes several horizontal diffusion times to become time dependent. The band of wavenumbers is almost wholly contained in the stable portion of the Busse balloon between the zig-zag and skew varicose instability boundaries. Figures 1(a) and 1(c) are stable patterns. I add one extra observation here from a numerical experiment in a rectangular cell (11.5×16) which Arter and I[6] have just carried out. We find that the finite box length (which limits the range of perturbation wavevectors) tends to push the skew varicose boundary to slightly larger wavenumbers.

In the third regime, $R > 4, 5R_c$, the pattern remains time dependent via repetitive nucleation of dislocation pairs due to what appear to be extra wavenumber production at sidewall foci. Although the dynamics is not periodic, it has a quasiperiod of about 20–40 vertical diffusion times. The distribution of pattern wavenumbers lies across the skew varicose instability boundary.

The challenge to the theoretician is to develop a theory that can allow one to understand the observed behavior and furthermore make predictions which prove to be accurate. The ideas I am about to present were developed several years ago in collaboration with Mike Cross.

[1]Editor's note: This refers to "Experiments on Bifurcations and One-Dimensional Patterns in Nonlinear Systems Far From Equilibrium," in *Lectures in the Sciences of Complexity*, edited by D. Stein, 175–224. Santa Fe Institute Studies in the Sciences of Complexity, Lect. Vol. I. Reading, MA: Addison-Wesley, 1989.

C. THE CROSS-NEWELL (CN) EQUATIONS

The Cross-Newell equations are

$$\tau(k)(\Theta_T + \vec{u} \cdot \vec{k}) + \nabla \cdot (\vec{k}B(k)) + \mu^4 J(k)\nabla^2\nabla \cdot \vec{k} = 0 \tag{5.4}$$

$$\Omega(R, A^2, k^2) = \mu^4 \left\{ \frac{A_T}{A}, \frac{(\vec{k} \cdot \nabla)^2 A}{A}, \cdots \right\} \tag{5.5}$$

and in the rigid boundary case, the mean drift velocity field $\vec{u}(X, Y, T)$ is given by the solenoidal part of the vector field $-\gamma \vec{k} \nabla \cdot (\vec{k}A^2)$, i.e.,

$$\vec{u} = \left(\frac{\partial \psi}{\partial Y}, -\frac{\partial \psi}{\partial X} \right)$$

and

$$-(\nabla \times \vec{u}) \cdot \hat{z} = \nabla^2 \psi = \gamma \hat{z} \cdot \nabla \times \vec{k}\nabla \cdot (\vec{k}A^2). \tag{5.6}$$

In these equations it is understood that $\nabla = (\frac{\partial}{\partial X}, \frac{\partial}{\partial Y}, 0)$ is effectively a two-dimensional gradient operator. The functions $\tau(k), B(k), J(k)$ are all related to the modulus k of the wavevector $\vec{k}$ and $A(k)$, the amplitude. For order-one amplitudes, the amplitude A is slaved to k by an algebraic relation

$$\Omega(R, A^2, k^2) = 0, \tag{5.7}$$

which in linear or nonlinear WKB theory would have been known as the eikonal or dispersion relation. In general, it would also contain the fast-time derivative of the phase, $\theta_t = -\omega(X, Y, T)$. Here there is no fast-time behavior in the pattern (I will briefly discuss the extra difficulties one encounters when there is) and Eq. (5.7) gives A^2 as a function of R and k^2. The reader might like to derive Eqs. (5.5)–(5.7) for the models:

I. $\left(\dfrac{\partial}{\partial t} - (\nabla^2 - 1) \right) (\nabla^2 - 1)^2\psi + (R - \psi\psi^* - \nu\psi\psi^*\nabla^2)\nabla^2\psi = 0.$

II. $\dfrac{\partial \psi}{\partial t} + (\nabla^2 + 1)^2\psi - R\psi + \psi^2\psi^* = 0.$

One finds

I. $\tau = k^2 A^2(1 - \nu k^2)(k^2 + 1)^2$, $\quad B = k^4 A^2 \dfrac{dA^2}{dk^2}(1 - \nu k^2)^2$, and

$\Omega(R, A^2, K^2) = R - A^2(1 - \nu k^2) - \dfrac{(k^2 + 1)^3}{k^2}.$

When $\nu = 0$, $\quad J = 3k^2(k^2 + 1)A^2.$

II. $\tau = A^2$, $\quad B = A^2 \dfrac{dA^2}{dk^2}$, $\quad J = A^2$ and $\Omega(R, A^2, k^2) = R - A^2 - (k^2 - 1)^2.$

In Eq. (5.7), the phenomenological constant γ, which can be worked out when R is close to R_c for the Oberbeck-Boussinesq equations, is intrinsically positive and inversely proportional to the Prandtl number. Although, in these two examples, the zero of B occurs at the same value of k as the maximum of A^2, that will not generally be true and indeed the onset of time-dependent patterns through the mechanism of the skew vector instability will depend very much on the fact that the wavenumber k_0 for which $B(k_0) = 0$ will be greater than the value which maximizes A^2. I emphasize that Eq. (5.6) was appended by hand, its form being guided by small amplitude theory. To date, the derivation of the phase, mean-drift equations for finite Prandtl numbers and order-one values of $R - R_c$ has yet to be done and provides nontrivial difficulties. (At this writing, two colleagues—Thierry Passot and Mohammad Souli—and I have developed the equations, but they have not been checked against experiment.)

The appropriate boundary condition on $\vec{k}$ is that the roll axis is perpendicular; i.e., $\vec{k} \cdot \hat{n} = 0$ where $\hat{n}$ is the outward normal to the boundary. When the higher-order derivative terms are included, I do not know what is the correct lateral wall condition, nor is it possible to demand that the total mean drift $\vec{u}$ is zero on the boundaries, and so auxiliary boundary layer fields may have to be added to satisfy these conditions.

D. THE SMALL AMPLITUDE LIMIT AND THE RECOVERY OF THE NWS EQUATIONS

Equations (5.4) and (5.5) correspond to the phase and amplitude parts of the NWS equation. Far from onset, when $R - R_c$ is order one, A is finite, the right-hand side of Eq. (5.5) is small, and A is determined algebraically. On the other hand, as $\epsilon^2 = (R - R_c)/R_c$ (or R if $R_c = 0$), becomes small, and then each of the terms in Ω on the left-hand side of Eq. (5.5) is small and of order ϵ^2. Further, when ϵ is small but still larger than μ^2, the dynamical variables can change over distance $\epsilon^{-1} \ll \Gamma$ in which case the right-hand side of Eq. (5.5) can be order ϵ^2, the same as each of the terms of the left-hand side of Eq. (5.5). In this limit, Eq. (5.5) becomes the amplitude part of the NWS equations and the amplitude is no longer slaved to the phase gradient by an algebraic equation. In addition, as the small ϵ limit is approached, $B(k)$ which is proportional to A^2 becomes order ϵ^2. Further, derivatives along the roll (which we will take to be the Y direction) are stronger than derivatives perpendicular to the roll and or order $\epsilon^{-1/2}$. Therefore, the last term in the phase equation $\mu^4 J \nabla^2 \nabla \cdot \vec{k}$ becomes $\epsilon^2 J \Theta_{YYYY}$ which combines together with $\nabla \cdot \vec{k} B$ (which is now also order ϵ^2) to give the phase part of the NWS equations. Finally, the Eq. (5.7) for the mean drift does not reduce to the Siggia-Zippelius calculation because in deriving Eq. (5.7) we assumed that X and Y derivatives were compatible. In order to have Eq. (5.7) reduce correctly we must add a term $-(1/2)\mu^2 \nabla^2 \cdot (A^* \nabla A - A \nabla A^*)$ to $(\nabla \cdot \vec{k} A^2)$.

E. PROPERTIES OF THE PHASE EQUATION IN THE INFINITE PRANDTL NUMBER LIMIT

In this section, we discuss properties of the solutions to the phase equations (5.4) which are consistent with experimental observation. Notable among these are: (i) the wavenumber selected by curved, and in particular circular, roll patches lies at the zig-zag or left stability boundary of the Busse balloon, (ii) the "almost" existence of a Lyapunov functional, (iii) the prediction that patterns cannot stabilize until $t = \tau_H \cdot (L/d)$, τ_H the horizontal diffusion time, (iv) the importance of larger "along the roll" derivatives and the resulting s-shaped roll patterns, and (v) the prediction of the shape of stationary dislocations. We begin by using the theory to recover the long-wave stability boundaries of the Busse balloon.

(I) THE BUSSE BALLOON AND THE NATURE OF THE LONG WAVE INSTABILITIES. If $\gamma = 0$, there is no mean drift and the phase equation (5.4) can be rewritten as ($\vec{k} = (m, n) = (\Theta_X, \Theta_Y)$)

$$\tau\Theta_T + \left(B + \frac{m^2}{k}\frac{dB}{dk}\right)\Theta_{XX} + \frac{2mn}{k}\frac{dB}{dk}\Theta_{XY} + \left(B + \frac{n^2}{k}\frac{dB}{dk}\right)\Theta_{YY} + O(\mu^4) = 0.$$

(5.8)

Depending on the coefficients, the part of the equation involving second spatial derivatives is (a) elliptic stable ($B < 0, (d/dk)kB < 0$), hyperbolic unstable with either (b) $B > 0, (d/dk)kB < 0$, (c) $B < 0, (d/dk)kB > 0$, or (d) elliptic unstable ($B > 0, (d/dk)kB > 0$). It is convenient to write Eq. (5.8) in a local coordinate system with X parallel and Y perpendicular to $\vec{k}$. We obtain

$$\Theta_T + \frac{1}{\tau}\left(\frac{d}{dk}kB\right)\Theta_{XX} + \frac{1}{\tau}B\Theta_{YY} = 0,$$

(5.9)

the phase diffusion equation of Pomeau and Mannville[52] except that in Eq. (5.9) the parallel and perpendicular diffusion coefficients are explicitly calculated. In this form, the nature of the instabilities are obvious. The typical graphs of A and kB are shown in Figure 10. The stable portion (k_c, k_E) defined the boundaries of the Busse balloon. The left border, $k = k_c$ where B is zero, is the boundary at which the rolls lose their resistance to bending and to the left of which one obtains the zig-zag instability. At the right border $k = k_E$ one finds the Eckhaus instability.

(II) DYNAMICS ON THE HORIZONTAL DIFFUSION TIME SCALE. We begin by noticing that the stationary solutions of the phase equation, given by $\nabla \cdot \vec{k}B = 0$, have the property that $k(B/l)$, where $l = \sqrt{\beta_x^2 + \beta_Y^2}$ and $\beta = $ constant are a family of orthogonal trajectories to the constant phase contours, is constant along each trajectory.[21] If the phase contours were circles, θ would be $r = \sqrt{X^2 + Y^2}$, β would

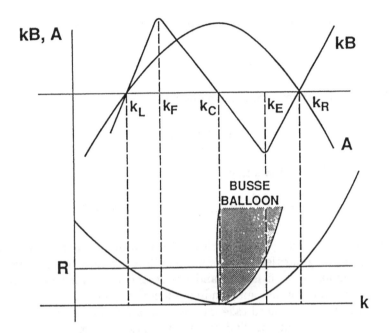

FIGURE 10 The functions kB and A and the Busse balloon.

be $\phi = \tan^{-1}(Y/X)$ in which case $l = 1/r$. In a rectangular box, the focus singularity usually resides in a corner. Since the equation $rkB = $ constant has to hold down to scales $r = \mu^2$, the constant is at most μ^2. Therefore, away from the focus, B is order μ^2 and k is within μ^2 of k_c everywhere in the circular patch. Since foci singularities are observed to occur widely (in both numerical and experimental situations), the wavenumber selection mechanism can have important consequences which we shall discuss later.

Stationary solutions for which the divergence of $\vec{k}B$ is zero are also important in situations when either the geometry or external influences are slowly changing. The function B depends on the external influences and, if they are changing slowly enough, the equation $\nabla \cdot \vec{k}B = 0$ still obtains. For example, let me illustrate the idea in an almost one-dimensional geometry, where R, in either model I or II, is slowly varying in X. Think of a gradually tapered Taylor-Couette column or a convection layer with nonuniform heating on the lower plate. From Eq. (5.4), kB is constant. But if R is subcritical anywhere in the interval, the constant must be zero (we know that the point where $R = R_c, k = k_c$ and thus B is zero) and hence the wavenumber of the pattern is uniquely determined. This is the main idea of the paper of Kramer et al.[38] An interesting calculation concerning the time dependence of the pattern in situations where the slowly varying geometrics select two different wavenumbers has been done by Rehberg and Riecke.[55,56]

We also want to mention another remarkable result. If we define

$$G(k) = -\frac{1}{2} \int^{K^2} B dk^2$$

and define F to be its integral over the box D

$$F = \iint_D G dX dY,$$ (5.10)

then a little calculation shows that

$$\frac{dF}{dT} = 0 \iint_D \tau(\Theta_T)^2 dX dY - \int_{\partial D} ds \Theta_T B \vec{k} \cdot \hat{n}$$ (5.11)

where ∂D is the boundary of D with outward normal $\hat{n}$, and includes contours surrounding any singularities and s is arc length along these boundaries. Now, on the natural boundary, $\vec{k} \cdot \hat{n}$ is zero and the second term vanishes. Also if ∂D is any part of a circular contour surrounding a focus singularity, B is zero. Therefore, the only contributions we need take into account are those due to dislocation-type defects. If v_d is the velocity of climb in the dislocation shown in Figure 11, i.e., the velocity in the direction which will decrease the number of rolls, then the second term in Eq. (5.11) is $2\pi v_d k B(k)$ where k is the local wavenumber. If the signs of v_d and B are everywhere opposite, i.e., if $v_d \propto (k - k_c)^\alpha$ (recall B is negative), then the total contribution of the defect motion to Eq. (5.11) is negative and F is indeed a *Lyapunov functional*. In certain situations, it would appear that stationary defects occur for values of $k = k_c$, namely, at the left boundary of the Busse balloon. In most cases, however, the sign of v_d is determined by the sign of $k - k_d$ where $k_d \neq k_c$[53] and in these circumstances, one cannot argue that F always decreases.

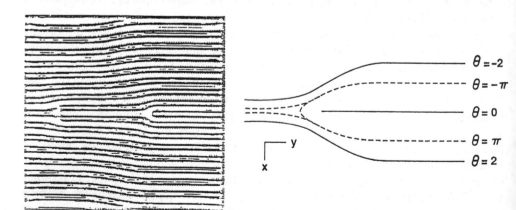

FIGURE 11 The shape of dislocations.

Nevertheless, on the horizontal diffusion time scale it would appear that the dominant pattern structures are circular or almost circular patches centered on focus singularities and in these patches the modulus of $\vec{k}$ is close to k_c and its direction is parallel to any sidewall. For experimental evidence of these facts, we refer the reader to Gollub,[29,30,31,34] and the numerical work of Greenside and Coughran[32] on the Swift-Hohenberg equation. This behavior is relatively independent of the Prandtl number, at least down to Prandtl numbers of order one. Therefore, we expect the wavenumber selection mechanism in circular roll patterns to be important.

(III) BEYOND THE HORIZONTAL DIFFUSION SCALE. It is clear from the previous discussion that patterns cannot settle down on the horizontal diffusion time scale because a plane box cannot be tiled smoothly with circular patches. Furthermore, because k is everywhere close to k_c, the position of the zig-zag instability, the rolls are at a wavelength where they lose their resistance to lateral bending ($B = 0$ in the diffusion equation (3.9)). Therefore, the derivative of the wavevector in the direction parallel to the roll axis can become larger until a balance between the $\nabla \cdot (kB)$ (which is now order μ^2) and $\mu^4 J(k)\nabla^2(\nabla \cdot \vec{k})$ terms is possible. This occurs when $\partial/\partial Y = O(\mu^{-1})$ or when the original scaling is $X = \mu^2 x, Y = \mu y$. This new dynamics takes place on a time scale $(L^2/\kappa) \times \mu^{-2}$, or the aspect ratio times the thermal diffusion time scale, and it is the earliest time on which one can expect patterns to become stationary if indeed they do. The prediction that this is the earliest time scale on which patterns can become stationary was one of the successes of Cross-Newell theory.

There are, therefore, three time scales of importance in these patterns. First, there is the vertical diffusion scale $\tau_v = d^2/\kappa$ on which a roll turnover occurs and which, incidentally, will give rise to an upper bound on the velocity of defects. Then there is the horizontal diffusion scale $\tau_H = L^2/\kappa$ on which the pattern evolution is well described by a macroscopic field equation, and on which time scale the wavevector pattern is dominated by circular patches surrounding foci singularities. Next there is the scale $\tau_L = \tau_H(L/d)$ on which scales of length $u^{-1} = \sqrt{Ld}$ along the rolls play an important role. What happens on this time in the high Prandtl number limit is well illustrated by Figures 2e and 2f. Several defects in the upper left (lower right) corner have glided to the right under the influence of the greater roll curvature (now proportional to μ rather than μ^2) and the outer rolls in the circular patch about the focus in the left lower (right upper) corner have become attached to the upper wall. They have taken the shape of an "s." Indeed Zaleski, Pomeau, and Pumir[72] have shown that the s-shaped rather than straight rolls are more likely to arise and, in addition, the roll axis intersects the boundary at an angle of order $((R - R_c/R_c)^{1/4}$ (here equivalent to μ) to the boundary normal. In lower Prandtl number situations, we shall see shortly that the effect of mean drift is to take the rolls away from the zig-zag instability boundary where they lose their resistance to lateral bending. In these cases, the behavior of the pattern on the τ_L scale is similar in some respects to high Prandtl number structure. The rolls in the center of the box or the cylinder have a tendency to straighten out and become parallel. Those attached to the "side" boundaries (the part of the wall almost parallel to the roll axes in the cylindrical geometry or the short sidewalls in the rectangular geometry) are connected through dislocations to small circular patches on the cylindrical geometry or grain boundaries in the rectangular geometry. (See Figures 1 and 2, respectively.)

Depending on how far R is from critical, the patterns on this time scale can become stationary or retain independently a slow time dependence. I will return to this point later.

(IV) AN EXACT SOLUTION FOR THE SHAPES OF STATIONARY DEFECTS. Finally, I mention an interesting exact solution[40,46] which mimics extremely well the shape of the stationary defects of the Swift-Hohenberg equation (model II which is considered a very good model for the Oberbeck-Boussinesq equations when the Prandtl number is infinite). (See Figure 11.) Assuming that k is close to k_c, we set (with $k_c = 1$)

$$\Theta(X, Y, T) = X + \mu^2 F\left(\frac{Y}{\mu\sqrt{2X}}\right) \quad X > 0$$
$$= X - \mu^2 F\left(\frac{Y}{\mu\sqrt{-2X}}\right) \quad X < 0 \tag{5.12}$$

into the phase equation, and obtain an o.d.e.

$$F'''' = 4\eta^2 F'' + 12\eta F' - 12\eta F' F'' - 8F'^2 + 6F'^2 F'' \tag{5.13}$$

for F as function of $\eta = y/\sqrt{2x}$ when written in the original coordinates. In Figure 12, we plot, after the dislocation has reached a stationary state, the numerically calculated values of the phase $\Theta - X$ at many points along the contours of constant $\eta = y/\sqrt{2x}$ so that for each value of η, there are several points. The dark line is the exact solution of Eq. (5.13) with the constant of integration chosen so that the total change in Θ from $y = -\infty$ to $y = +\infty$ is π.

It is encouraging that one can find the shape of the stationary dislocations as an exact, singular solution (Θ has a jump of 2π across $x = 0$ for large negative y), of the phase equation.

F. THE PHASE EQUATION AT FINITE PRANDTL NUMBERS; THE EFFECTS OF MEAN DRIFT

In this section, we look at the coupled Eqs. (5.4) and (5.6) and the importance of mean drift. The key new features are:

1. The decreased importance of the zig-zag instability which moves leftward from a value of k for which $B(k) = 0$ to a value from which $B(k) = \gamma \tau k^2 A^2$.

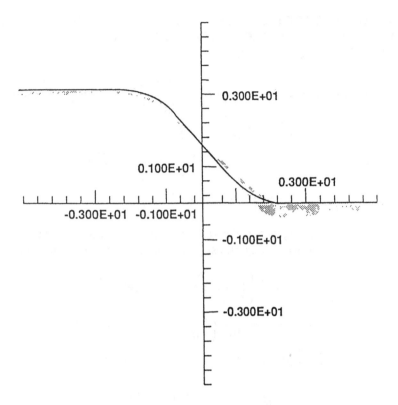

FIGURE 12   Numerical verification of the self-similar solution.

2. The appearance of new instability, the *skew varicose* instability, at values of k less than k_E, the Eckhaus boundary, which occurs when the mean drift advection of wavenumber overcomes the stabilizing influence of cross-roll diffusion $((d/dk)(kB) < 0)$.

3. The role of mean drift in creating stable off-center patches of almost circular rolls surrounding a focus singularity. In circular boxes, where the focus singularity is on the sidewall, the outermost rolls of the almost circular patch have their virtual centers farther outside the boundary than the innermost rolls, thereby creating a compression of the outermost rolls. When this compression becomes severe, the skew varicose instability occurs and nucleates defect pairs which travel to the walls.

4. For certain parameter values, the mean drift effect causes the focus singularity to become a continuous source of wavenumber. This leads to a continuous compression of the outermost rolls in the patch past the skew varicose instability boundary and the continuous nucleation of defects, resulting in a time-dependent and aperiodic pattern.

(I) THE DEFORMATION OF THE BUSSE BALLOON DUE TO MEAN DRIFTS; THE SKEW VARICOSE INSTABILITY AND THE NUCLEATION OF DEFECTS. We now explore the full equations including mean drift

$$\Theta_T + \psi_Y \Theta_X - \psi_X \Theta_Y + \frac{1}{\tau}\left\{\left(B + \frac{m^2}{k}\frac{dB}{dk}\right)\Theta_{XX} + 2\frac{mn}{k}\frac{dB}{dk}\Theta_{XY}\right.$$
$$\left. + \left(B + \frac{n^2}{k}\frac{dB}{dk}\right)\Theta_{YY}\right\} = 0 \tag{5.14}$$

$$\psi_{XX} + \psi_{YY} = \gamma\left(\frac{\partial}{\partial X}n - \frac{\partial}{\partial Y}m\right)\left\{\left(A^2 + \frac{m^2}{k}\frac{dA^2}{dk}\right)\Theta_{XX}\right.$$
$$\left. + 2\frac{mn}{k}\frac{dA^2}{dk}\Theta_{XY} + \left(A^2 + \frac{n^2}{k}\frac{dA^2}{dk}\right)\Theta_{YY}\right\}. \tag{5.15}$$

The stability of the straight parallel roll solution $\Theta = k_0 X$ is treated by setting $\Theta = k_0 X + \Theta', \psi = \psi'$ (dropping the superscripts and the subscript zero on k) to obtain

$$\Theta_T + k\psi_Y + \tau^{-1}\left\{\left(\frac{d}{dk}kB\right)\Theta_{XX} + B\Theta_{YY}\right\} = 0$$

$$\psi_{XX} + \psi_{YY} = -\gamma k\frac{\partial}{\partial Y}\left\{\left(\frac{d}{dk}kA^2\right)\Theta_{XX} + A^2\Theta_{YY}\right\}.$$

If $\Theta, \psi \propto \exp(i\vec{K} \cdot \vec{X} + \sigma T), \vec{K} = K(\cos\rho, \sin\rho)$, we obtain

$$\tau\frac{\sigma}{K^2} = \alpha + \beta\sin^2\rho - \delta\sin^4\rho \tag{5.16}$$

when $\alpha = (d/dk)kB, \beta = -(d/dk)kB + B - \gamma k^2\tau(d/dk)kA^2$, and $\delta = -\gamma k^3\tau(dA^2/dk)$. Note the following.

1. When $\rho = 0$, i.e., Y dependence is ignored, we recover the Eckhaus instability at the border where $(d/dk)kB$ becomes positive.
2. When $\rho = \pi/2$, we recover a modified version of the zig-zag instability with growth rate

$$\tau\frac{\sigma}{K^2} = B - \gamma\tau k^2 A^2. \tag{5.17}$$

The zig-zag boundary is now defined, not by $B = 0$ but by $B = \gamma\tau k^2 A^2$. Hence one effect of mean drift is to force the zig-zag instability boundary leftward. This results in both the left Eckhaus and left cross-roll instabilities becoming more important.
3. The maximum growth occurs for a value of ρ given by $\sin^2\rho = \beta/2\delta$ for which $\tau(\sigma/K^2)$ is $\alpha + (\beta^2/2\delta)$. This instability is the *skew-varicose instability*. Observe that in order for the skew-varicose instability boundary k_{sv} to lie inside that of the Eckhaus instability, $\delta = -\gamma k^3\tau(A^2)'$ must be positive. This means that the k value at which A^2 is maximum must be less than k_{sv}. Moreover, since β must be positive, it is necessary that $-\gamma k^2\tau(kA^2)' > kB'$ which is certainly

satisfied if the maximum of kA^2 lies to the left of k_{sv}, the right-hand side is negative. We will see later that the parameter

$$-\frac{\gamma k^2 \tau (kA^2)'}{kB'} = s \qquad (5.18)$$

is very important. A necessary condition for the skew-varicose instability to occur, and the negation of which is a sufficient condition for it not to is,

$$s < 1. \qquad (5.19)$$

We shall be returning to the importance of this instability when we discuss circular roll patterns in 5F(ii) and the onset of time dependence in 5F(iii). In anticipation of the results of these sections, let us rewrite Eq. (5.16) when k is chosen so that B is zero and B' negative. Then, the growth rate is

$$-\frac{\tau \sigma}{kB'} \geq -K_x^2 - sK_y^2. \qquad (5.20)$$

In circular rolls, K_x will measure the radial and K_y (proportional to curvature) the angular wavenumbers. It is also important to notice that the skew-varicose instability acts to increase the wavenumber at some locations along the roll so that at these locations the instability is more strongly driven and, therefore, the nonlinear feedback is positive and serves to continue the deformation.

The Busse balloon for model I, with the choice of $\gamma = 20$ and $\nu = 0$, is shown in Figure 13. See also Figure 14, which depicts the likely form of the stability balloon as a function of Prandtl number.

While the skew varicose instability may be thought of as the modification of the large wavenumber Eckhaus instability, the resulting finite amplitude behaviors of the two instabilities are significantly different. The Eckhaus instability removes a roll pair using the X variations by strongly modulating the envelope. In the infinite horizontal geometry, the amplitude will locally reach zero at some point and the roll pair at this X location will disappear. In the case of too few rolls, a similar scenario obtains except, in this instance, a roll pair is born in the low amplitude region. In a finite geometry, the subtraction or addition of a roll pair always occurs at the boundaries where the large amplitude modulation is already present due to boundary conditions. In the case of too many rolls, the roll nearest either end wall becomes compressed laterally and the flow near to the wall becomes very slow. In the near wall region, this roll continues to shrink and it is absorbed by its neighbor which expands. Thereafter, a slow readjustment takes place until all rolls have approximately the same width.[5]

On the other hand, the skew varicose instability exhibits a markedly different behavior.

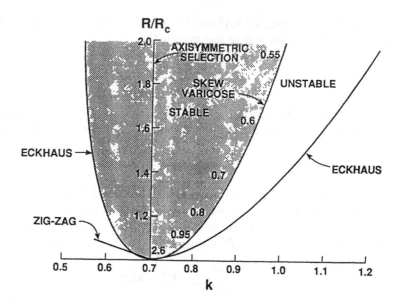

FIGURE 13   The Busse balloon for model I with $\gamma = 20, \nu = 0$. Numbers adjacent to the skew varicose line give the value of $\tan \rho$ of maximum growth rate.

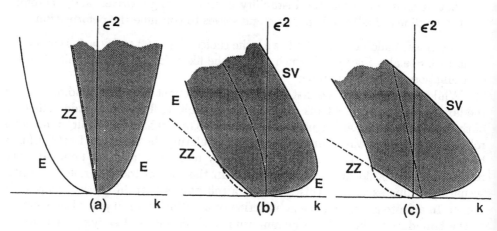

FIGURE 14   Suggestions for likely form of stability balloons for convection (hatched region is stable). (a) No drift (large Prandtl numbers); (b) and (c) with drift. The instability lines are E = Eckhaus, S.V. = skew varicose, and Z.Z. = zig-zag. The dotted line is the wavenumber selected by axisymmetric focus singularities.

Since it is difficult to modulate the amplitude to zero at a particular X for all Y, the pattern relieves the stress of an extra roll pair by nucleating two dislocations. The nucleation process is the finite amplitude state of the skew varicose instability. I describe a numerical experiment done by Arter using FLOW 3-D and reported in detail in Arter and Newell.[6] The box has horizontal dimensions 11.5 by 16 (the depth is one) and the Prandtl number is 2.5. The Rayleigh number is 8500 and the starting conditions are such that any roll compression will lead to a crossing of the skew varicose instability boundary of the Busse balloon. The velocity field is zero on all boundaries and the sidewalls are taken to be thermal insulators. At early time, a single defect was observed to nucleate near the boundary and its effect was to compress the main roll pattern near $x = 0$ at $y = 8$. Figure 15 records the zero contours of the u_y field in the region $0 \leq x \leq 8$, $3 \leq y \leq 13$ of the plane $z = 2/3$ at a sequence of times ($t = 1$ is the vertical diffusion time), $t = 6.2$ through $t = 10$. Observe that the compression of the central roll pattern, due to the wall nucleated defect (upper right corner), reaches critical near $x = 0$ at $t = 6.8$ through distortions of the straight roll pattern which clearly resemble the skew varicose instability. The roll E shrinks away from the wall, its flow field fades, and the flow fields of D and F merge near the wall. Almost immediately D is pinched off at $x = 2$ so that the second defect is the term minus of D and E and lies on the boundary between C and F. The defect then propagates to the right with a velocity of approximately unity.

We suggest that the finite amplitude evolution of the skew varicose instability is such that the X (along the roll) wavenumber increases and causes the total wavenumber to pass through the linear stability boundary, at which point the amplitude of convection is zero and the phase can undergo a discontinuous change. The result is the formation of a singularity. Again the shape of the dislocation suggests that the phase contours $\Theta - k_0 Y$ near the defect lie on parabola $X - X_0 \propto \pm\sqrt{Y - v_d T}$. The reader will recall that in describing the shapes of

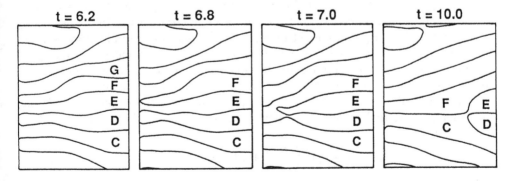

FIGURE 15   The nucleation of a defect after a skew varicose instability. Numerical simulation of full equations with real boundary conditions.

dislocations in the infinite Prandtl number case, the larger wavenumbers, which occur along the roll, were due to the loss of resistance to lateral bending caused by a proximity to the zig-zag instability boundary. In the structure described here, it is proximity to the right or skew varicose boundary that has increased the bending along the roll. It therefore would be a worthwhile exercise to look for solutions of Eqs. (5.4) and (5.6) which have this structure. The mean drift velocity field $\vec{u}$ at the defect should have its velocity there.

The reader might like to read Arter and Newell[6] where more details of the numerical calculations, including the mean drift fields, are given.

(II) A FAMILY OF EXACT, SINGULAR SOLUTIONS IN WHICH THE FOCUS SINGULAR-ITIES OF CIRCULAR ROLL PATTERNS ARE MOVED OFF CENTER BY DIPOLE-LIKE MEAN DRIFT FIELDS.

We have mentioned already that in the infinite Prandtl number limit, the wavenumber of rolls in the path is selected to be zero of $B(k)$. Even in the finite Prandtl number case, if the rolls are circular, no mean drift is produced and again the wavenumber is selected. However, suppose the box is cylindrical and the wavenumber chosen by $B = 0$ is not a wavenumber that precisely fits in the box. How is this situation resolved? We describe below a solution which partly addresses this problem and which is consistent with the observations of Croquette and Pocheau[51] and Ahlers, Cannell, and Steinberg.[4] These authors have noted that instead of the roll pattern being axisymmetric about a focus at the center, there is a noticable shift of the focus which breaks the axisymmetric symmetry. The phase contours appear to be circles whose center shift $(0, 0) \rightarrow (0, -D(r, T))$ becomes smaller the farther the phase contour is from the center; i.e., $D(r, T) \rightarrow 0$ as $r \rightarrow \infty$. Motivated by this observation, I look for solutions to Eqs. (5.4) and (5.6) in cylindrical coordinates (r, θ) in the form

$$\Theta = k_0 r + k_0 D(r, T) \sin \theta \qquad (5.21)$$

where $B(k_0) = 0$, reflecting the idea that the constant phase contours are circles about the point $x = 0, y = -D(r, T)$. Alternatively, one may take $D(0, T)$ to be zero and $-D(r, T)$ increasing in r so that the roll wavenumber increases in the negative y direction. This latter picture might be more convenient, for example, if we consider a segment of an off-center circular roll patch surrounding a focus singularity on a sidewall boundary (see Figure 16) in which the outermost rolls appear to be circles with a virtual center at $y = +D(r, T)$. The following calculation holds for both pictures. We consider D to be small, and linearize the equations in D and its derivatives. A little calculation shows that the right-hand side of Eq. (5.6) is ($\prime = \partial/\partial r$).

$$-\frac{1}{r^2}\gamma k^2 \left(\frac{d}{dk}(kA^2)\right)_0 (rD')' \cos\theta$$

(a)   (b)

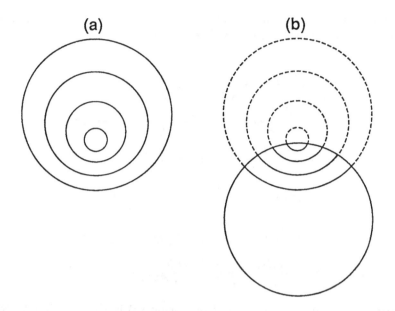

FIGURE 16   (a) The (greatly exaggerated) off-center shift of circular rolls. (b) Circular rolls surrounding a focus in a cylindrical box.

which is the negative of the vorticity ζ. The corresponding stream function is

$$\psi = \gamma k_0^2 (k_0 A_0^2)' \left(r \int_r^\infty \frac{D'}{r^2} dr \right) \cos \theta . \tag{5.22}$$

We determine $D(r, T)$ by substituting Eqs. (5.21) and (5.22) in the phase equation (5.4), whence we obtain

$$-\frac{\tau}{k_0 B_0'} \frac{\partial D'}{\partial T} = \frac{\partial^2 D'}{\partial r^2} + \frac{1}{r} \frac{\partial D'}{\partial r} - \frac{1+s}{r^2} D' \tag{5.23}$$

and we have used the definitions (5.18) for $-\gamma k_0^2 \tau_0 (k_0 A_0^2)' = s k_0 B_0'$. The effect of mean drift is contained only in the last term on the right-hand side of Eq. (5.23). Recall that k_0 is that value for which B is zero and at which $B_0' < 0$. Normally this value will be less than that value required to trigger the skew varicose instability and therefore we can look for stationary solutions to Eq. (5.23) which decay as $r \to \infty$. We find

$$D' = \frac{d_0}{r^{\sqrt{1+s}}} \tag{5.24}$$

as long as $1 + s > 0$. If $s < -1$, an instability which we call the focus instability will be triggered and I will return to this point when we consider time-dependent patterns. The free constant in Eq. (5.24) is chosen to be negative because $D(r)$ decays from a positive value to zero. These solutions correspond to the following fields

$$\Theta = k_0 r \left(1 + \frac{d_0}{r\sqrt{1+s}} \frac{1}{1 - \sqrt{1+s}} \sin\theta \right),$$

$$\vec{k} = k_0 \left(1 - \frac{d_0}{r\sqrt{1+s}} \sin\theta \right) \hat{r} - \frac{d_0}{r\sqrt{1+s}} \frac{1}{1 - \sqrt{1+s}} \cos\hat{\theta},$$

$$\psi = -\gamma k_0 (k_0 A_0^2)' \frac{d_o}{r\sqrt{1+s}} \cdot \frac{1}{\sqrt{1+s}+1} \cos\theta,$$

$$u_r = \frac{1}{r} \frac{\partial\psi}{\partial\theta} = \gamma k_0 (k_0 A_0^2)' \frac{d_0}{r\sqrt{1+s}+1} \frac{1}{\sqrt{1-s}+1} \sin\theta, \qquad (5.25)$$

$$u_\theta = -\frac{\partial\psi}{\partial r} = -\gamma k_0 (k_0 A_0^2)' \frac{d_0\sqrt{1+s}}{r\sqrt{1+s}+1} \frac{1}{\sqrt{1+s}+1} \cos\theta, \text{ and}$$

$$\zeta = -\nabla^2\psi = \frac{-k_0 B_0'}{\tau_0} \frac{s^2 d_0 \cos\theta}{(\sqrt{1+s}+1)r\sqrt{1+s}+2}.$$

The vertical vorticity, ζ, is always positive (negative) in the right-hand (left-hand) plane ($x \neq 0$) for $1+s > 0$. The vorticity field is dipolar and the flow sweeps the fluid in the direction of the most compressed rolls. The largest (smallest) wavenumber occurs on the negative (positive) y axis, $\theta = (-\pi/2)(\pi/2)$.

It should be emphasized that I have only discussed the far field behavior. I assume that the degeneracy involved in the position of the umbilicus and in d_0 is removed when one considers the nonlinear terms in D, but I have not done this calculation. In the following section 5F(III), I will discuss what happens as k increases to the point where $1 + s = 0$.

In his thesis,[51] Pocheau has also carried out these computations.

(III) ROUTES TO TIME DEPENDENCE; THE ROLES OF DEFECTS AND FOCUS SINGULARITIES.

An analysis of the equation for the dynamics of the phase variable due to slow inhomogeneities in the wavevector k for the model discussed here and other models leads us to the following speculations concerning the onset of time dependence in large Rayleigh-Bernard cells.

In the case of equations with a smooth gradient expansion—corresponding to infinite Prandtl number in convection—the only long wavelength instabilities are the Eckhaus and zig-zag instabilities. The presence of focus singularities, forced on the system by the tendency of rolls to approach a sidewall normally, selects a unique wavenumber over much of the cell. This wavenumber lies at the boundary of the zig-zag instability (Figure 14a). Although it has been speculated that this marginal stability may be important in the onset of turbulence in large cells, no detailed mechanism has been put forward. In any case, experiments at large Prandtl numbers do not see turbulence close to the threshold, but only after an instability of short length which occurs at many times the threshold Rayleigh number.

At lower Prandtl numbers, mean drift effects severely alter this picture. Now, no unique wavenumber is necessarily selected by focus singularities. If, however, we use the value selected by axisymmetric foci as a guide to the trend in the wavenumber, we may suggest two possibilities. In the first, Figure 14b, the selected wavenumber always resides within the stability balloon for long wavelength disturbances. The onset of time dependence will presumably develop either when the selected wavenumber intersects a short wavelength instability, or at lower Rayleigh numbers in defected regions of the cell, where our analysis does not apply. A second likely possibility is that the selected wavenumber intersects the skew varicose instability.[21,29,30] Let us suppose that the k in 5F(II) is chosen so as to satisfy the circular roll selection criterion, namely, $B = 0$. We can interpret that calculation as a linear stability analysis of a patch of circular rolls surrounding a focus singularity. If, in Eq. (5.23), $1 + s = \nu^2 > 0$, and $D' = \exp(k_0, (B'_0/\tau_0)\sigma^2 T))f(r)$, then $f(r)$ satisfies $f'' + (1/r)f' + (\sigma^2 - (\nu^2/r^2))f = 0$, for which the solution, bounded for all r, is $J_\nu(\sigma r)$. If the circular box is finite of radius R, then σ belongs to a discrete spectrum given by $J_\nu(\sigma R) = 0$. If R is infinite, σ is continuous. The important thing, however, is that σ is real, σ^2 is positive and (remember $B'_0 < 0$) disturbances are damped. On the other hand, when $1 + s$ changes sign, disturbances will grow and the basic solution $\Theta = k_0 r$ is unstable. To see what happens, let us write the phase equation as

$$-\frac{\tau_0}{k_0^2 B'_0}\frac{\partial \Theta}{\partial T} = \frac{\tau_0}{k_0^2 B'_0}k_0 u_r + \frac{1}{r}\frac{\partial}{\partial r}(rD')\sin\theta$$

where u_r is the radial velocity given by

$$u_r = -\gamma k_0^2 (k_0 A_0^2)'\left(\int_r^\infty \frac{D'}{r^2}dr\right)\sin\theta.$$

Let us look at the increase of phase $\Delta\Theta = \Theta(r + \Delta r) - \Theta(r)$ (the increase in radial wavenumber between r and $r + \Delta r$),

$$-\frac{\tau_0}{k_0^2 B'_o}\frac{\partial \Delta\Theta}{\partial T} = \frac{\tau_0}{k_0^2 B'_0}[u_r]_r^{r+\Delta r} + \left[\frac{\partial}{\partial r}rD'\right]_r^{r+\Delta r}\sin\theta. \qquad (5.26)$$

The second term on the right-hand side is diffusive when $B'_0 < 0$ ($k_0 < k_E$) and acts to lessen $\Delta\Theta$ and lower the radial wavenumber. The advection of wavenumber by the mean flow, however, can either add or subtract wavenumbers. This term is

$$\frac{\tau_0}{k_0^2 B'_0}\left(\gamma k_0^2 (k_0 A_0^2)'\frac{\Delta r D'}{r^2}\sin\theta\right). \qquad (5.27)$$

Consider Figure 16(b) which is a reproduction of a figure with the same number from Heutmaker and Gollub at a value of $\epsilon^2 = ((R/R_c)/R_c) = 3.84$ on the unstable side of the skew varicose instability boundary. If one completes the circular rolls about the focus, it is clear that this picture corresponds to the situation in 5F(II) with the umbilicus displaced to $y = -D$. Hence D' and $\sin\theta$ are negative. Therefore, once the ratio $(\gamma k_0^2 \tau_0 (k_0 A_0^2)')/k_0 B_0'$ becomes positive (s becomes negative), *the mean flow makes the focus act like a source of new rolls.* This is exactly what Heutmaker and Gollub observe. For $s < -1$, the mean flow advection overcomes the stabilizing role of diffusion. Note that at these values the velocity field points away from the focus and bends toward the $\theta = -(\pi/2)$ axis. New rolls are continually created at the focus singularity and the rolls in the outermost part of the circular patch are compressed past the skew varicose instability wavelength. Defects are continually nucleated, the pattern remains time dependent, and there is no reason to expect that this time dependence will be periodic. Defects can move quickly with velocities comparable with the vertical diffusion time. In its direction of motion, a defect is a domain wall between rising and falling fluid and its speed is only limited by how quickly a region of falling fluid can reverse its polarity.

It should be remembered that the arguments I have just made use model equations which I expect will be qualitatively relevant to the dynamics of real fluid convection patterns. It is important to do the real calculations, however, in order to gain a more definite understanding of what quantities B, A^2, and kA^2 stand for. For example, it might be useful if one could interpret the zeros of B, the maxima of A and kA^2 in the k, R parameter space in terms of the locus of the roll wavelength which maximize quantities such as heat flux, total dissipation or other less obvious functionals of the flow field which emphasize the importance of mean drift. Although an outline of what needs to be done is given in Cross and Newell,[21] this important calculation has not yet been carried out.

What are the effects of the finite aspect ratio and the particular geometric configurations? Numerical work by Arter and this author[6] for water convecting in an 11.5×16 roll rectangular box suggests clearly that the finite aspect ratio inhibits the skew varicose instability (k_{sv} is increased) because it limits the length of disturbances along the roll. The geometric configuration does not seem to play a crucial role at the high Rayleigh numbers. On the other hand Ahlers and Behringer[2] and Heutmaker and Gollub[34] have observed chaotic motion over long time scales at Rayleigh numbers less than $1.2 R_c$. In square boxes of the same aspect ratio, the latter authors have found the stable states do exist but are realized only after very long times. As a first guess, it is reasonable to speculate that the different geometries account for the different behavior. One might suggest two consequences of the cylindrical geometry. The first is that the nonaxisymmetric pattern expected is free to rotate: perturbations associated with the random slow rotation of the whole pattern may lead to the chaotic heat flux measure. A second is that the inconsistency between the conditions of normal boundary conditions at a closed boundary and a fixed roll diameter is now present locally, rather than at isolated corners. It is conceivable that this local inconsistency may lead to a dynamic state. Note that the former suggestion relies on the global symmetry, whereas the latter depends only on the local curvature. Investigating the presence of noise near threshold in other curved boxes (e.g., ellipses) would discriminate between these two ideas.

On the other hand, our numerical experiments[6] in a rectangular box suggest
that the time dependence may have another origin. For Rayleigh numbers of 2000,
there is clear evidence for instability of a regular 14-roll pattern after 12 vertical diffusion times. The configuration should be stable to the zig-zag instability, however,
it turns out that the rolls adjacent to the walls are slightly larger and are subject to
this instability, but the disturbance is confined to the walls. The disturbance helps
initiate a cross-roll instability at the boundary and the outcome is a gain boundary at each wall. Cross, Tesauro, and Greenside[23] have suggested that the effect
of cross rolls is to introduce another selection mechanism for the wavenumber in
the bulk different from the one chosen by a purely two-dimensional configuration.
Numerical work by these authors suggests that the resulting competition causes at
least a periodic and perhaps an aperiodic time dependence.

Finally, I want to mention the importance of defects in controlling turbulent
behavior. So far, in most of the numerical and real experiments, the starting fields
were fairly smooth either as a result of applying a smooth seed field (in the former,
in Arter and Newell,[6] we began many runs by constraining the flow to be two-
dimensional for the initial period) or by slowly raising the Rayleigh number (in
the latter) so that relatively large patches of compatible roll structures can evolve.
In these situations, the number of defects in the turbulent region is minimal and
the dynamics appears to be low-dimensional. However, if one suddenly raises the
Rayleigh number, there is less propagation of convection from one unstable location
to another and many convection patches are independently triggered. As a result,
the size of the typical roll patch is smaller and there are many more defects. At
subturbulent levels of the stress parameter, most of the defects travel quickly to the
walls and are destroyed leaving the kind of patterns seen in Figure 1. At values of
the Rayleigh number above the turbulence threshold, one can expect the defects to
be continuously nucleated and the number of defects remains large. Since defects are
excellent decorrelators of the field variable at neighboring spatial sites, one expects,
in these cases, that the dynamics remains large-dimensional. We suggest, therefore,
that the initial excitation of many defects in the pattern at turbulent values of
the stress parameter can cause the system to bypass the usual onset scenarios
associated with low- to moderate-dimension transitions. This scenario has often
been emphasized and demonstrated by Ribotta.[57]

G. WHEN THE UNDERLYING PATTERN IS PERIODIC IN THE SPATIAL AND TEMPORAL VARIABLES

The analysis proceeds as before (see Howard and Kopell,[35] and Bernoff[7]) with

$$w = f\left(\theta = \frac{1}{\epsilon}\Theta(X = \epsilon x, Y = \epsilon y, T = \epsilon t), \ A(X, Y, T)\right) \qquad (5.28)$$

except that now $\omega = -\theta_t = \Theta_T$ is an additional variable which will appear in the eikonal or dispersion relation

$$\Omega(\omega, k, R, A^2) = 0. \tag{5.29}$$

Conservation of wavenumber gives the additional relation

$$\vec{k}_t = \nabla\omega = 0. \tag{5.30}$$

When diffusive, e.g., $\nabla\tau^{-1}\nabla \cdot (\vec{k}B)$, and mean drift terms are added at a higher order, Eq. (5.30) will be the phase equation. If we consider Eq. (5.29) as giving ω in terms of k, then Eq. (5.30) is a nonlinear p.d.e. for $\vec{k}$ whose dominant terms at leading order give a first-order nonlinear p.d.e. for the wavevector $\vec{k}$. Depending on initial conditions, this equation can have shock-like solutions and, although the resulting discontinuities in $\vec{k}$ can be mediated and smoothed by both diffusive- and dispersive-like terms which appear in Eq. (5.30) at higher order, one now has to deal with the propagation of these shock layers which divide regions of slowly varying $\vec{k}$. Nevertheless, it is still possible in very simple cases to handle these complications, although it should be clear that the problem presents a severe challenge. I would suggest that it might be worthwhile to extend the results of Bernoff[7] to the two-dimensional CGL equation in the high Prandtl number situation when mean drifts can be ignored. Such an analysis might help provide some description of the break-up of two-dimensional patterns in convection in binary mixtures.

6. CONCLUDING REMARKS

It seems that the experimentalists have put the ball in the theoreticians' court. The challenges include:

1. Find the Cross-Newell equations for an Oberbeck-Boussinesq fluid using real boundary conditions and observe how the functionals corresponding to B and kA^2 behave as the parameters change. (This is now done.)
2. Explain the curious continuous time dependence of patterns at Rayleigh numbers less than $1.2R_c$.
3. Explain the stable configurations found for $1.2R_c < R < 4.5R_c$. Ignoring rotational and other symmetries, how many solutions are there for a fixed R? For example, both the configuration 1A and a configuration very like 1C are stable states for $R = 2.61R_c$. What determines their bases of attraction? Is there a functional which acts like a Lyapunov functional and which can guide one's thinking?
4. Understand better the reasons for and the interplay among the various wavenumber selection mechanisms, circular rolls, $k_c \ s \cdot t \ B(k_c) = 0$, stationary

defects k_d $s \cdot t$ $v_d(k_d) = 0$, grain boundaries, slowly varying external influences, etc.

5. Create a consistent macroscopic field theory which contains both a smooth component (the wavevector field is almost everywhere smooth) and a singular or particle-like component consisting of point defects (dislocations, foci) and line defects (grain boundaries).
6. How does one mathematically find the equation for mean drift as a solvability condition? (This is now understood.)
7. Construct an efficient numerical scheme for handling Oberbeck-Boussinesq fluids in boxes of aspect ratios up to 20, for times long with respect to the horizontal diffusion time. This might then be used to do a signal analysis on the dynamics to determine the various properties of the attractor, particularly when $R < 1.2R_c$ and $R > 4.5R_c$.
8. Understand the football states seen in binary mixture convection. These should be accessible to analysis by the envelope equation as they lie on a low-amplitude branch which bifurcates from the purely conductive solutions.
9. Understand whether the velocity of fronts joining stable but turbulent solutions to unstable quiescent states is determined using the same ideas when the stable solution is stationary.
10. Make predictions concerning the nature of binary mixture convection in boxes which are wide in all directions.

As this article was going to press, we (Thierry Possot, Mohommad Soull, and myself) have been able to derive the phase diffusion and mean drift equations for the full Oberbeck-Boussinesq equations. We are able to reproduce the borders of the Busse balloon exactly, predict the wavenumber selected by circular patterns (the zero of $B(k)$) which prediction agrees with experimental observation, calculate the value of the Rayleigh number at which the focus instability (where $s + 1 = 0$) occurs and, in general, use the theory to suggest a scenario for the onset of time dependence which is consistent with experimentally observed behavior. This work will be published shortly.

REFERENCES

1. Ahlers, G. Private communication, 1988.
2. Ahlers, G., and R. Behringer. *Phys. Rev. Lett.* **40** (1978): 712.
3. Ahlers, G., V. Steinberg, and D. S. Cannell. *Phys. Rev. Lett.* **54** (1985): 1373.
4. Ahlers, G., D. S. Cannell, and R. Heinricks. *Phys. Rev. Lett.* **A35** (1987): 2761.
5. Arter, W., A. Bernoff, and A. C. Newell. "Wavenumber Selection of Convection Rolls in a Box." *Physics of Fluids* **30(12)** (1987): 3840–3842.
6. Arter, W., and A. C. Newell. "Numerical Simulation of Rayleigh-Benard Convection in Shallow Tanks." *Physics of Fluids* **31** (1988): 2474–2485.
7. Bernoff, A. *Physica* **30D** (1988): 363–391.
8. Berge, P. *Chaos and Order in Nature*, edited by H. Haken. Berlin: Springer-Verlag, 1981.
9. Bleistein, N., and R. Handlesman. *Asymptotic Expansions of Integrals*. New York: Holt, Reinhart, and Winston, 1971.
10. Bolton, E. W., F. H. Busse, and R. M. Clever. *J. Fluid Mech.* **164** (1966): 469.
11. Brand, H. R., P. S. Lomdahl, and A. C. Newell. *Physica* **22D** (1986): 345.
12. Brand, H. R., P. S. Lomdahl, and A. C. Newell. *Phys. Lett.* **118A** (1986): 67.
13. Bretherton, C., and E. Spiegel. *Phys. Lett.* **A96** (1983): 152.
14. Busse, F. H. *Rep. Prog. Phys.* **41** (1978): 1929.
15. Busse, F. H. "Hydrodynamic Instabilities." *Topics in Applied Physics* **45** (1985).
16. Busse, F. H., and R. M. Clever. *J. Fluid Mech.* **91** (1979): 391.
17. Chandrasekhar, S. *Hydrodynamic and Hydromagnetic Stability*. Oxford: Oxford University Press, 1961.
18. Coullet, P., C. Elphick, and D. Repaux. *Propagation in Systems Far from Equilibrium* (Proceedings of Les Houches Workshop, March 1987), eds. J. E. Wesfried, H. R. Brand, P. Manneville, G. Albinet, and N. Boccara. Berlin, Heidelberg: Springer-Verlag, 1988.
19. Coullet, P, C. Elphick, J. Lega, and L. Gill. *Phys. Rev. Lett.* **59** (1987): 884–888.
20. Cross, M. C., P. G. Daniels, P. C. Hohenberg, and E. D. Siggia. *J. Fluid Mech.* **127** (1983): 255.
21. Cross, M. C., and A. C. Newell. *Physica* **10D** (1984): 299.
22. Cross, M. C. *Phys. Rev. Lett.* **57** (1986): 2935.
23. Cross, M. C., G. Tesauro, and H. Greenside. *Physica D* (1988): to be published.
24. Croquette, V., M. Mony, and F. Schosseler. *J. Phys. (Paris)* **44** (1983): 293.
25. Croquette, V., and A. Pocheau. *J. Phys. (Paris)* **45** (1984): 35.
26. Dee, G., and J. Langer *Phys. Rev. Lett.* **50** (1983): 383.

27. Doering, C. R., J. D. Gibbon, D. D. Holm, and B. Nicoleanko. *Nonlinearity* (1988): to appear.
28. Erdelyi, A. *Asymptotic Expansions* Dover, 1956.
29. Gollub, J. P., and J. F. Steinman. *Phys. Rev. Lett.* **47** (1981): 505.
30. Gollub, J. P., A. R. McCarrier, and J. F. Steinman. *J. Fluid Mech.* **125** (1982): 259.
31. Gollub, J. P., and A. R. McCarrier. *Phys. Rev. A* **26** (1982): 3470.
32. Greenside, H. S., and W. M. Coughran. *Phys Rev. A* **30** (1986): 398.
33. Heutmaker, M. S. P. N. Frankel, and J. P. Gollub. *Phys. Rev. Lett.* **54** (1985): 1369.
34. Heutmaker, M. S., and J. P. Gollub. *Phys. Rev. A* **35** (1987): 242.
35. Howard, L. N., and N. Koppel. *Stud. in Appl. Math.* **56** (1977): 95.
36. Keefe, L. *Stud in Appl. Math.* **73** (1985): 91.
37. Kolodner, P., R. W. Wadden, A. Passner, and C. M. Surko. *J. Fluid Mech.* **163** (1986): 195.
38. Kramer, L., E. Ben Jacob, H. Brand, and M. C. Cross. *Phys. Rev. Lett.* **49** (1982): 1891.
39. Lange, C. G., and A. C. Newell. *SIAM J. of Appl. Math.* **27** (1974): 441.
40. Meiron, D., and A. C. Newell. *Phys. Lett.* **113A** (1985): 289.
41. Moon, M. T., P. Huerre, and L. Redekopp. *Physica* **7D** (1983): 135.
42. Newell, A. C., and J. A. Whitehead. "Finite Bandwidth, Finite Amplitude Convection." *J. Fluid Mech.* **38** (1969): 179.
43. Newell, A. C., and J. A. Whitehead. "Review of the Finite Bandwidth Concept." In *Proc. IUTAM Conference on Instabilities in Continuous Systems*, edited by H. Leipholz, 284–289. Berlin: Springer-Verlag, 1971.
44. Newell, A. C. *Am. Math. Soc.* **15** (1974): 157.
45. Newell, A. C. *Pattern Formation and Pattern Recognition*, 244. Berlin: Springer-Verlag, 1979.
46. Newell, A. C. *Lecture Notes in Num. App. Anal.* **5** (1982): 205.
47. Newell, A. C. "Dynamics of Pattens: A Survey." In *Propagation in Systems Far from Equilibrium, Proceedings of Les Houches Workshop*, March 1987, edited by J. E. Wesfried, H. R. Brand, P. Manneville, G. Albinet and N. Boccara. Berlin, Heidelberg: Springer-Verlag, 1987.
48. Nozaki, K., and N. Bekki. *Phys. Rev. Lett.* **51** (1983): 2171.
49. Nozaki, K., and N. Bekki. *J. Phys. Soc. Jap.* **53** (1984): 581.
50. Olver, F. W. J. *Asymptotics and Special Functions*. New York: Academic Press, 1971.
51. Pocheau, A. Thesis, 1987.
52. Pomeau, Y., and P. Manneville. *J. Physique Lett.* **40** (1979): 609.
53. Pomeau, Y., and P. Manneville *J. Phys. (Paris)* **42** (1981): 1067.
54. Pomeau, Y., S., Zakeski, and P. Manneville. *Phys. Rev. A* **27** (1983): 2710.
55. Rehberg, I., and H. Riecke. "Physics of Structure Formation." In *Proceedings of Conf. Tubingen*. Berlin: Springer-Verlag, 1986.

56. Rehberg, I., and F. H. Busse. In *Propagation in Systems Far from Equilibrium, Proceedings of Les Houches Workshop*, March 1987, edited by J. E. Wesfried, H. R. Brand, P. Manneville, G. Albinet, and N. Boccara. Berlin, Heidelberg: Springer-Verlag, 1988.
57. Ribotta, R., and A. Joets. *J. Physique* **478** (1986): 595–606.
58. Ribotta, R., A. Joets, and X. D. Yang. In *Propagation in Systems Far from Equilibrium, Proceedings of Les Houches Workshop*, March 1987, edited by. J. E. Wesfried, H. R. Brand, P. Manneville, G. Albinet, and N. Boccara. Berlin, Heidelberg: Springer-Verlag, 1988.
59. Segel, L. A. *J. Fluid Mech.* **38** (1969): 203.
60. Siggia, E. D., and A. Zippelius. *Phys. Rev. A* **24** (1982): 1036.
61. Steinberg, V., G. Ahlers, and D. S. Cannell. *Physica Scripta* **32** (1985): 534.
62. Steinberg, V., and E. Moses. *Phys. Rev. A* **34** (1986): 693.
63. Stewartson, K., and J. T. Stuart. *J. Fluid Mech.* **48** (1971): 529.
64. Stuart, J. T. *J. Fluid Mech.* **9** (1960): 353.
65. Stuart, J. T., and R. DiPrima. *Proc. R. Soc. Lond. A* **362** (1978): 27.
66. Swinney, H., and R. DiPrima. "Hydrodynamic Instabilities." *Topics in Applied Physics* **45** (1985): 31.
67. Walden, R. W., P. Kolodner, A. Passner, and C. M. Surko. *Phys. Rev. Lett.* **55** (1985): 496.
68. Watson, J. *J. Fluid Mech.* **9** (1960): 371.
69. Wesfried, J. E., H. R. Brand, P. Manneville, G. Albinet, and N. Boccara, eds. *Propagation in Systems Far from Equilibrium, Proceedings of Les Houches Workshop*, March 1987. Berlin, Heidelberg: Springer-Verlag, 1988.
70. Whitham, G. B. *J. Fluid Mech.* **44** (1970): 373.
71. Whitham, G. B. *Linear and Nonlinear Waves.* New York: Wiley Interscience, 1975.
72. Zaleski, S., Y. Pomeau, and S. Pumir. *Phys. Rev. A* **29** (1984): 366.

H. F. Nijhout
Department of Zoology, Duke University, Durham, NC 27706

Pattern Formation in Biological Systems

This chapter originally appeared in *1991 Lectures in Complex Systems*, edited by L. Nadel and D. Stein, 159–187. Santa Fe Institute Studies in the Sciences of Complexity, Lect. Vol. IV. Reading, MA: Addison-Wesley, 1992. Reprinted by permission.

Pattern formation refers to the processes in development by which ordered structures arise within an initially homogeneous or unstructured system. Understanding these processes is absolutely essential for understanding regulatory mechanism in development. It is also essential for understanding the developmental origin of biological form, and ultimately, for understanding morphological evolution. In practice, pattern formation refers to things like the processes in embryos that determine where gastrulation will occur, or the processes that define where bones will condense in the mesenchyme of a developing limb, how many there will be, their shape, and their position in relation to each other. Or in plants, where leaves will form on the stem of a plant, and what shape those leaves will have.

Here we will be particularly concerned with processes of pattern formation that occur quite late in animal development, in particular, the development of pigment patterns. Pigment patterns have several advantages as model systems in which to study the principles of pattern formation. First, color patterns are almost always two-dimensional, so they can be studied on the plane without having to use projections or collapse dimensions. This makes them far easier to deal with than three-dimensional processes in development, and makes color patterns particularly attractive for computer modeling, because the whole pattern can be represented on the two-dimensional computer screen. Second, since they develop relatively late, the processes that give rise to the pattern occur in a system that is usually macroscopic and, therefore, more easily manipulated experimentally than are early embryos. Third, since pattern is manifested as the local synthesis of pigment, it is easy to detect. Fourth, since the chemical nature and biosynthetic pathways of most pigments are known, it is in principle possible to fully understand all control pathways in the system at the chemical and molecular level. Finally, it is in systems like color patterns, where all the molecular and biochemical steps are in principle knowable and understandable, that we have the best chance of uncovering the full sequence of events that links genotype and phenotype, something that has yet to be done for any morphological event.

THREE MECHANISMS

The processes that result in local specialization of structure and function can be formally subdivided into two distinctive kinds: those that involve cell migration and mechanical interactions among cells (such as traction and differential adhesivity), and those that involve chemical prepatterning.[14,19] Murray[14] points out that the two mechanisms are quite different because in chemical prepatterning the chemical pattern precedes morphogenesis, while in patterning by mechanochemical cell-cell interactions, morphogenesis is the immediate consequence of the patterning process. There are a few examples of patterning systems that are purely one or the other (the formation of butterfly wing patterns, which we will deal with below, is one of them), but in most cases both mechanisms seem to operate, such as when a chemical gradient allows migrating cells to aggregate and interact.

Among the best studied examples of cell movement-mediated patterning are aggregation and fruiting body formation in the slime mold *Dictyostelium*, and the formation of bones in the developing limbs of vertebrates. In *Dictyostelium* we have one of the very few cases in which we actually know what the chemical morphogen is whose gradient stimulates the initial aggregation. Here the aggregation signal is cyclic AMP (cAMP), which is secreted by isolated cells when they run out of food. When other cells perceive this signal, they are attracted to its source and migrate up the cAMP gradient. Such migrating cells also begin to secrete cAMP themselves, and a complex set of interactions ensues that transiently gives rise to interesting cell aggregation patterns and eventually results in the clumped aggregates. While aggregating, the population of cells does exhibit spatial patterns of spiral waves very similar to those seen in the Belousov-Zhabotinski reaction, and in many models using cellular automata (see figures in Winfree,[32] Tomchik and Devreotes,[27] and Murray[14]).

Patterned bone formation in the mesenchyme of developing vertebrate limbs has been studied in a variety of contexts. Perturbation experiments have revealed complex interactions that have been modeled conceptually as the well-known clock face model of French et al.[8] and Bryant et al.,[2] and mechanistically as a traction-aggregation mechanism by Oster et al.[20] Evolutionary morphologists have been particularly interested in the development of bone patterns in vertebrate limbs because the well-established homologies among the bones, and the extensive historical pattern of their transformation preserved in the fossil record, makes this one of the most attractive (and tractable) systems in which to study the interplay of developmental and evolutionary processes in the shaping of biological form.[9,21]

Pigment patterns arise by the same two mechanisms, of cell movement and chemical prepatterning and, in mollusk shells, by a third distinctive mechanism that involves a complex interplay between the tissue of the mantle and the shell as it is secreted. In vertebrates, the pigment pattern of the skin is produced by melanophores, which are cells that produce the black/brown pigment, melanin. Melanophores arise from the neural crest (along the dorsal midline) early in embryonic development and from there migrate across the body surface.[5,29,30] The color patterns of fish, frogs, zebras, giraffes, and leopards are therefore the consequence of the migration and patterned accumulation of pigment-producing cells.

In insect color patterns, the mechanism is quite different. The insect epidermis is only one cell layer thick, and the cells are attached to the overlying cuticle most of the time. Cell migration and cell rearrangement are therefore generally impossible. All patterning in the epidermis must therefore take place by mechanisms of cell-to-cell communication. The cells of the insect epidermis are interconnected by gap junctions and are thus coupled electrically and are potentially coupled by diffusion. Signals can thus be transmitted across substantial distances and control over this communication can be exercised by modulating the number and distribution of gap junctions that are open at any one time. Pigment patterns are thus the result of local cell differentiation in a static monolayer of cells. Formation of the pattern does not involve cell migration, nor is the pattern subsequently modified by cell rearrangement.

In the shells of gastropods (snails) and bivalves (clams), the color pattern is laid down as the shell is secreted. The pigment of the pattern resides not in cells but in the nonliving shell. In contrast to the two previous cases, pattern formation in shells is essentially a one-dimensional process. During growth the mantle adds material to the leading edge of the shell and at the same time secretes pigments at appropriate locations to produce species-characteristic color patterns (stripes, spots, zig-zags, etc.). The mantle is a motile organ and moves frequently relative to the margin of the shell as the animal locomotes, rests, and hides. Consequently, shell deposition is not continuous but shows both regular and erratic periods of growth and rest. The mantle ultimately controls where and when the shell will grow, and also where exactly pigment will be deposited. The pigment pattern is thus the result of the behavior of the whole mantle and of the way it interacts with the growing edge of the shell.

The color patterns of vertebrates, of insects, and of mollusk shells thus come about by three fundamentally different mechanisms. Theoretical work has shown, however, that the essence of these three pattern-forming processes can be captured by very similar sets of mathematical equations.[7,14,19] This suggests that the principles involved in each process could be fundamentally similar even though the actual mechanisms are not. In almost all cases, lateral inhibition (short-range activation coupled with long-range inhibition of a particular event) provides the organizing mechanism, and, while systems may differ in the exact means by which effective activation and inhibition is achieved (e.g., cooperativity, autocatalysis, or positive feedback, versus catabolism, interference, or competition), the final spatial results of the process are similar if not identical.

In the sections that follow we will assume, for the sake of simplicity, that chemical prepatterning is the process at work because such a process can be modeled without having to take account of the movement of cells relative to one another. In addition, we will assume a perfectly two-dimensional system. Thus, what follows will apply, strictly speaking, only to pattern formation in the insect integument. As we will see, these assumptions produce a rich, complex, and largely nonintuitive world of patterns that begs further exploration, both experimental and theoretical.

DIFFUSION

In biological systems, convection (usually via a circulatory system) and diffusion (within cells and, via gap junctions, between cells) provide the most common means of chemical communication within and among cells and tissues. Convection is generally used for long-range transport and appears to play no role in any of the pattern formation systems that have been studied so far. Thus, to understand patterning, we need to understand the mechanism and consequences of diffusion.

Diffusion comes about by the random movement of particles produced by thermal agitation. The mathematics of diffusion has been widely studied, and the reader is referred to the text by Crank[4] for the fundamentals, and to Carslaw and Jaeger[3] for a more elaborate treatment of special cases. In one spatial dimension, the diffusion equation is usually written as:

$$\frac{\partial c}{\partial t} = D\nabla^2 c \qquad (1)$$

where c is the concentration of the diffusion substance, and D is the diffusion coefficient. On macroscopic scales diffusion is a slow process. The dimension of the diffusion coefficient, D, can be used to get an idea of the rate of diffusion. If the diffusion coefficient is expressed in the units cm^2/sec, then the average time (in seconds) it takes for a particle to diffuse through a given distance, d (in centimeters), is approximately d^2/D.[6] Moderately large biochemical molecules diffuse through the cytoplasm of a cell with $D = 10^{-7}$. Such a molecule would take an average of $(10^{-4})^2/10^{-7} = 10$ seconds, to diffuse across the diameter of a typical 10 micron cell. The average distance over which diffusion acts within a given period of time is proportional to $(Dt)^{1/2}$.[6] Even though diffusion is an inherently slow process it does clearly provide a relatively effective means of communication over the small distances (usually 1 mm or less) and time periods (hours to days) that are relevant to most developmental systems.

Diffusion-dependent processes can also exert their effect rapidly and over much larger distances if they are coupled to some amplifying machinery. The diffusion of a large charged molecule (say, $D = 10^{-7}$ cm^2/sec) across a cell membrane can rapidly change the local balance of charge, and cause the diffusion of small ions toward or away from the area. If the small ions have a diffusion coefficient of, say, 10^{-5}, and act as intermediate messengers, then the rate of "signal" propagation caused by diffusion of the large molecule would have been amplified 100-fold. The propagation of an action potential, which is basically a cytochemical cascade mechanism, is a well-known example of the amplification of a diffusible signal.

EVOKING AN EFFECT: THRESHOLDS

Simple linear diffusion from a source into a medium, or from a source to a sink, sets up a gradient in the concentration of the diffusing substance. The concentration at a particular point (p) along such a gradient carries information. It can be used to estimate both the distance between p and the source (or the sink), and the time since diffusion began. For the purposes of pattern formation, the former, the estimation of position within a diffusion field, is the more interesting and useful one. In the simplest case of pattern formation, diffusion from a point source sets up a gradient of a chemical across an otherwise homogeneous developmental field, and some novel developmental event is caused to occur wherever the concentration of the gradient is above (or below) some critical value. As we will see below, the eyespots in the wing patterns of butterflies are produced by just such a simple mechanism. Changes in the threshold, and changes in the shape of the gradient can both alter the dimension and position of the "pattern" within the total field. The formal requirements and consequences of pattern formation by such simple gradient systems have been explored by Lewis Wolpert[34] in his "Theory of Positional Information."

On this view, the problem of pattern formation is twofold: first, how to establish a source for the diffusing signal, and second, how to retrieve the information in the diffusion gradient. The first of these problems is by far the most difficult one, and we will take it up below. The second problem can be rephrased to ask: how do you set up a threshold so that the continuously distributed gradient in one substance (the diffusing signal) is translated into a sharply discontinuous and stable change of some developmental or biochemical event?

Lewis et al.[11] have developed an elegant model for a threshold mechanism. They note that most threshold models assume an allosteric enzyme whose activity is a sigmoidal function of substrate concentration (Figure 1). The problem with such a model is that along a gradient of substrate concentration the transition from the inactive form to the active form of the enzyme is gradual and occurs over a relatively long distance (Figure 1). Increasing the number of cooperating subunits in the enzyme increases the steepness of the sigmoid transition and thus sharpens the "threshold" to some degree. Allosteric enzymes generally, however, have no more than four subunits, and that puts a practical limit to the refinement of a threshold by this means. Lewis et al.[11] suggested a modification of the allosteric model to include also a linear degradation term. They suggest the following structure. Suppose a gene G, which produces a product g, that stimulates its own synthesis by positive feedback at a rate that is a sigmoidal function of its concentration $(K_a g^2 / k_b + g^2)$, and that g breaks down at a rate proportional to its concentration $(-k_c g)$. Suppose further that the synthesis of g is also stimulated by a signal molecule S, at a rate that is linearly proportional to the concentration of S. This gives the following relationship:

$$\frac{dg}{dt} = k_1 S + \frac{k_2 g^2}{k_3 + g^2} - k_4 g \qquad (2)$$

which is shown graphically in Figure 2. The graph of the rate of production of g is in effect an inclined sigmoid curve whose position is controlled by the value of S. When S is small, the reaction has three steady states, two of which are stable (Figure 2). If the system starts with gene G off, and thus with no g present, the concentration of g will tend toward its low steady state. Small and moderate perturbations in its concentration will always cause g to return to this low steady state. However, if the concentration of S goes up, the level of the curve rises and there is eventually only one steady state of g (Figure 2), much higher than the previous one. Thus, if S increases gradually, there will be a sudden transition in the concentration of g from its low to its high steady state. A smooth and continuous change in the concentration of S thus results in an abrupt switch in the concentration of g. This gives us, then, a mechanism for a sharp threshold in the control variable, S, with no intermediate values between the extremes of the response variable, g.

An additional interesting and useful feature of this model is that it has a kind of "memory" because, once g has switched to its higher steady state, it will stay there even if S subsequently declines or disappears. Thus we have essentially a mechanism for the irreversible activation of a gene. If such a gene controls, for instance, the synthesis of pigment-forming enzymes, then we have a mechanism for producing a patch of pigment wherever the concentration of S is above the threshold defined by Eq. (2).

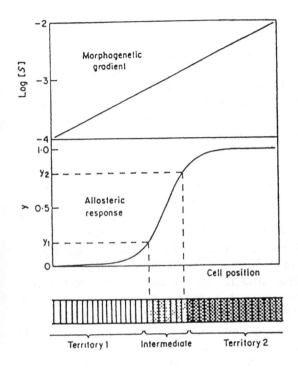

FIGURE 1 Allosteric model for a threshold. The concentration gradient in S activates an allosteric enzyme that obeys the Hill equation. The degree of saturation of the enzyme, y, that corresponds to various points along the gradient is shown in the lower graph. The threshold provided by this mechanism is not sharp and the transition can extend across many cells. Reprinted by permission of the publisher from "Thresholds in Development" by J. Lewis et al., *J. Theor. Biol.* **65**, 579–590. Copyright © 1977 by Academic Press Inc. (London).

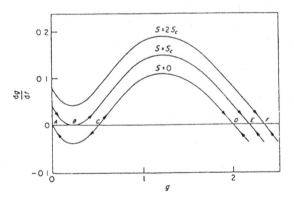

FIGURE 2 Curves produced by Eq. (2) for three values of the signal substance, S. As S gradually increases, the number of stable states abruptly falls from two to one.

REACTION DIFFUSION

Pattern formation by diffusion gradients requires at the very minimum the existence of a source of the diffusing chemical. If pattern regulation is important, then a sink is also essential, so that all intermediate values of the gradient are always present within the developmental field. It should be clear that this requirement for a source (and a sink) in effect pushes the problem of pattern formation back one step, and the issue becomes one of determining what causes the sources and sinks to be where they are.

Though unsatisfying from a mathematical point of view, such potentially infinite regressions in control mechanisms are biologically reasonable and probably the rule rather than the exception. Development is, after all, a complex network of causal connections in which any process works correctly only if all the preceding ones did (at least within certain tolerances). There are, however, certain conditions under which a stable pattern can emerge in an initially homogeneous and randomly perturbed field without the need for initial sources or organizing centers. The conditions under which this can occur were discovered by Turing[28] and this discovery constitutes one of the major advances ever made in the theory of biological development. Turing[28] showed that the steady-state condition of certain kinds of biochemical reactions can be made spatially unstable if at least two of the reactants are able to diffuse. In other words, if the reactants are free to diffuse, then it is possible for them to become stably patterned into areas of high and areas of low steady-state concentrations. On first sight this is a nonintuitive result, because one generally thinks of diffusion as having a homogenizing effect. Under certain conditions, however, diffusion can act to amplify spatial waves of certain critical frequencies. The mathematics behind this process were outlined by Turing and have been more fully explored by many other authors since. Particularly readable accounts of the theory and the conditions under which such diffusive instabilities arise in chemical reaction systems are given by Segel and Jackson[24] and by Edelstein-Keshet,[6] and a more technical treatise with many examples is given by Murray.[14] The most elaborate exploration of the consequences and possible uses of one class of these *reaction-diffusion* mechanisms is given by Meinhardt.[12]

The conditions necessary for chemical pattern formation in reaction-diffusion systems are given by Edelstein-Keshet[6] as follows:

1. There must be at least two chemical species.
2. These chemicals must affect each other's rate of production and/or breakdown in particular ways.
3. These chemicals must also have different diffusion coefficients.

The general equation system for reaction diffusion is:

$$\frac{\partial A}{\partial t} = F(A, B) + D_A \nabla^2 A,$$

$$\frac{\partial B}{\partial t} = G(A, B) + D_B \nabla^2 B,$$

in which $F(A, B)$ and $G(A, B)$ define the reaction equations for the two interacting chemical species.

Most mechanisms for chemical patterning produce a set of conditions that are referred to as *lateral inhibition*. What this means is that one of the chemicals, usually called the *activator,* has a low diffusion coefficient and exerts its influence over a fairly short range while the other, called the *inhibitor,* has a much higher diffusion coefficient and thus exerts its effect over a much longer range. The term is derived from physiology where similar short-range activation, long-range inhibition systems are common, and particularly well studied in the retina where lateral inhibition is in part responsible for the detection of edges and patterns.

Three reaction-diffusion systems have achieved particular popularity for problems in developmental biology and biological pattern formation. The model of Schnakenberg[22] is one of the simplest systems that exhibits chemical pattern formation. Its reaction dynamics are given by:

$$
\begin{aligned}
F(A, B) &= k_1 - k_2 A + k_3 A^2 B\,, \\
G(A, B) &= k_4 - k_3 A^2 B\,.
\end{aligned}
\tag{3}
$$

The lateral inhibition system of Meinhardt[12] is the one whose behavior has been studied most extensively:

$$
\begin{aligned}
F(A, B) &= k_1 - k_2 A + \frac{k_3 A^2}{B}\,, \\
G(A, B) &= k_4 A^2 - k_5 B\,.
\end{aligned}
\tag{4}
$$

The reaction system of Thomas,[26] while more complicated than the preceding two, has the virtue that it is the only system that is empirical, based on real chemistry. It involves three reactants as follows:

$$
\begin{aligned}
F(A, B) &= k_1 - k_2 A - H(A, B)\,, \\
G(A, B) &= k_3 - k_4 B - H(A, B)\,, \\
H(A, B) &= \frac{k_5 AB}{k_6 + k_7 A + k_8 A^2}\,.
\end{aligned}
\tag{5}
$$

For many purposes it is convenient to express equations such as these in a nondimensional form. One reason is that nondimensionalization always reduces the number of parameters in the model, which simplifies the analysis of the scope of the model. Another is that it removes the units of measurement and thus allows one to examine the effects of scale more effectively.[14,23] Murray[14] suggests the following general nondimensional form for reaction-diffusion systems:

$$
\begin{aligned}
u_t &= \gamma f(u, v) + \nabla^2 u\,, \\
v_t &= \gamma g(u, v) + d\nabla^2 v\,.
\end{aligned}
\tag{6}
$$

With the appropriate scaling, the reaction dynamics for the three systems mentioned above can be rewritten as follows:

$$
\begin{aligned}
f(u, v) &= a - u + u^2 v\,, \\
g(u, v) &= b - u^2 v\,,
\end{aligned}
\tag{7}
$$

for the Schnakenberg system;

$$f(u, v) = a - bu + \frac{u^2}{v},$$
$$g(u, v) = u^2 - v,$$

(8)

for the Meinhardt system; and

$$f(u, v) = a - u - h(u, v),$$
$$g(uv) = a(b - v) - h(u, v),$$
$$h(u, v) = \frac{puv}{1 + u + Ku^2},$$

(9)

for the Thomas system.

The parameter d in Eq. (6) corresponds to the *ratio* of the diffusion coefficients of inhibitor and activator, while the parameter γ represents the scale of the system. Murray[14] suggests that γ is proportional to the area of the system, for two-dimensional diffusion. γ can also represent the strength of the reaction term relative to the diffusion term. An increase in γ can be offset by a decrease in d. The advantage of having a single variable that can represent the scale of the system is that the consequences of pattern formation in a growing system can be easily studied, and predictions can be made about the differences in pattern that would be produced when the same mechanism acts in developmental fields of different sizes. Both features are of interest to developmental biologists who perforce deal with many systems that undergo growth during the period of study.

The advantages of nondimensionalized systems in facilitating studies on the effects of scaling are offset, for the biologist at least, by the fact that other biologically important parameters (such as the reaction constants for the synthesis and breakdown of specific chemical species) become inaccessible to manipulation. Since such reaction constants provide the only direct link to the genome (genes code for enzymes, whose activity is represented by the reaction constant), it becomes virtually impossible to study the effects of single gene alterations. Thus biologists interested in exploring the potential of gradualistic accumulations of small genetic changes to cause gradualistic (or discontinuous) morphological change will need to work with fully dimensional forms of a system.

In addition to the general conditions for chemical pattern formation mentioned above, there are several specific conditions that must be met. These are treated in detail and with several examples by Segel and Jackson,[24] Edelstein-Keshet,[6] and Murray.[14] Only the summary conclusions will be given here. The form of the null clines (the graphs of $dx/dt = 0$) of the reactions gives essential information on whether diffusive instability is in principle possible. The character of the crossover point of the two null clines (the system's steady state) is critical: the activator and inhibitor must both have positive slopes or both have negative slopes at steady state, and, in either case, the slope of the inhibitor must be steeper than that of the activator.[6] The null clines for the nondimensional forms of the three reaction-diffusion equations listed above are given in Murray.[14]

Whether or not a system with null clines of the required shapes will exhibit diffusive instability depends critically on the values of the parameters, and these are specific to each system. Murray[13] has worked out the parameter space for the nondimensionalized forms of the three reaction systems listed above, and has shown that they are surprisingly narrow. In almost all cases parameters must be chosen with considerable precision and the choice of one parameter value places significant constraints on the possible values of the remaining parameters. Once the parameter space for a given system is known, however, it can form the basis for the numerical exploration of its pattern-forming properties.

The analysis of the general behavior of a reaction-diffusion system is not a trivial matter. Because reaction-diffusion systems involve coupled nonlinear equations, they usually cannot be solved analytically and their behavior must be studied by numerical simulation. It is, however, possible to get a general idea of how a particular system behaves by studying perturbations near the steady state of a linearized system (see Edelstein-Keshet[6] for the description of a method). Such a linear theory approach can predict the number of modes that will form after random perturbation of a field of given dimensions. Arcuri and Murray[1] have used linear theory to predict the pattern generated by the nondimensionalized Thomas system in one space dimension (Figure 3(a)). The theory predicts a regular increase in the number of modes as either d or γ increases (Figure 3(a)). Numerical simulation of the full nondimensional Thomas system, however, gives somewhat different results (Figure 3(b)). Odd modes appear to be favored over even modes, something which linear theory does not predict. The solution space for modes 2 and 6 is particularly small in the full nonlinear system.

Arcuri and Murray[1] also calculated how the modes of a Thomas system would behave in a growing field. As the field grows, it can support a progressively larger number of modes. Existing modes appear initially to split in two, which would result in a doubling of the number of modes. But this is not what happens. Instead, at a critical point the system appears to become unstable and reorganizes so that only a single mode is added. Only in some cases is more than one mode added as the field grows, which is consistent with the behavior shown in Figure 3(b), where many odd numbered modes are adjacent.

In two dimensions the succession of modes is more complicated, and is critically dependent on both the chemistry of the system and the geometry of the field. In a rectangular field two perpendicular sets of waves are possible, and as the scale of the field increases, more waves can be fitted along both axes. The succession of modes for a general reaction-diffusion system on a rectangular field with no-flux boundaries has been studied by Edelstein-Keshet,[6] and is shown in Figure 4 for a field of dimension 1×2. The quantity E^2 in Figure 4 has the following correspondence:

$$E^2 = m^2 + n^2/\gamma^2 \tag{10}$$

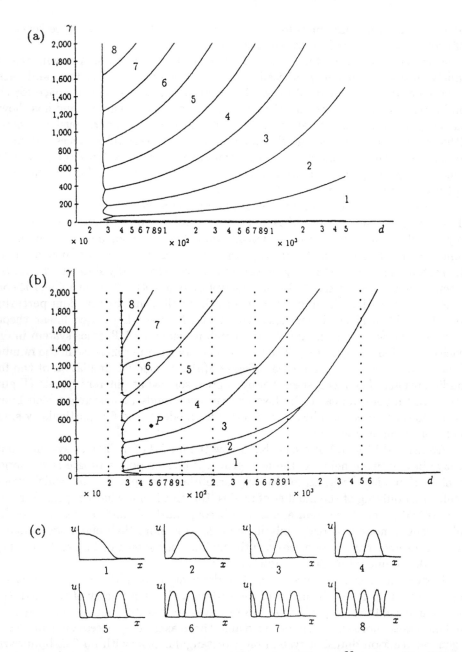

FIGURE 3 Solution space for the nondimensionalized Thomas[26] system (Eq. (9)) in one spatial dimension, with no-flux boundary conditions. (a) Modes for various values of d and γ obtained from linear theory. (b) Solution space obtained by simulation of the full nonlinear system. (c) Spatial distribution of morphogen concentration (continued)

FIGURE 3 (continued) for several of the regions indicated in (b). Reprinted by permission of the publisher from "Pattern Sensitivity to Boundary and Initial Conditions in Reaction-Diffusion Models" by P. Arcuri and J. D. Murray, *J. Math. Biol.* 24 141–165. Copyright © 1986 by Springer-Verlag.

where m and n are integers that represent the number of wavelengths parallel to the x and y axis, respectively, and γ is the dimension of the field parallel to the y axis divided by the dimension parallel to the x axis. The succession of modes is then given by the sequence of values of m and n, ranked in order of increasing values of E^2.[6] Figure 5 illustrates the patterns that correspond to several of these modes. In a real system E^2 can be derived as a function of the area of the field and the ranges of activation and inhibition, and thus the succession of modes shown in Figures 4 and 5 can be the consequence of gradual changes in any of these three parameters. Of course, fields of different shapes may have a different succession of modes, determined also by the value of γ.

In circular and elliptical fields, the succession of modes is also different. Kauffman[10] has calculated the mode progression for a general reaction-diffusion system on an elliptical domain of increasing size and showed that the succession of nodal lines on such a growing field was very similar to the succession of compartmental boundaries that form in the wing imaginal disks of *Drosophila*. It may therefore be that the progressive compartmentalization of the *Drosophila* imaginal disks is the simple and spontaneous consequence of a reaction-diffusion system operating on a growing domain. It is interesting to modes that the succession of modes is also similar to the succession of nodes in vibrating circular and elliptical plates. Xu et al.[35] have shown that the vibrational modes of plates of more complex shapes also corresponds generally to the pattern boundaries produced by reaction-diffusion systems on similar-shaped fields. Murray[14] has noted that the initial stages of chemical pattern formation by reaction-diffusion poses the same mathematical eigenvalue problem as that describing the vibration of thin plates. Thus, assuming equivalent boundary conditions can be established on vibrating plates, we may have here an analog model of pattern formation by a general reaction-diffusion system that solves for the pattern almost instantaneously and would therefore afford a fast and efficient way of exploring patterning in complex geometries.

A DISCRETE MODEL OF PATTERN FORMATION BY LATERAL INHIBITION

Young[36] has demonstrated that instead of using continuous partial differential equations to describe pattern formation by reaction diffusion, it is possible to obtain

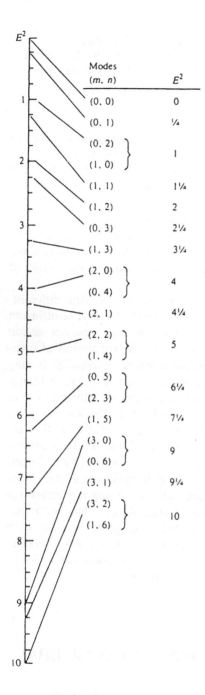

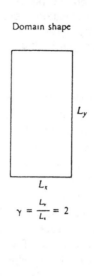

Domain shape

$$\gamma = \frac{L_y}{L_x} = 2$$

FIGURE 4   Progression of modes of a typical reaction-diffusion system in two dimensions, with increasing values of E^2. The modes m and n correspond to the number of wave peaks supported in the x and y direction, respectively, as the domain size increases, or as the range of the inhibitor decreases. Reprinted by permission of the publisher from *Mathematical Models in Biology* by L. Edelstein-Keshet. Copyright © 1988 by Random House Publishers.

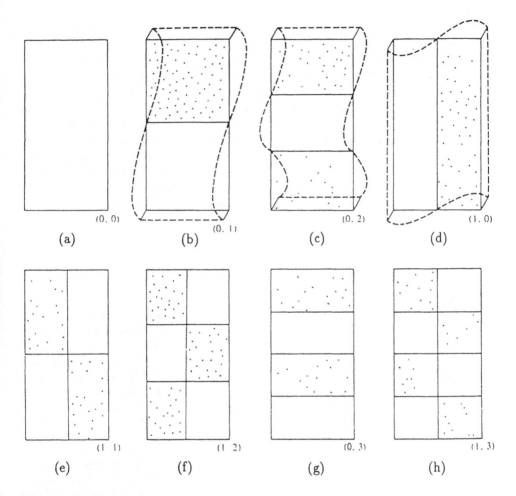

FIGURE 5   Examples of the first seven two-dimensional patterns predicted for a typical reaction-diffusion system under the conditions described in Figure 4. Reprinted by permission of the publisher from *Mathematical Models in Biology* by L. Edelstein-Keshet. Copyright © 1988 by Random House Publishers.

equivalent results with a completely discrete model that captures the essence of lateral inhibition but does not require solution of the diffusion equation. Young's theory is modeled on the one proposed by Swindale[25] for explaining patterns in the visual cortex of the brain.

Young[36] models the combined effect of a short-range activator and a long-range inhibitor by assuming that around each "source" cell there are two concentric circular regions: an inner one where there is a constant positive value of some control parameter, and an outer one where there is a constant but negative value of the same parameter (Figure 6). This condition corresponds to the short-range activation and long-range inhibition of a lateral inhibition model, the principal difference being that in reaction-diffusion systems the "activity" of the activator and inhibitor decline gradually with distance from the center of activation.

The Young mechanism produces spots or irregular stripes, depending on the ratio of activator and inhibitor levels (Figure 7). Small values produce spots, while large values of the ratio produce stripes. The size of these pattern elements is determined by the range of the activator, while their spacing is determined in large measure by the range of the inhibition. One of the chief advantages of the Young method is that the patterns form and stabilize after only three or four iterations. This mechanism can therefore produce patterns far more rapidly than one that depends on the numerical simulation of diffusion (which may require thousands of iterations).

(a)

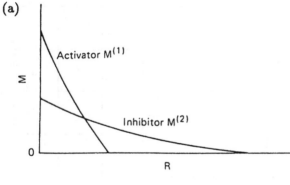

(b)

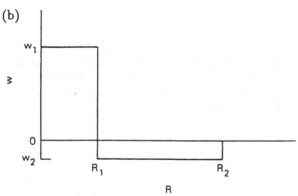

FIGURE 6 Discrete lateral-inhibition model of Young.[36] (a) The typical lateral inhibition system with continuously variable values of activator and inhibitor. (b) Young's model with discrete and spatially constant values for activator and inhibitor. Each differentiated cell exerts a constant short-range activating effect (w_1) and a constant long-range inhibitory effect (w_2) on its neighbors. Reprinted by permission of the publisher from "A Local Activator-Inhibitor Model of Vertebrate Skin Patterns" by D. A. Young, *Math. Biosci.* **72**, 51–58. Copyright © 1984 by Elsevier Science Publishing Co., Inc.

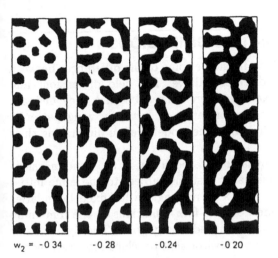

FIGURE 7   Patterns produced by
the Young model for different values
of w_2 (while w_1 is held constant at
1), after random activation of some
cells. Reprinted by permission of the
publisher from "A Local Activator-
Inhibitor Model of Vertebrate Skin
Patterns" by D. A. Young, *Math.
Biosci.* **72,** 51–58. Copyright ©
1984 by Elsevier Science Publishing
Co., Inc.

$w_2 = -0\ 34$ $-0\ 28$ -0.24 $-0\ 20$

RANDOM AND NONRANDOM PATTERNS

The patterns produced by the Young mechanism (Figure 7) illustrate one of the
limitations of the standard approach to the simulation of pattern formation. When
patterning is initiated by random perturbation of the steady state (as is usually
done to study the general properties of a given reaction-diffusion mechanism), then
the pattern produced is also random. These patterns thus mimic the stripes on the
coats of zebras, or the spotting patterns of cheetahs, leopards, and giraffes, all of
which are random and characterize the individual-like fingerprints. Randomness is,
in fact, the hallmark of vertebrate color patterns, and of certain developmental patterns such as the interdigitating ocular dominance stripes in the vertebrate visual
cortex. The vast majority of patterns in development, however, are regular and are
reproduced identically from individual to individual. To obtain regularity and re-
peatability, it is necessary to define the boundary conditions and initial conditions
by a nonrandom mechanism. The trick in modeling pattern formation in development is to find a nonarbitrary means of defining initial and boundary conditions.
This generally requires substantial knowledge of the developmental biology of the
system under study. Thus, while reaction-diffusion mechanisms can make patterns
that look remarkably like those seen in nature, we can only accept a given pattern
and mechanism as representing nature in a significant and meaningful way if it is
backed up by a body of experimental evidence that gives us confidence that we
have applied the correct boundary conditions.

RESULTS OF SIMULATIONS IN TWO DIMENSIONS

We use numerical simulation methods to illustrate some of the differences between the three reaction-diffusion schemes discussed above. The field dimensions and boundary conditions used in these examples were chosen because they define a problem of biological interest, namely the formation of butterfly wing patterns. We will first, however, examine the behavior of the models before illustrating their application to a biological problem.

Figures 8, 9, and 10 illustrate the behaviors, respectively, of the nondimensionalized Schnakenberg, Thomas, and Meinhardt systems subject to the same initial and boundary conditions. The field is a (1×2) rectangle, with fixed boundaries on one short side and the two long sides, and no-flux conditions at the remaining short side. Initial conditions were the unperturbed steady state. The figures show the near steady state concentration of the activator that develops after setting the fixed boundaries to 1.1 times the initial steady state. Each panel explores the d/γ parameter space. It will be recalled that an increase in the parameter γ can be interpreted as an increase in the size of the field, while an increase in parameter d represents an increase in the range of the inhibitor. The patterns produced by fixed boundary conditions on all four sides can be visualized by reflection on the horizontal midline of each figure.

It is obvious that the three mechanisms produce dramatically different patterns. The Thomas and Schnakenberg systems produce mostly linear patterns, while the Meinhardt mechanism stabilizes as point patterns. The patterns produced by the Thomas and Schnakenberg systems differ considerably in detail. The Thomas patterns are relatively simple lines, while the Schnakenberg patterns tend to develop bulges and isolated islands of activator concentration. It is possible to get an idea of the sensitivity of these systems to variation in parameters and field size by noting the changes in pattern that are associated with, say, a 10% change in d or γ. On the whole, variation of this magnitude has relatively little effect on the pattern.

It is evident that the three reaction-diffusion systems are far from equivalent, even though linear theory predicts the same general behavior for all three systems. The details of the patterns produced by each system, and the characteristic differences between them, can only be uncovered by simulation. This means that there is no way of using the information in Figures 8 to 10 to predict how these three systems will behave under different boundary conditions. We can be assured that each will produce characteristically different patterns, but their form cannot be predicted without simulation.

Both the one-dimensional simulations of Arcuri and Murray[1] and the two-dimensional simulations shown above illustrate that the full nonlinear systems produce patterns whose details differ significantly (and often dramatically) from those predicted by linear theory. For most developmental systems the *details* of the pattern are more important than its general features, and this means that each

γ

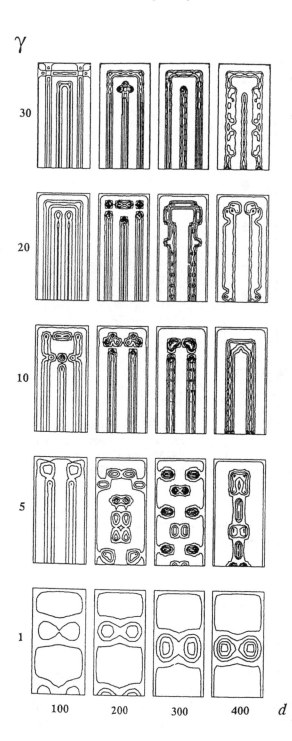

FIGURE 8 Patterns produced for various values of d and γ by the Schnakenberg system (Eq. (7)) in two spatial dimensions.

γ

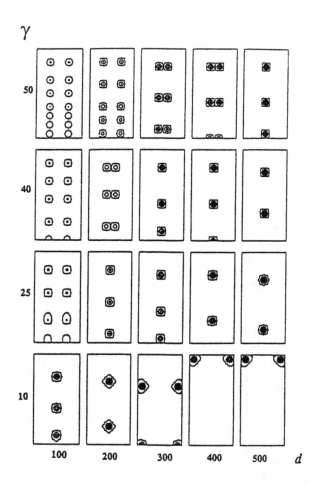

FIGURE 9 Patterns produced for various values of d and γ by the Meinhardt system (Eq. (8)) in two spatial dimensions.

biological problem in which reaction diffusion is believed to play a role must be studied by full simulation of the nonlinear system.

CELLULAR AUTOMATA

We conclude the general section on pattern formation with a brief discussion of the usefulness of cellular automata for simulating pattern formation in development. In their pure form, cellular automata are points in space which can take on one of two values (0 or 1) depending on the values of other such points in their neighborhood. The rules of a cellular automaton determine how the values of neighbors are interpreted. With relatively simple rules operating on such binary automata, it is

γ

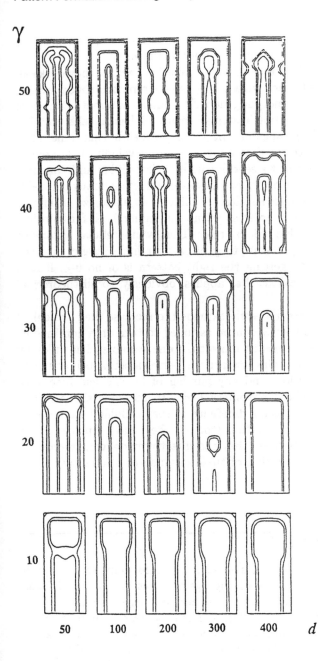

FIGURE 10   Patterns produced for various values of d and γ by the Thomas system (Eq. (9)) in two spatial dimensions.

possible to produce a vast array of complicated patterns that have fascinated mathematicians and biologists for nearly a decade (e.g., Wolfram[33]). Such automata have been used, among others, to simulate the color patterns on mollusk shells, and the branching pattens of algae. Spiral waves, such as those of the Belousov-Zhabotinski reaction, and interdigitating patterns, resembling ocular dominance stripes, are particularly easy to mimic and emerge from a variety of automata.

Cellular automata are attractive for biological simulation because they evoke an immediate image of biological cells, each with a fairly simple repertoire of behaviors, but collectively capable of complex morphogenesis.[33] Cellular automata can serve as models of biological pattern-formation systems because biological cells, too, behave by interacting only with their immediate neighbors, while obeying some set of internal "rules." The complex patterns that appear during development are emergent properties of the interaction of those rules with their cellular and chemical environment.[17] Many theoretical biologists are, however, reluctant to accept cellular automata models because the formal rules are difficult to analogize to known biological processes, and because there exist as yet no general methods for translating biological interactions into a table of local rules. Thus, while cellular automata can produce biologically realistic patterns, they often offer little insight into the biological process. In other words, getting the right pattern is of no use, if it is obtained for the wrong reason (a caveat that applies, obviously, to all theoretical modeling in biology).

Cellular automata can, however, be easily extended to increase their biological realism. Each point (or cell) in an array can be assumed to take on a continuous, range of values, and can possess values in more than one variable. The rules by which these values change can reflect the interactions between cells, such as receptor binding, competition, or diffusion, and any number of biochemical reactions. Clearly, with such extensions cellular automata begin to resemble the methods used for numerical simulation. The main difference is that cellular automata do not attempt to model a differential equation (though they may). Such complex automata are useful for biologists because they can directly model communication between cells, and they allow examination of the consequences of qualitative and quantitative rules of interaction.

SIMULATION AND MIMICRY

Cellular automata, like reaction-diffusion systems, are useful only to the extent that they give insights into the biology of the system that is being simulated. In this regard it is perhaps useful to make a distinction between simulation and mimicry. In simulation the theoretical model grasps and accurately summarizes the principles behind the process being simulated, while in mimicry the model is wrong even though it produces the right kind of pattern. Mimicry in theoretical modeling commits what statisticians would call a type 2 error: accepting a false hypothesis, or in this case, getting the right answer for the wrong reason.

Unfortunately, much modeling in theoretical developmental biology appears at present to be mimicry. In developmental modeling it is easy to get the right kinds of pattern for the wrong reason because certain categories of biologically reasonable patterns (zebra stripes, ocular dominance stripes, sea shell patterns) emerge readily from a variety of reaction-diffusion and cellular automata models. In most modeled systems, we simply do not know enough about the developmental physiology to make sensible choices between alternative models, and, even when we can imagine only one model mechanism, we cannot be sure it has captured the essence of the underlying process.

In order for a model to be biologically useful, it must obviously incorporate as much information as possible about the developmental physiology of the system. But that is generally not sufficient. In order to have reasonable assurance that a model has captured the essence of a process, it must produce a pattern whose *details* resemble those of the morphology being modeled, it must also reproduce in its dynamics reasonable portions of the *ontogenetic transformation* that the real pattern undergoes, and because morphological evolution is gradualistic, it must be able to produce by simple changes of parameter values (and not by adding more terms to the model) a range of *diversity* of the pattern identical to that found to occur in nature. Few models meet these expectations.

PATTERN FORMATION ON BUTTERFLY WINGS

Here we briefly discuss pattern formation on the wings of butterflies as a concrete example of color pattern formation because it is one of the few systems that meets the expectations of physiology, detail, ontogeny, and diversity, mentioned above. It has the added advantage that the patterns are strictly two dimensional, exhibit an evolved system of homologous elements with transformations across the thousands of species of butterflies, and can be easily modeled without having to collapse any dimensions. This system has provided a variety of insights into the way in which developmental processes change during morphological evolution.[18]

The color patterns of butterflies are all variants on a theme of homologies called the nymphalid ground plan (Figure 11). The entire diversity of color patterns comesabout through the selective expression and modification of the individual pattern elements that make up the ground plan. The wing pattern is compartmentalized into two developmentally independent systems. First, the overall pattern is divided into three parallel symmetry systems: the basal symmetry system (elements **b** and **c**, in Figure 11), the central symmetry system (elements **d** and **f**), and the border symmetry system (elements **g** and **i**). In the centers of the latter two systems, there are two additional pattern elements, the discal spot (element **e**) and the border ocelli (element **h**). Second, the development of the elements of these symmetry systems within a given wing cell is uncoupled from that in adjoining wing cells. As a consequence of this developmental isolation, each element of the pattern has been free to evolve morphologically with nearly complete independence from the other pattern elements. The overall wing pattern is thus a mosaic of

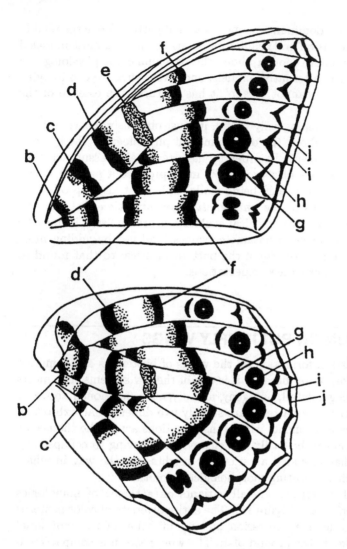

FIGURE 11 The nymphalid ground plan. This is a diagrammatic representation of the general distribution of pattern elements (labeled *b–j*) on the wings of butterflies. The pattern elements are arranged in serially homologous series that repeat from wing cell to wing cell. (From Nijhout[18]; reprinted by permission of the author.)

semi-independent pattern elements that can be modified and arranged on the wing surface to provide a variety of optical effects, ranging from camouflage to mimicry.[18]

The presumptive evolution of the nymphalid ground plan is illustrated diagrammatically in Figure 12. The ancestor is believed to have had a simple pattern with a single symmetry system, as is found in many species of moths today. Evolution of complexity progressed by the addition of more symmetry systems (Figures 12(b)–(d)), possibly by a system that sets up an increasing number of standing waves on the wing. The number of symmetry systems became stabilized at three, and each gradually evolved a distinctive morphology (Figures 12(e)–(h)), probably

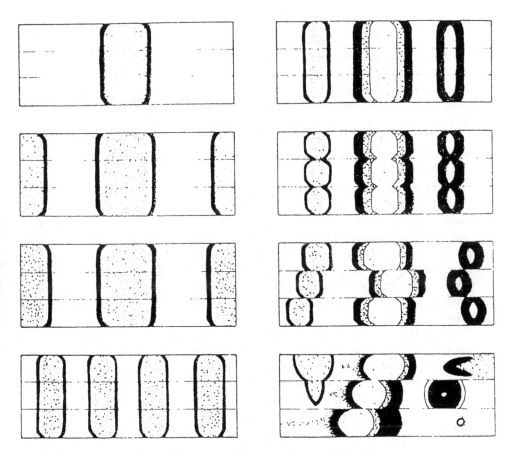

FIGURE 12 Hypothetical evolution of the nymphalid ground plan from ancestors with a few simple uncompartmentalized symmetry systems.

due to the evolution of a proximo-distal gradient or discontinuity in some variables that interact with the wave pattern. In the immediate ancestors of the butterflies, the wing veins became boundaries to pattern formation and the pattern became compartmentalized to each wing cell (Figures 12(f)–(h)). With this developmental isolation the pattern elements in each wing cell became free to diverge both in position (Figures 12(f) and (g)) and morphology (Figures 12(g) and (h)).

The developmental compartmentalization of the wing pattern greatly facilitates its modeling, because each pattern element in each wing cell can be modeled separately without having to worry about possible interactions with distant patterns. Nijhout[16,18] has shown that a relatively simple model can account for nearly the entire diversity of shapes of pattern elements that are found among the thousands of species of butterflies. The model generates the pattern in two steps, in accordance with what is known about the developmental physiology of pattern formation in this system. The first step establishes a system of line and point sources of a diffusible substance, and the second step establishes the pattern as a simple threshold on the diffusion gradients produced by those sources.

The distribution of diffusion sources (and barriers to diffusion) in real butterfly wings is known from experimental perturbation studies and from studies of the comparative morphology of normal and aberrant patterns.[18] When activated singly or in pairs, this distribution of sources (Figure 13) has been shown, by simulation, to be capable of producing nearly the entire diversity of pattern shapes found in the butterflies.

Sources in the exact locations shown in Figure 13, are readily produced by the Meinhardt[12] lateral inhibition system, and by no other reaction-diffusion system that has been examined so far.[16] The Meinhardt system produces the right patterns, but only when provided with fixed boundary conditions for the activator on three of the four sides of the rectangle that simulates a wing cell. These are the three locations of the wing veins around a typical wing cell. The wing veins afford the only means by which material can enter or leave the developing wing, and provide reasonable physical constant-level sources for materials, which are modeled as fixed boundaries.

Perhaps the most important feature of the Meinhardt lateral inhibition system implemented in this way is the dynamic progression of source distributions it produces as the reaction-diffusion progresses (Figure 14). This progression of sources produces patterns that closely resemble the diversity of color patterns seen among closely related species in several genera of butterflies. Diversity of this type in essence constitutes a heterochrony. This example illustrates that the most interesting feature of reaction-diffusion systems, from a biological perspective, is probably not the steady-state patterns to which a system tends, but the dynamic progression of patterns well before the steady state is reached. Development, like most of biology,

Point Sources   Line Sources

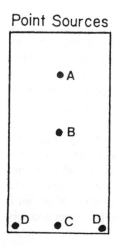
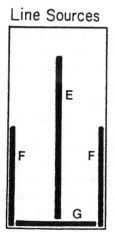

FIGURE 13   Distribution of sources (or sinks) that can produce nearly the entire diversity of patterns found in the wing cells of butterflies. The rectangular field represents a single wing cell in which veins make up the two long side boundaries and the top boundary. (From Nijhout[18]; reprinted by permission of the author.)

is not an equilibrium phenomenon. Dynamically changing patterns like those of evolving reaction-diffusion systems may provide useful models for the progression of determinative processes during development.

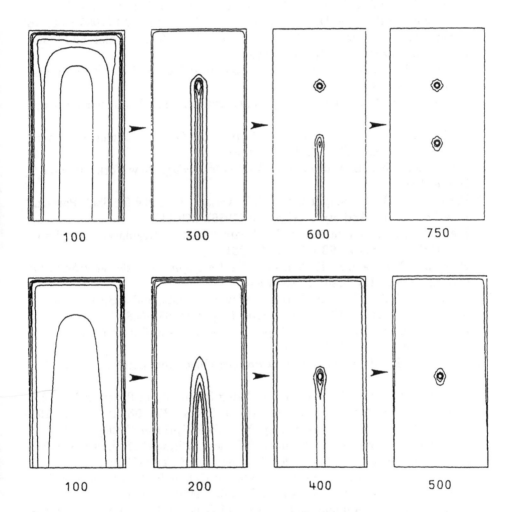

FIGURE 14 The lateral inhibition model of Meinhardt (Eq. (4)) can produce the diversity of source distribution shown in Figure 14 by varying boundary conditions and reaction constants. The series shown is a typical time sequence of activator concentration which gradually transforms from a high ridge to a series of point sources on the wing-cell midline. (From Nijhout[18]; reprinted by permission of the author.)

REFERENCES

1. Arcuri, P., and J. D. Murray. "Pattern Sensitivity to Boundary and Initial Conditions in Reaction-Diffusion Models." *J. Math. Biol.* **24** (1986): 141–165.
2. Bryant, S. V., V. French, and P. J. Bryant. "Distal Regeneration and Symmetry." *Science* **212** (1981): 993–1002.
3. Carslaw, H. S., and J. C. Jaeger. *Conduction of Heat in Solids.* Oxford: Oxford University Press, 1959.
4. Crank, J. *The Mathematics of Diffusion.* Oxford: Oxford University Press, 1975.
5. DuShane, G. P. "An Experimental Study of the Origin of Pigment Cells in Amphibia." *J. Exp. Zool.* **72** (1935): 1–31.
6. Edelstein-Keshet, L. *Mathematical Models in Biology.* New York: Random House, 1988.
7. Ermentrout, B., J. Campbell, and G. F. Oster. "A Model for Shell Patterns Based on Neural Activity." *Veliger* **28** (1986): 369–388.
8. French, V., P. J. Bryant, and S. V. Bryant. "Pattern Regulation in Epimorphic Fields." *Science* **193** (1976): 969–981.
9. Hinchliffe, J. R., and D. R. Johnson. *The Development of the Vertebrate Limb.* Oxford: Oxford University Press, 1980.
10. Kauffman, S. A. "Chemical Patterns, Compartments, and a Binary Epigenetic Code in *Drosophila.*" *Amer. Zool.* **17** (1977): 631–648.
11. Lewis, J., J. M. Slack, and L. Wolpert. "Thresholds in Development." *J. Theor. Biol.* **65** (1977): 579–590.
12. Meinhardt, H. *Models of Biological Pattern Formation.* London: Academic Press, 1982.
13. Murray, J. D. "Parameter Space for Turing Instability in Reaction-Diffusion Mechanisms: A Comparison of Models." *J. Theor. Biol.* **98** (1982): 143–163.
14. Murray, J. D. *Mathematical Biology.* New York: Springer Verlag, 1989.
15. Newman, S. A., and W. D. Comper. "'Generic' Physical Mechanisms of Morphogenesis and Pattern Formation." *Development* **110** (1990): 1–18.
16. Nijhout, H. F. "A Comprehensive Model for Colour Pattern Formation in Butterflies." *Proc. Roy. Soc. London* B **239** (1990): 81–113.
17. Nijhout, H. F. "Metaphors and the Role of Genes in Development." *BioEssays* **12** (1990): 441–446.
18. Nijhout, H. F. *The Development and Evolution of Butterfly Wing Patterns.* Washington, DC: Smithsonian Institution Press, 1991.
19. Oster, G. F., and J. D. Murray. "Pattern Formation Models and Developmental Constraints." *J. Exp. Zool.* **251** (1989): 186–202.
20. Oster, G. F., J. D. Murray, and P. K. Maini. "A Model for Chondrogenic Condensation in the Developing Limb: The Role of Extracellular Matrix and Tractions." *J. Embryol. Exp. Morphol.* **89** (1985): 93–112.

21. Oster, G. F., N. Shubin, J. D. Murray, and P. Alberch. "Evolution and Morphogenetic Rules. The Shape of the Vertebrate Limb in Ontogeny and Phylogeny." *Evolution* **45** (1988): 862–884.
22. Schnakenberg, J. "Simple Chemical Reaction Systems with Limit Cycle Behavior." *J. Theor. Biol.* **81** (1979): 389–400.
23. Segel, L. A. *Modeling Dynamic Phenomena in Molecular and Cellular Biology.* Cambridge: Cambridge University Press, 1984.
24. Segel, L. A., and J. L. Jackson. "Dissipative Structures: An Explanation and an Ecological Example." *J. Theor. Biol.* **37** (1972): 545–559.
25. Swindale, N. V. "A Model for the Formation of Ocular Dominance Stripes." *Proc. Roy. Soc. London* **B 208** (1980): 243–264.
26. Thomas, D. "Artificial Enzyme Membranes, Transport, Memory, and Oscillatory Phenomena." In *Analysis and Control of Immobilized Enzyme Systems*, edited by D. Thomas and and J.-P. Kervenez. New York: Springer Verlag, 1975.
27. Tomchik, K. J., and P. N. Devreotes. "Adenosine 3′,5′-Monophosphate Waves in *Dictyostelium discoideum*: A Demonstration by Isotope Dilution-Fluorography." *Science* **212** (1981): 443–446.
28. Turing, A. M. "The Chemical Basis of Morphogenesis." *Phil. Trans. Roy. Soc. London* **B 237** (1952): 37–72.
29. Twitty, V. C. "The Developmental Analysis of Specific Pigment Patterns." *J. Exp. Zool.* **100** (1945): 141–178.
30. Twitty, V. C. *Of Scientists and Salamanders.* New York: Freeman, 1966.
31. Waddington, C. H., and R. J. Cowe. "Computer Simulation of Molluscan Pigmentation Pattern." *J. Theor. Biol.* **25** (1969): 219–225.
32. Winfree, A. T. *The Geometry of Biological Time.* New York: Springer Verlag, 1980.
33. Wolfram, S. "Cellular Automata as Models of Complexity." *Nature* **311** (1984): 419–424.
34. Wolpert, L. "Positional Information and Pattern Formation." *Curr. Top. Dev. Biol.* **6** (1971): 183–224.
35. Xu, Y., C. M. Vest, and J. D. Murray. "Holographic Interferometry Used to Demonstrate a Theory of Pattern Formation in Animal Coats." *Appl. Optics* **22** (1983): 3479–3483.
36. Young, D. A. "A Local Activator-Inhibitor Model of Vertebrate Skin Patterns." *Math. Biosci.* **72** (1984): 51–58.

Carla J. Shatz
Department of Neurobiology, Stanford University School of Medicine, Stanford, CA 94305

Impulse Activity and the Patterning of Connections During CNS Development

Reprinted from *Neuron* **5** (1990): 745–756. Permission granted by Cell Press.

How are the highly ordered sets of axonal connections so characteristic of organization in the adult vertebrate central nervous system formed during development? Many problems must be solved to achieve such precise wiring: axons must grow along the correct pathways and must select their appropriate target(s). Even once the process of target selection is complete, however, the many axons that comprise a particular projection must still arrange themselves in an orderly and highly stereotyped pattern, typically one in which nearest-neighbor relations are preserved so that the terminal arbors of neighboring projection neurons are also neighbors within the target. Here, I would like to consider the process by which this final patterning of neuronal connections comes about during development. Studies of the vertebrate visual system, reviewed here, have provided extensive evidence in favor of the hypothesis that an activity-dependent competition between axonal inputs for common postsynaptic neurons is responsible in good part for the establishment of orderly sets of connections.

COMPETITION IN THE FORMATION OF OCULAR DOMINANCE COLUMNS IN THE MAMMALIAN PRIMARY VISUAL CORTEX

Many insights into developmental mechanisms underlying the formation of orderly connections have come from studies of the mammalian visual system, in which the clear-cut patterning of connections is exemplified in the highly topographic ordering of projections and strict segregation of inputs from the two eyes at successive levels of visual information processing (for reviews, see Rodieck[54] and Sherman and Spear[68]). Ganglion cell axons from each eye project to the lateral geniculate nucleus (LGN) on both sides of the brain. However, within the LGN, axons from the two eyes terminate in a set of separate, alternating eye-specific layers that are strictly monocular[22] (see Figure 1). Neurons in the LGN project, in turn, to layer 4 of the primary visual cortex where, again, axons are segregated according to eye of origin into alternating monocularly innervated patches that represent the system of ocular dominance columns within cortical layer 4.[23,26,59,60]

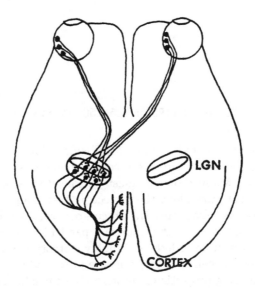

FIGURE 1   A simplified diagram of the mammalian visual pathways. Only connections from each eye to the left side of the brain are shown. Retinal ganglion cell axons from the two eyes travel to the lateral geniculate nucleus (LGN) of the thamalus, where their terminals are segregated in separate eye-specific layers. The axons of neighboring retinal ganglion cells within each eye terminate in neighboring regions within the appropriate layers, establishing a topographically ordered map. LGN neurons, in turn, project to layer 4 of the primary visual cortex where again axonal terminal arbors of LGN neurons representing the two eyes are segregated into alternating ocular dominance patches.

E 40   E 46   E 53   E 63

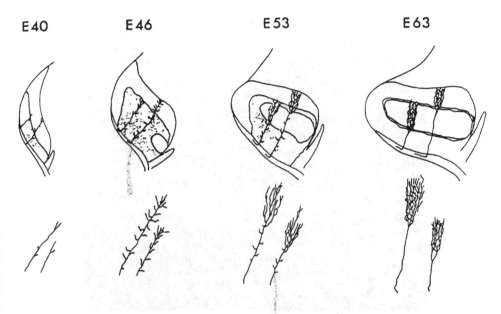

FIGURE 2 Summary of the prenatal development of the eye-specific layers in the
cat's LGN. Shaded areas indicate regions within the LGN simultaneously occupied by
ganglion cell axons from the two eyes at different times in development, as derived
by the anterograde transport of intraocularly injected tracers. Stick figures show the
appearance of representative ganglion cell axons from the ipsilateral (shorter axons
at each age) and contralateral (longer) eyes, based on studies of the morphology of
individual axons filled with horseradish peroxidase *in vitro* (see Shatz,[67] for more
details; reproduced with permission from Shatz[65]). The eye-specific layers emerge
as retinal ganglion cell axons, withdraw delicate sidebranches from inappropriate
regions, and elaborate complex terminal arbors within appropriate regions of the LGN.
E=embryonic age; gestation in the cat is 65 days.

Remarkably, neither the layers within the LGN nor the columns within the cortex are present initially during development (for reviews, see Sretavan and Shatz,[69]
Shatz,[67] and Miller and Stryker[42]). When retinal ganglion cell axons from the two
eyes first grow into the LGN, they are intermixed with each other throughout a good
portion of the nucleus; the eye-specific layers emerge as axons from the two eyes
gradually remodel by withdrawing modest branches from inappropriate territory
and growing extensive terminal arbors within appropriate territory[69] (Figure 2).
Physiological studies *in vitro*[62] and electron microscopic examination of identified
retinal ganglion cell axons[8,67] suggest that this remodeling is accompanied by the
reorganization of synapses from the two eyes such that initial binocular convergence
is replaced by monocular inputs.

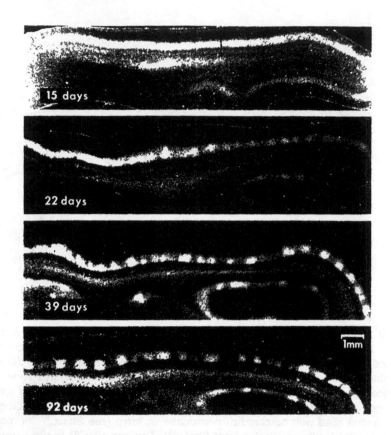

FIGURE 3 The postnatal development of the ocular dominance patches within layer 4 of the primary visual cortex of the cat is summarized. The location of LGN axons is monitored by means of the transneuronal transport through the LGN of radioactively labeled material (which appears white in these darkfield photographs) injected into one eye. The adult pattern of layer 4 labeling—patches separated by gaps of roughly equal size—can be seen by 92 days postnatal. However, at 2 weeks postnatal, the pattern of labeling within layer 4 is continuous, indicating that LGN axons representing the two eyes are intermixed with each other. (Based on experiments presented in LeVay et al.[34])

Ocular dominance columns in layer 4 form from extensively intermixed LGN inputs representing the two eyes (Figure 3), presumably also by a process of axonal remodeling and synapse elimination. At present, little is known about the exact morphological details because few individual axons have been successfully labeled for study, but microelectrode recordings have shown that initially the majority of neurons in cortical layer 4 receive functional inputs from LGN afferents representing

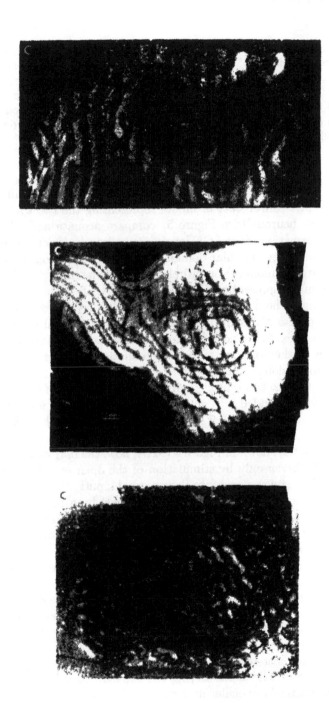

FIGURE 4   The effects of monocular eye closure at birth on the subsequent (adult) organization of the ocular dominance columns in layer 4 of the monkey visual cortex, as revealed by the transneuronal transport method (see Figure 3). (a) The normal tangential organization of LGN afferents within layer 4 into alternating (continued.)

FIGURE 4 (continued.) stripes of equal width representing the injected and uninjected eye. (b) The representation of the open eye within layer 4 following monocular deprivation—LGN axons occupy most of layer 4, with only small unlabeled regions remaining for the LGN axons representing the closed eye. (c) The pattern of transneuronal labeling resulting from injection of the closed eye is complementary to that shown in (b), indicating a shrinkage of territory devoted to the representation of the closed eye within layer 4. Reprinted with permission from Hubel et al.[26]

both eyes.[34,35] Thus, here too, ocular segregation emerges from an initial condition of functional synaptic convergence of inputs representing the two eyes onto common (layer 4 cortical) neurons (See Figure 5: compare neonate and adult). In higher mammals, the formation of the LGN layers occurs largely, if not entirely, prenatally and precedes the onset of ocular dominance column formation within the cortex, which occurs largely (monkey) or entirely (cat) postnatally.[34,35,47,61]

How do inputs representing the two eyes segregate from each other to form layers or columns? The first clues came from the pioneering studies of Hubel and Wiesel on the effects of visual deprivation on the functional organization of the primary visual cortex. In the normal adult visual cortex, the majority of neurons are binocular: that is, they respond to visual stimulation of either eye. Even binocular neurons tend to be dominated by one eye or the other, and as mentioned above, layer 4 neurons tend to be exclusively driven by stimulation of one eye only so that the cortex is evenly divided into ocular dominance columns for both eyes.[23,26,60] However, if one eye is deprived of vision by closing the eyelids at birth for several days to weeks, the ocular dominance distribution of neurons in visual cortex is drastically shifted: as shown in Figure 5 (MD), now, the majority (90%) of neurons are monocularly driven only by stimulation of the open eye.[25,26] (Neurons in the retina and LGN remain responsive to their normal inputs.[78]) The physiological shift in ocular dominance within the cortex is paralleled by a profound change in the anatomical organization of LGN axons within layer 4: LGN axons representing the open eye now occupy most of layer 4, while those representing the closed eye are relegated to very small patches[26] (see Figure 4).

The observation that the wiring of LGN axons and the eye preference of cortical neurons can be influenced by early visual experience sets the stage for the idea that *use* of the visual system is required for its normal development and for the maintenance of its connections, at least during an early period of susceptibility called the "Critical Period."[25,35] But how might abnormal use, such as the occlusion of one eye, result in such profound changes in connectivity at the level of the visual cortex? The most reasonable explanation is that a use-dependent synaptic competition between LGN axons serving the two eyes for layer 4 neurons normally drives the formation of the ocular dominance columns during the critical period.

Consequently, unequal use caused by monocular deprivation could bias the outcome in favor of the open eye. Many lines of evidence support the suggestion that

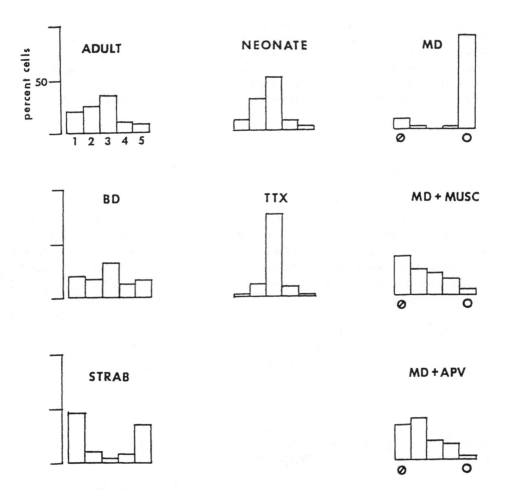

FIGURE 5 Idealized summary diagram of the effects of various manipulations that alter the pattern or levels of visually driven activity on the ocular dominance of visual cortical neurons as assessed physiologically. In the normal adult cortex (ADULT), the majority of neurons are binocularly driven, with a roughly even distribution of neurons representing each eye (group 1 = neurons responding exclusively from the right eye; group 2 = neurons responding predominantly to the right eye but some also from the left eye; group 3 = neurons responding equally to the two eyes; group 4 = neurons responding predominantly to the left eye; and group 5 = neurons responding exclusively to the left eye).[60,73] The majority of group 1 and 5 neurons are found within layer 4. In neonates, there are very few monocularly driven neurons, presumably since inputs from the two eyes are extensively intermixed even within layer 4.[34] If the right eye is closed at birth (Ø), then an ocular dominance shift in favor of the left eye (O) occurs with long-term monocular deprivation (MD).[25,60] However, if MD is combined with cortical infusion of muscimol (MD+MUSC)[52] during the critical period, then the shift in favor of the open eye is prevented close to the infusion site and a shift in favor of the closed (continued)

FIGURE 5   (continued) eye results. Binocular deprivation (BD)[73,79] during the cortical period, however, does not have an obvious effect on cortical ocular dominance, whereas intraocular injections of TTX during the same period retains, or possibly exaggerates, the highly binocular distribution present in neonates[73] (cf. TTX and Neonates). In contrast, alternating monocular deprivation or strabismus (STRAB) causes a complete loss of binocular neurons within the cortex.[77] See text for further details.

competitive interactions are involved. Binocular deprivation leaves the ocular dominance of cortical neurons unaltered (Figure 5: BD),[79] although neurons eventuallydo not respond briskly to visual stimulation. In a clever experiment, Guillery[19] demonstrated that competitive interactions are not only present, but must occur locally within the cortex, between LGN axons subserving corresponding regions of the visual field. He sutured one eye closed and then made just a small lesion in the open eye, destroying a localized group of ganglion cells there. As a consequence, the effects of monocular deprivation were manifested everywhere *except* within the small region receiving LGN axons representing the lesioned area of the open eye and the corresponding region of the closed eye. Thus equal use of the two eyes during the critical period subserves competitive interactions whose outcome is manifested in the even distribution of ocular dominance columns. Similar competitive interactions are thought to operate even earlier in development to drive the formation of LGN layers, as discussed more fully at the conclusion of this article.

THE ROLE OF PATTERNED NEURAL ACTIVITY IN COMPETITIVE INTERACTIONS

Signalling by neurons is, of course, via action potentials and synaptic transmission; hence, the effects of visual experience on cortical organization must be a consequence of alterations in either the level or patterning (or both) of neural activity within the visual pathways. The most graphic demonstration that this must be the case comes from experiments in which the inputs from both eyes are completely silenced by injecting Tetrodotoxin (TTX), a blocker of the sodium channel, for several weeks postnatally during the critical period and then examining the consequences on the formation of ocular dominance columns in layer 4.[73] Intraocular application of TTX conveniently silences the entire pathway from retina to cortex since there is very little spontaneously generated activity in central visual pathways in the absence of the eyes.[73] Segregation of LGN axons into patches within cortical layer 4 was prevented completely, and neurons in layer 4, normally monocularly driven, were instead binocularly driven (Figure 5: TTX), reminiscent of the initial period of normal postnatal development.[34,35] Indeed, at present it is not known whether the effect of the TTX treatment is to simply arrest development or to permit continued but undirected growth of LGN axon terminals within layer 4. Examination of axonal morphology should eventually clarify this issue. Analogous results are obtained when cortical activity alone is blocked (both pre- and postsynaptically), by infusing TTX locally via an osmotic minipump[51]: such treatment, when performed during the critical period, prevents the shift in cortical ocular dominance produced by monocular eye closure.

These experiments indicate that neural activity is necessary for ocular dominance columns to form during development (and for them to be perturbed with monocular deprivation), but they do not reveal how an activity-dependent signal might permit the selection of appropriate inputs from each eye to generate the segregated pattern characteristic of the adult geniculocortical projection. Experiments in which the use of the two eyes remains equal, but is never synchronous, provide some clues. During the critical period, if artificial strabismus is produced by cutting the extraocular muscles of one eye, thereby disrupting normal eye alignment, or if the eyes are closed alternately so that the total amount of vision received by each eye is the same, but vision is never binocular, then essentially every neuron in the primary visual cortex becomes exclusively monocularly innervated, with cells of like ocular dominance grouped into entirely "monocular" columns (see Figure 5: AMD) (recall that in normal animals, only layer 4 is monocular).[24,77] These results suggest that information concerning the relative *timing* of activity in the two eyes is somehow used to distinguish inputs at the cortical level: asynchrony leads to ocular segregation; synchrony maintains binocularity.

The conclusion that the formation of ocular dominance columns is influenced by the timing and patterning of neuronal activity within the retinae is underscored by the results of an experiment by Stryker and Strickland[72] in which retinal activity was first blocked by intraocular injections of TTX, but then experimentally controlled by electrically stimulating the optic nerves either synchronously or asynchronously. Synchronous stimulation of the two nerves prevented the formation of ocular dominance columns, whereas asynchronous stimulation permitted them to form. The only difference between the two experiments was the timing of stimulation, thereby demonstrating directly that the patterning of neural activity provides sufficient information for ocular segregation to occur, at least at the level of the primary visual cortex.

The above considerations can also explain why ocular dominance columns can develop even when animals are binocularly deprived or reared in the dark during the critical period.[73] In the absence of visual stimulation, ganglion cells in the mammalian retina of adults[36,53] and even in fetal animals[18,39] fire action potentials spontaneously. Such spontaneous firing could supply activity-dependent cues provided that ganglion cell firing in the two eyes is asynchronous.[43,82]

Before examining further how the timing and patterning of impulse activity might lead to segregation of geniculocortical afferents, it is worth considering briefly why an alternative hypothesis for the formation of segregated inputs, one that invokes the existence of eye-specific molecular labels within the cortex, is at odds with most experimental observations. First, geniculocortical axons segregate to form ocular dominance columns whose precise locations within the visual cortex are unpredictable, although the global arrangement of the columns is similar from one animal to the next.[26] Second, blockade of neural activity within the eyes prevents segregation of geniculocortical axons (themselves not directly affected by TTX). Thus, if eye-specific labels were present within the cortex, they should still have been recognized by LGN axons. Moreover, such markers should operate to form columns regardless of the *pattern* of electrical stimulation of the optic nerves (synchronous vs. asynchronous). Third, there is no obvious tendency for axons representing the right or left eyes to be grouped together prior to segregation[34,63]; indeed, activity-dependent models of ocular dominance column formation can easily produce segregated inputs from an initially randomly intermixed condition (see Miller et al.,[43] for more details). It should be noted that the absence of such labels with respect to eye of origin in no way argues against the existence of specific molecular cues that could initially guide axons to their appropriate targets (LGN, visual cortex) during development, or that could help to establish coarse retinotopic projections within these targets. By analogy with studies in lower vertebrates (for review see Udin and Fawcett[76]), such cues are highly likely to be present in the mammalian CNS as well. However, once axons reach their correct target and establish a coarse topographic projection, activity-dependent interactions could provide the major cues necessary for segregation.

Finally, a set of creative experiments performed in the amphibian visual system also argues against the presence of intrinsic eye-specific labels within the postsynaptic targets of retinal ganglion cells. In amphibians during larval development, the projections from retinal ganglion cells to their principal target, the optic tectum, are entirely crossed. Consequently, each tectum receives a map from the whole contralateral retina. The map is topographically orderly, such that the axons of neighboring retinal ganglion cells terminate in neighboring regions of the optic tectum. In frogs, it is possible to perform experimental manipulations in the embryo to transplant an extra eye onto one side of the head. Axons from both the normal and transplanted eyes are then capable of growing into the optic tectum, artificially creating a competitive situation. Constantine-Paton and her colleagues have shown that axons from both eyes segregate into eye-specific stripes reminiscent of the stripe-like pattern of the mammalian geniculocortical projection[32] (see Figure 6(a)). Thus, segregation of eye-specific inputs can occur in an experimentally manipulated system that normally never forms a segregated projection and therefore is highly unlikely to contain intrinsic eye-specific labels within the postsynaptic target. Moreover, blockade of action potential activity with TTX causes ganglion cell axons from the two eyes to desegregate[6,40,50] (see Figure 6(b)). In amphibia, connections between retina and tectum continue to grow throughout larval and early postmetarnorphic life, in a process involving the continual reshaping of synaptic connections.[14,49] These experiments suggest an analogous conclusion,

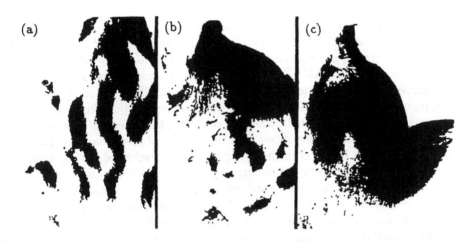

FIGURE 6 The organization of ganglion cell axon projections to a dually innervated optic tectum in three-eyed frogs, as revealed in tectal wholemounts by injecting one of the two eyes with horseradish peroxidase. (a) Axons from the two eyes segregate into alternating stripes reminiscent of the system of mammalian ocular dominance columns in cortical layer 4. Desegregation occurs when either TTX (not shown) or APV ((b) after 2.5 weeks; (c) after 4 weeks treatment) is infused into the tectum. Modified, with permission, from Cline et al.[9]

that the maintenance of segregated inputs in these three-eyed frogs is a dynamic ongoing process that requires neural activity (presumably asychronous) in the two eyes.

CELLULAR CORRELATES OF ACTIVITY-DEPENDENT COMPETITION

The finding that the synchronous activation of afferents prevents them from segregating, while asynchronous activation promotes segregation, indicates that the timing of presynaptic activity is crucial to the process. Studies also suggest that involvement of the postsynaptic cell is necessary. For example, in the mammalian visual cortex, when visual stimulation through one eye is paired simultaneously with postsynaptic depolarization produced by extracellular stimulation, the strength of inputs from the stimulated eye can be enhanced in some cells from minutes to hours.[17] The effect is quite variable and is more frequently produced in young animals during the critical period than in adults; nevertheless, this experiment serves to illustrate the point that coincidence of pre- with postsynaptic activity can, at least under certain circumstances, enhance visually driven inputs.

Manipulations that block postsynaptic activity exclusively can also alter the outcome of competition in the visual system. Reiter and Stryker[52] have shown that when cortical neurons are silenced during the critical period by the intracortical infusion of muscimol, a GABA-A receptor agonist, monocular eye closure has surprising consequences for the inputs from the two eyes: within the silenced region of cortex, inputs from the *closed* eye come to dominate over inputs from the open eye (see Figure 5: MD and muscimol), whereas, of course, the reverse is true outside the silenced zone. This observation shows that the activity of postsynaptic cortical cells is highly likely to be involved in the synaptic reorganization occurring during the critical period, since the same patterning of presynaptic activity produces different outcomes depending on the state of activation of the postsynaptic cell.

The requirement for the participation of both pre- and postsynaptic partners in activity-dependent rearrangements, and the fact that coincident activation can strengthen coactivated inputs, is consistent with the idea that a Hebb rule may govern the process of synapse rearrangements during ocular dominance column development in mammals (and in three-eyed frogs) (for review see Brown et al.[7]; see Kossel et al.[30] for an alternate view). Hebb[21] suggested that when pre- and postsynaptic neurons are coactivated, their synaptic connections are strengthened, whereas connections are weakened with the lack of coincident activation. In this context, the muscimol experiment described above[52] is also consistent with a Hebb rule in the sense that the levels of presynaptic activity in geniculocortical axons representing the closed eye are better matched to the silenced postsynaptic neurons than those inputs representing the open eye. Thus, the correlated firing of nearby ganglion cells within one eye, and the lack of synchronous firing of ganglion cells in the other eye could provide appropriate signals to produce the regional strengthening and weakening of synaptic inputs needed for segregation to take place.

These activity-dependent properties of visual cortical synapses during the critical period are very reminiscent of some of the well-known characteristics of synapses in the adult mammalian hippocampus that are capable of undergoing long-term potentiation (LTP): that is, a long lasting increase in synaptic strength produced with the appropriate matching of pre- and postsynaptic activation.[7,45] In the CA1 region of the hippocampus, many lines of experimentation indicate that activation of the NMDA receptor (N-methhyl-D-aspartate) on postsynaptic neurons by means of the presynaptic release of glutamate is required for LTP.[80] The consequent strengthening of synaptic transmission appears to be due at least in part to a presynaptic change: an increase in transmitter release from the presynaptic terminals.[5,37,81] The wealth of information on LTP, and its similarities with activity-dependent development, has prompted many recent experiments in the visual system designed to learn whether the two forms of synaptic change share similar cellular mechanisms. Of course, it should be noted that in at least one respect, the two must differ ultimately in that during development major structural changes occur not only in individual synapses but also in the overall morphology of presynaptic terminals, since some terminals are actually eliminated while others are newly formed. Moreover the physiological properties of developing synapses are very different from those of adult, suggesting that the parameters for patterned activity to produce synaptic change may also differ.

A major question is whether NMDA-receptor activation is necessary for developmental plasticity. This is a reasonable question to pose since glutamate is thought to be the excitatory neurotransmitter released by retinal ganglion cells in all vertebrates[27,31] and also by LGN neurons in mammals.[20] The most compelling evidence in favor of the specific involvement of NMDA-receptors in activity-dependent development comes from recent studies of the retinotectal system in fish and frogs. For instance, in three-eyed frogs, the ocular dominance stripes desegregate in the presence of the NMDA receptor antagonist APV (2-amino-5-phosphonovaleric acid),[9] suggesting that activation of this receptor is necessary for the maintenance of segregated inputs (see Figures 6(b) and (c)).

NMDA receptor activation is also apparently necessary for the maintenance of two other activity-dependent processes known to occur in the retinotectal system. The first is in the refinement of topographic projections from retina to tectum that occurs during regeneration of the optic nerve in goldfish. Ganglion cell axons can establish coarse topographic projections even when activity is blocked with TTX, presumably because activity-independent molecular cues are unaltered.[12,16] However, the fine-tuning of axon terminal arbors necessary for the re-establishment of highly refined connections is prevented.[41,56] In this case, topographic fine-tuning would be expected to occur if the activity of neighboring retinal ganglion cells was highly correlated, while that of distant ganglion cells was not—a situation naturally produced with visual stimulation. Consistent with this suggestion, rearing animals in stroboscopic light, which causes all ganglion cells to fire in near synchrony, prevents the fine-tuning of topography during regeneration.[57] Recent experiments have demonstrated that infusion of APV also blocks the fine-tuning of the retinotectal map.[58] Moreover, Schmidt has demonstrated that during the period of map refinement following optic nerve regeneration, low-frequency electrical stimulation of the optic nerve causes long-term potentiation of the postsynaptic tectal response which is also blocked by APV.

Another example demonstrating the involvement of NMDA receptors involves the process by which binocular neurons are normally created and maintained in the frog optic tectum. Although the optic tectum only receives direct input from the retinal ganglion cells in the opposite eye, an indirect pathway from one tectum to the other via a relay nucleus, the isthmo-tectal nucleus, does convey input from the other eye to create binocular neurons. Here too, the maintenance of the binocular map is activity dependent, as demonstrated by the fact that rotation of one eye in its orbit leads to a systematic and anatomically demonstrable rewiring of isthmo-tectal connections so as to preserve ocular correspondence.[75] The rewiring induced by eye rotation is prevented by infusion of APV into the optic tectum.[55] An essential finding in these studies is that the levels of APV necessary to prevent the activity-dependent rearrangements apparently do not block appreciably retinotectal synaptic transmission or the excitability of the postsynaptic neuron.[9,58] Thus, the APV treatment does not act like TTX to block neural activity generally, but more likely acts specifically to prevent whatever cascade of events is triggered by NMDA receptor activation.

The specific involvement of the NMDA receptor in the synaptic alterations occurring during the critical period in the mammalian visual cortex is more controversial, but there is no doubt that NMDA receptors are present throughout (but also after) the relevant times in the cat visual system. Physiological studies of cortical neurons demonstrate that both their spontaneous firing and their responses to visual stimulation can be decreased by APV, and that lower doses of APV are needed in younger animals.[15,44,74] Moreover, Fox et al.[15] found that there is a systematic change in the laminar distribution of responsiveness to iontophoretic application of APV with age: in neonates, neurons in all cortical layers are sensitive to APV, whereas by the end of the critical period, the visually evoked responses of neurons in the deeper cortical layers (layers 4, 5, and 6) are not affected by APV iontophoresis. Thus, the changing susceptibility of cortical neurons in layer 4 to APV application is generally correlated with the period in which segregation of the geniculocortical afferents occurs. However, the fact that the superficial cortical layers remain highly sensitive to NMDA receptor blockade after the critical period draws to a close is difficult to reconcile with a simple view for the participation of NMDA receptors in the events of activity-dependent segregation and visual cortical plasticity.

If NMDA receptors are to contribute to the mechanism underlying synaptic rearrangements during the critical period, then pharmacological blockade of the receptor might be expected to prevent the segregation of LGN axons into ocular dominance patches within layer 4 in a fashion analogous to that found for the desegregation of stripes in three-eyed frogs. At present, this possibility has not been investigated in the mammalian visual system. A correlate, that receptor blockade might prevent the shift in ocular dominance toward the open eye caused by monocular eye closure, has been studied by using minipumps to infuse APV into the cat visual cortex during the critical period.[4,28] Within the infusion zone, a shift towards the open eye was prevented and, in fact, there was an unanticipated shift in favor of the *closed* eye—reminiscent of the results obtained in a similar experiment decribed above in which muscimol[52] was infused in order to silence selectively the postsynaptic cortical neurons without also blocking presynaptic afferent inputs. At first glance, then, these results would seem to conform nicely to the hypothesis that NMDA receptors play a specific role in activity-dependent cortical development and plasticity. Unfortunately, the alternative interpretation exists, namely that APV acts in a nonselective fashion to block postsynaptic activity, much as muscimol does; that is, current flowing through an NMDA-gated channel is not exclusively a "plasticity" signal. This alternate interpretation seems quite likely in view of the results of the iontophoresis experiments described above demonstrating that activation of NMDA receptors is necessary for cortical neurons to respond normally to visual stimulation. Thus, at present, it is not possible to draw strict parallels between the requirement for NMDA receptor activation in hippocampal LTP and an analagous role in visual cortical plasticity during development.

Even if a specific role for the NMDA receptor during development of the visual cortex is eventually clearly established, the synaptic basis for its mode of action remains to be elucidated. Clues come from experiments performed on rat visual cortical slices *in vitro* which suggest that synaptic connections can undergo LTP following appropriate tetanic stimulation of the white matter (which contains the incoming LGN axons), both during neonatal life and in adulthood.[1,29,30,46] LTP is much more difficult to induce in cortical neurons than in the hippocampus (only about 30% of all recorded neurons demonstrate the phenomenon in cortex), and, in fact, frequently requires a concommitant reduction in local inhibitory influences by application of a GABA antagonist; nevertheless, as in the hippocampus, LTP can be blocked consistently by iontophoresis of APV.[1] However, unlike the hippocampus, the circuitry of the cortex makes it difficult to stimulate an isolated excitatory pathway in order to separate monosynaptic from polysynaptic inputs. Thus, while the LTP studied in hippocampus clearly involves a change in the efficacy of a single type of excitatory synapse, what is called LTP in cortical slices may involve a mixture of several effects, both excitatory and inhibitory. Nevertheless, these observations raise the possibility that a cascade of physiological and biochemical events similar to those known to occur during hippocampal LTP might also take place during activity-dependent strengthening of visual cortical connections during development.

In the formation of ocular dominance columns, both normally during development and when perturbed by abnormal visual experience, some connections are strengthened, but others must be weakened and likely even eliminated in order for neurons in layer 4 to become monocularly driven. While a mechanism such as LTP could help to explain synaptic strengthening in the visual cortex, what about the reverse? A recent experiment by Artola et al.[2] suggests that it may be possible to produce a weakening, or long-term depression (LTD), of synaptic transmission in neurons in slices of rat visual cortex. These authors suggest that a level of membrane depolarization above resting level but below the greater level required for the induction of LTP can produce LTD in active synapses. (A similar phenomenon has been described in the hippocampus by Stanton and Sejnowski.[71]) Moreover, Artola et al. report that LTD can be produced even in the presence of APV, consistent with the idea that activation of an NMDA-gated channel is not involved. This result may help to explain why monocular deprivation combined with cortically infused muscimol[52] or APV[4] causes an ocular dominance shift in favor of the closed eye within the infusion zone. Perhaps in the presence of these agents, activation of inputs from the open eye brings cortical neurons only to a level of membrane potential critical for LTD, consequently weakening those inputs. While these experiments provide a convenient conceptual framework for thinking about how activity-dependent synaptic change may occur during visual cortical development, it will be essential first to understand these effects *in vitro* at the level of single identified synapses and next to demonstrate that similar alterations in synaptic efficacy indeed take place *in vivo* during the critical period as a consequence of natural visual stimulation.

GENERALITY OF ACTIVITY-DEPENDENT DEVELOPMENT IN THE CENTRAL NERVOUS SYSTEM

The experiments discussed thus far provide compelling evidence in favor of the idea that activity-dependent competitive interactions in the visual system can account for the establishment of highly segregated and topographically ordered sets of connections during ocular dominance column development in mammals, and in the regeneration and maintenance of retinotectal connections in lower vertebrates. Moreover, a variety of new experiments has begun to draw exciting parallels between the cellular bases for these events and those thought to underlie LTP in the hippocampus. A common thread in all these examples is that synaptic change can be produced by the appropriate patterning of presynaptic activity and its conjunction with postsynaptic activity. In the hippocampus, when these requirements are met, evidence suggests that the resulting alterations may subserve memory and learning.[7] In the postnatal visual system, they subserve synaptic rearrangements that are generally dependent upon visual stimulation in order to provide the presynaptic correlations in neural activity necessary to preserve topographic relations, and the asynchrony required for ocular segregation.

Studies of the development of connections between retinal ganglion cells and their target neurons in the LGN suggest that structured activity may even play a role long before vision is possible. As mentioned early in this article, in the adult, ganglion cell axons from the two eyes project to each LGN, where they terminate in strictly segregated eye-specific layers. These layers are not present initially in development but rather emerge as retinal ganglion cell axons from the two eyes remodel their terminals[67] (see Figure 2). In the cat and monkey visual system, the period during which the layers form is entirely prenatal. It begins before all photoreceptor cells become postmitotic and is complete before photoreceptor outer segments are present.[13] Nevertheless, many lines of evidence suggest that here too segregation comes about by a process of activity-dependent synaptic competition. The idea that competitive interactions of some form might govern layer formation originates with observations that removal of one eye during development permits axons from the other eye to occupy the entire LGN.[48,69] Hints that the competition might be mediated by synaptic interactions comes from physiological observations that individual LGN neurons initially receive binocular inputs when the optic nerves are electrically stimulated *in vitro*[62] and that retinal ganglion cell axons from one eye can make synaptic contacts in regions later exclusively innervated by axons from the other eye.[8,67] These observations provide evidence to suggest that synaptic remodelling accompanies the formation of the eye-specific layers in the LGN.

What might be the source of activity-dependent signals during these early times in development when vision not possible? The most likely source is the spontaneously generated activity of retinal ganglion cells. In a technically remarkable experiment, Galli and Maffei[18] succeeded in making microelectrode recordings from fetal rat retinal ganglion cells *in vivo* and found that they fired spontaneously, sometimes correlated with each other when several cells were recorded together on the same electrode.[36] Recently it has been possible to examine the spatial and temporal pattern of firing of up to 100 retinal ganglion cells simultaneously by removing fetal and neonatal retinae and recording *in vitro* using a multielectrode array; results show that even in the absence of photoreceptor function, ganglion cells fire in a very stereotyped bursting pattern, with neighboring cells firing in near synchrony.[39] These two experiments together provide evidence that the spontaneous activity of retinal ganglion cells may have the appropriate spatiotemporal patterning to provide necessary activity-dependent cues for the formation of topographically ordered and segregated inputs to the LGN and other central visual targets of ganglion cell axons.

If spontaneous activity does play a role in the segregation of retinal ganglion cell axons into the eye-specific layers within the LGN, then blockade of such activity should prevent the formation of the layers. Minipump infusions of TTX into the thalamus of fetal cats, indeed, block layer formation[66] and correspondingly perturb the branching pattern of individual retinal ganglion cell axons so that branches are no longer restricted to appropriate zones within the LGN.[70] Indeed, the effects of TTX on the shapes of retinal ganglion cell axons in the cat are remarkably similar to its effects on ganglion cell axons in the optic tectum of three-eyed frogs,[50] as shown in Figure 7. However, a criticism of the results is that TTX may have acted in a non-specific fashion to cause unregulated growth of the axons.[10] Definitive proof that this is not the case requires an experiment analogous to that performed by Stryker and Strickland,[72] in which the patterning of neural activity is specifically perturbed. This should be possible in future, when the mechanisms for the generation of synchronous bursting among retinal ganglion cells are better understood. Meanwhile, it should be noted that ganglion cell axon growth is not entirely unregulated in the presence of TTX: the axons are still capable of detecting and stopping their growth at the LGN boundaries.

The results of the experiments described above permit an important generalization concerning the universality of activity-dependent synaptic interactions. During normal development, such interactions may be driven not only by the normal pattern of use (e.g., visually evoked activity), but even earlier before vision begins by patterned spontaneously generated activity. This suggestion raises the possiblility that spontaneously generated activity elsewhere in the CNS during development may play a similar role in establishing orderly sets of connections. If so, then the synaptic changes produced by activity-dependent interactions early in development may be at one end of a continuum of synaptic change, the other end of which are the use-dependent alterations in synaptic strength associated with learning and memory. Although the changes occurring during development require major anatomical restructuring of axons whereas those occurring during learning and memory are more likely to be confined to individual synapses,[3] evidence presented here suggests that the two types of change may not be all that different in terms of cellular mechanisms. Future experiments will reveal the extent to which the two areas of investigation converge, and whether there are familarities at the molecular level as well. The existence of similar mechanisms could represent an extremely elegant solution to the complex problem of establishing and maintaining specific synaptic connections throughout life.

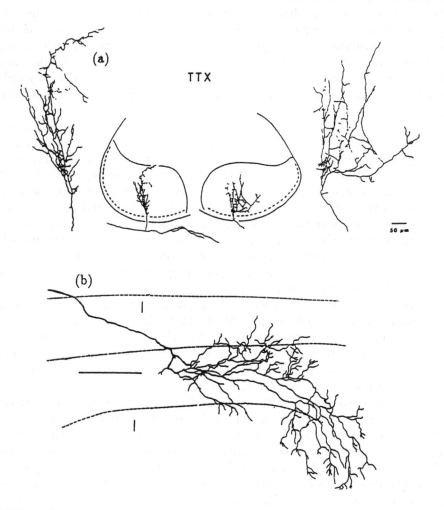

FIGURE 7   A comparison of the morphology of retinal ganglion cell axons in the fetal cat at E57 (a) and the three-eyed frog (b) following TTX treatment. In both cases, the terminal arbors of the axons are not as restricted as usual: in fetal cats, retinal ganglion cell axons normally have terminal arbors that branch only in the inner or outer half of the LGN rather than throughout (compare with Figure 2). In three-eyed frogs, the arbors are usually restricted to one stripe and do not cross stripe boundaries (indicated by dashed lines). Adapted from Sretavan et al.,[70] and Reh and Constantine-Paton.[50]

REFERENCES

1. Artola, A., and W. Singer. "Long-Term Potentiation and NMDA Receptors in Rat Visual Cortex." *Nature* **330** (1987): 649-652.
2. Artola, A., S. Brocher, and W. Singer. "Different Voltage-Dependent Thresholds for Inducing Long-Term Depression and Long-Term Potentiation in Slices of Rat Visual Cortex." *Nature* **347** (1990): 69–72.
3. Bailey, C. H., and M. Chen. "Structural Plasticity at Identified Synapses During Long-Term Memory in *Aplysia*." *J. Neurobiol.* **20** (1989): 356–372.
4. Bear, M. F., A. Kleinschmidt, Q. Gu, and W. Singer. "Disruption of Experience-Dependent Synaptic Modifications in Striate Cortex by Infusion of an NMDA Receptor Antagonist." *J. Neurosci.* **10** (1990): 909–925.
5. Bekkers, J. M., and C. F. Stevens. "Presynaptic Mechanism for Long-Term Potentiation in the Hippocampus." *Nature* **346** (1990): 724–729.
6. Boss, V. C., and J. T. Schmidt. "Activity and the Formation of Ocular Dominance Patches in Dually Innervated Tectum of Goldfish." *J. Neurosci.* **4** (1984): 2891–2905.
7. Brown, T. H., E. W. Kairiss, and C. Keenan. "Hebbian Synapses: Mechanisms and Algorithms." *Ann. Rev. Neurosci.* **13** (1990): 475–511.
8. Campbell, G., and C. J. Shatz. "Synapses Formed by Identified Retinogeniculate Axons During the Segregation of Eye Input." *J. Neurosci.* **12** (1992): 1847–1858.
9. Cline, H. T., E. A. Debski, and M. Constantine-Paton. "NMDA Receptor Antagonist Desegregates Eye-Specific Stripes." *PNAS* **84** (1987): 4342–4345.
10. Cohan, C. S., and S. B. Kater. "Suppression of Neurite Elongation and Growth Cone Motility by Electrical Activity." *Science* **232** (1986): 1638–1640.
11. Constantine-Paton, M., H. T. Cline, and E. Debski. "Patterned Activity, Synaptic Convergence, and the NMDA Receptor in Developing Visual Pathways." *Ann. Rev. Neurosci.* **13** (1990): 129–154.
12. Cox, E. C., B. Muller, and F. Bonhoeffer. "Axonal Guidance in the Chick Visual System: Posterior Tectal Membranes Induce Collapse of Growth Cones from the Temporal Retina." *Neuron* **2** (1990): 31–37.
13. Donovan, A. "Postnatal Development of the Cat Retina." *Exp. Eye Res.* **5** (1966): 249–254.
14. Easter, S. S., Jr., and C. A. G. Stuermer. "An Evaluation of the Hypothesis of Shifting Terminals in the Goldfish Optic Tectum." *J. Neurosci.* **4** (1984): 1052–1063.
15. Fox, K. H. Sato, and N. W. Daw. "The Location and Function of NMDA Receptors in Cat and Kitten Visual Cortex." *J. Neurosci.* **9** (1989): 2443–2454.
16. Fraser, S. E., and D. H. Perkel. "Competitive and Positional Cues in the Patterning of Nerve Conditions." *J. Neurobiol.* **21** (1990): 51–72.

17. Fregnac, Y., D. Schulz, S. Thorpe, and E. Bienenstock. "A Cellular Analogue of Visual Cortical Plasticity." *Nature* **333** (1988): 367–370.
18. Galli, L., and L. Maffei. "Spontaneous Impulse Activity of Rat Retinal Ganglion Cells in Prenatal Life." *Science* **242** (1988): 90–91.
19. Guillery, R. W. "Binocular Competition in the Control of Geniculate Cell Growth." *J. Como. Neurol.* **144** (1972): 117–130.
20. Hagihara, K., T. Tsumoto, H. Sato, and Y. Hata. "Actions of Excitatory Amino Acid Antagonists on Geniculo-Cortical Transmission in the Cat's Visual Cortex." *Exp. Brain Res.* **69** (1988): 407–416.
21. Hebb, D. O. *The Organization of Behavior.* New York: John Wiley & Sons, 1949.
22. Hickey, T. L., and R. W. Guillery. "An Autoradiographic Study of Retinogeniculate Pathways in the Cat and Fox." *J. Comp. Neurol.* **156** (1974): 239–254.
23. Hubel, D. H., and T. N. Wiesel. "Shape and Arrangement of Columns in Cat's Striate Cortex." *J. Physiol.* **165** (1963): 559–568.
24. Hubel, D. H., and T. N. Wiesel. "Binocular Interaction in Striate Cortex of Kittens Reared with Artificial Squint." *J. Neurophysiol.* **28** (1965): 1041–1059
25. Hubel, D. H., and T. N. Wiesel. "The Period of Susceptibility to the Physiological Effects of Unilateral Eye Closure in Kittens." *J. Physiol.* **206** (1970): 419–436.
26. Hubel, D. H., T. N. Wiesel, and S. LeVay. "Plasticity of Ocular Dominance Columns in the Monkey Striate Cortex." *Phil. Trans. R. Soc. Lond. B* **278** (1977): 377–409.
27. Kemp, J. A., and A. M. Sillito. "The Mature of the Excitatory Transmitter Mediating X and Y Cell Inputs to the Cat Dorsal Lateral Geniculate Nucleus." *J. Physiol.* **323** (1982): 377–391.
28. Kleinschmidt, A., M. Bear, and W. Singer. "Blockade of NMDA Receptors Disrupts Experience-Dependent Modifications in Kitten Striate Cortex." *Science* **238** (1987): 355–358.
29. Komatsu, Y., K. Fujii, J. Maeda, H. Sakaguchi, and K. Toyama. "Long-Term Potentiation of Synaptic Transmission in Kitten Visual Cortex." *J. Neurophysiol.* **59** (1988): 124–141.
30. Kossel, A., T. Bonhoeffer, and J. Bolz. "Non-Hebbian Synapses in Rat Visual Cortex." *Neuroreport* **1** (1990): 115–118.
31. Langdon, R. B., and J. A. Freeman. "Pharmacology of Retinotectal Transmission in Goldfish: Effects of Nicotinic Ligands, Strychnine and Kynurenic Acid." *J. Neurosci.* **7** (1987): 760–773.
32. Law, M. I., and M. Constantine-Paton. "Anatomy and Physiology of Experimentally Produced Striped Tecta." *J. Neurosci.* **1** (1981): 741–759.
33. LeVay, S., D. H. Hubel, and T. N. Wiesel. "The Pattern of Ocular Dominance Columns in Macaque Visual Cortex Revealed by a Reduced Silver Stain." *J. Comp. Neurol.* **159** (1975): 559–576.

34. LeVay, S., M. P. Stryker, and C. J. Shatz. "Ocular Dominance Columns and Their Development in Layer IV of the Cat's Visual Cortex." *J. Comp. Neurol.* **179** (1978): 223–244.

35. LeVay, S., T. N. Wiesel, and D. H. Hubel. "The Development of Ocular Dominance Columns in Normal and Visually Deprived Monkeys." *J. Comp. Neurol.* **191** (1980): 1–51.

36. Maffei, L., and L. Galli-Resta. "Correlation in the Discharges of Neighboring Rat Retinal Ganglion Cells During Prenatal Life." *PNAS* **87** (1990): 2861–2864.

37. Malinow, R., and R. W. Tsien. "Presynaptic Enhancement Shown by Whole-Cell Recordings of Long-Term Potentiation in Hippocampal Slices." *Nature* **346** (1990): 177–180.

38. Mastronarde, D. N. "Correlated Firing of Cat Retinal Ganglion Cells. I. Spontaneously Active Inputs to X and Y Cells." *J. Neurophysiol.* **49** (1983): 303–324.

39. Meister, M., R. O. L. Wong, D. A. Baylor, and C. J. Shatz. "Synchronous Bursts of Action Potentials in Ganglion Cells of the Developing Mammalian Retina." *Science* **252** (1991): 939–943.

40. Meyer, R. L. "Tetrodotoxin Blocks the Formation of Ocular Dominance Columns in the Goldfish." *Science* **218** (1982): 589–591.

41. Meyer, R. L. "Tetrodotoxin Inhibits the Formation of Refined Retinotopograpphyin Goldfish." *Dev. Brain Res.* **6** (1983): 293–298.

42. Miller, K. D., and M. P. Stryker. "Development of Ocular Dominance Columns: Mechanisms and Models." In *Connectionist Modeling and Brain Function: The Developing Interface,* edited by S. J. Hanson and C. R. Olson, 255–305. Cambridge, MA: MIT Press, 1990.

43. Miller, K. D., J. B. Keller, and M. P. Stryker. "Ocular Dominance Column Development: Analysis and Simulation." *Science* **245** (1989) 605–615.

44. Miller, K. D., B. Chapman, and M. P. Stryker. "Visual Responses in Adult Cat Visual Cortex Depend on N-Methyl-D-Aspartate Receptors." *PNAS* **856** (1989): 5183–5187.

45. Nicoll, R. A., J. A. Kauer, and R. C. Malenka. "The Current Excitement in Long-Term Potentiation." *Neuron* **1** 97–103.

46. Perkins, A. T., and T. J. Teyler. "A Critical Period for Long-Term Potentiation in the Developing Rat Visual Cortex." *Brain Res.* **439** (1988): 222–229.

47. Rakic, P. "Prenatal Development of the Visual System in the Rhesus Monkey." *Phil Trans. R. Soc. Lond. B* **278** (1977): 245–260.

48. Rakic, P. "Mechanism of Ocular Dominance Segregation in the Lateral Geniculate Nucleus: Competitive Elimination Hypothesis." *TINS* **9** (1986): 11–15.

49. Reh, T. A., and M. Constantine-Paton. "Retinal Ganglion Cells Change Their Projection Sites During Larval Development of *Rana pipiens.*" *J. Neurosci.* **4** (1983): 442–457.

50. Reh, T. A., and M. Constantine-Paton. Eye-Specific Segregation Requires Neural Activity in Three-Eyed *Rana pipiens*." *J. Neurosci.* **5** (1985): 1132–1143.

51. Reiter, H. O., D. M. Waitzman, and M. P. Stryker. "Cortical Activity Blockade Prevents Ocular Dominance Plasticity in the Kitten Visual Cortex." *Exp. Brain Res.* **65** (1986): 182–188.

52. Reiter, H. O., and M. P. Stryker. "Neural Plasticity Without Postsynaptic Action Potentials: Less-Active Inputs Become Dominant when Kitten Visual Cortical Cells are Pharmacologically Inhibited." *PNAS* **85** (1988): 3620–3627.

53. Rodieck, R. W., and P. S. Smith. "Slow Dark Discharge Rhythms of Cat Retinal Ganglion Cells." *J. Neurophysiol.* **29** (1966): 942–953.

54. Rodieck, R. W. "Visual Pathways." *Ann. Rev. Neurosci.* **2** (1979): 193–255.

55. Scherer, W. S., and S. B. Udin. "N-Methyl-D-Aspartate Antagonists Prevent Interaction of Binocular Maps in *Xenopus* Tectum." *J. Neurosci.* **9** (1989): 3837–3843.

56. Schmidt, J. T., and D. L. Edwards. "Activity Sharpens the Map During the Regeneration of the Retinotectal Projection in Goldfish." *Brain Res.* **209** (1983): 29–39.

57. Schmidt, J. T., and L. E. Eisele. "Stroboscopic Illumination and Dark Rearing Block the Sharpening of the Regenerated Retinotectal Map in Goldfish." *Neuroscience* **14** (1985): 535–546.

58. Schmidt, J. T. "Long-Term Potentiation and Activity-Dependent Retinotopic Sharpening in the Regenerating Retinotectal Projection of Goldfish: Common Sensitive Period and Sensitivity to NMDA Blockers." *J. Neurosci.* **10** (1990): 233–246.

59. Shatz, C. J., S. Lindstrom, and T. N. Wiesel. "The Distribution of Afferents Representing the Right and Left Eyes in the Cat's Visual Cortex." *Brain Res.* **131** (1978): 103–116.

60. Shatz, C. J., and M. P. Stryker. "Ocular Dominance in Layer IV of the Cat's Visual Cortex and the Effects of Monocular Deprivation." *J. Physiol.* **281** (1978): 267–283.

61. Shatz, C. J. "The Prenatal Development of the Cat's Retinogeniculate Pathway." *J. Neurosci.* **3** (1983): 482–499.

62. Shatz, C. J., and P. A. Kirkwood. "Prenatal Development of Functional Connections in the Cat's Retinogeniculate Pathway." *J. Neurosci.* **4** (1984): 1378–1397.

63. Shatz, C. J., and M. B. Luskin. "The Relationship Between Geniculocortical Afferents and Their Cortical Target Cells During Development of the Cat's Primary Visual Cortex." *J. Neurosci.* **6** (1986): 3655–3668.

64. Shatz, C. J., and D. W. Sretavan. "Interactions Between Ganglion Cells During the Development of the Mammalian Visual System." *Ann. Rev. Neurosci.* **9** (1986): 171–207.

65. Shatz, C. J. "The Role of Function in the Prenatal Development of Retino-geniculate Connections." In *Cellular Thalamac Mechanisms*, edited by M. Bentivoglio and R. Spreafico, 435–446. New York: Elsevier, 1988.
66. Shatz, C. J., and M. P. Stryker. "Prenatal Tetrodotoxin Infusion Blocks Segregation of Retinogeniculate Afferents." *Science* **242** (1988): 87–89.
67. Shatz, C. J. "Competitive Interactions Between Retinal Ganglion Cells During Prenatal Development." *J. Neurobiol.* **21** (1990): 197–211.
68. Sherman, S. M., and P. D. Spear. "Organization of the Visual Pathways in Normal and Visually Deprived Cats." *Physiol. Rev.* **62** (1982): 738–855.
69. Sretavan, D. W., and C. J. Shatz. "Prenatal Development of Retinal Ganglion Cell Axons: Segregation into Eye-Specific Layers." *J. Neurosci.* **6** (1986): 234–251.
70. Sretavan, D. W., C. J. Shatz, and M. P. Stryker. "Modification of Retinal Ganglion Cell Axon Morphology by Prenatal Infusion of Tetrodotoxin." *Nature* **336** (1988): 468–471.
71. Stanton, P. K., and T. J. Sejnowski. "Associative Long-Term Depression in the Hippocampus: Induction of Synaptic Plasticity by Hebbian Covariance." *Nature* **339** (1989): 215–218.
72. Stryker, M. P., and S. L. Strickland. "Physiological Segregation of Ocular Dominance Columns Depends on the Pattern of Afferent Electrical Activity." *Invest. Ophthalmol. and Vis. Sci. (suppl.)* **25** (1984): 278.
73. Stryker, M. P., and W. Harris. "Binocular Impulse Blockade Prevents the Formation of Ocular Dominance Columns in Cat Visual Cortex." *J. Neurosci.* **6** (1986): 2117–2133.
74. Tsumoto, T., H. Hagihara, H. Sato, and Y. Hata. "NMDA Receptors in the Visual Cortex of Young Kittens are More Effective than Those of Adult Cats." *Nature* **327** (1987): 513–514.
75. Udin, S. B. "Abnormal Visual Input Leads to Development of Abnormal Axon Trajectories in Frogs." *Nature* **301** (1983): 336–338.
76. Udin, S. B., and J. W. Fawcett. "Formation of Topographic Maps." *Ann. Rev. Neurosci.* **11** (1990): 289–237.
77. Van Sluyters, R. C., and F. B. Levitt. "Experimental Strabismus in the Kitten." *J. Neurophysiol.* **43** (1980): 686–699.
78. Wiesel, T. N., and D. E. Hubel. "Effects of Visual Deprivation on Morphology and Physiology of Cells in the Cat's Lateral Geniculate Body." *J. Neurophysiol.* **26** (1963): 978–993.
79. Wiesel, T. N., and D. H. Hubel. "Comparison of the Effects of Unilateral and Bilateral Eye Closure on Cortical Unit Responses in Kittens." *J. Neurophysiol.* **28** (1965): 1029–1040.
80. Wigstrom, H., and B. Gustafsson. "On Long-Lasting Potentiation in the Hippocampus: A Proposed Mechanism for Its Dependence on Coincident Pre- and Post-Synaptic Activity." *Acta Physiol. Scand.* **123** (1985): 519–522.

81. Williams, J. H., M. L. Errington, M. A. Lynch, and T. V. P. Bliss. "Arachidonic Acid Induces a Long-Term Activity-Dependent Enhancement of Synaptic Transmission in the Hippocampus." *Nature* **341** (1989): 739–742.
82. Willshaw, D. J., and C. Von der Marlsberg. "How Patterned Neural Connections Can Be Set Up by Self-Organization." *Proc. R. Soc. Lond. (Biol.)* **194** (1976): 431–445.

Wing Yim Tam
Physics Department, University of Arizona, Tucson, AZ 85721

Pattern Formation in Chemical Systems: Roles of Open Reactors

This chapter originally appeared in *1990 Lectures in Complex Systems*, edited by L. Nadel and D. Stein, 99–123. Santa Fe Institute Studies in the Sciences of Complexity, Lect. Vol. III. Reading, MA: Addison-Wesley, 1991. Reprinted by permission.

INTRODUCTION

Pattern formation is the most fascinating subject in nature.[37] Patterns range from banded structures in rocks to beautiful skin coatings in animals. The formation of these structures is often chemical in origin. The discovery of oscillating chemical reactions early in this century has accelerated the study of chemical pattern formation, which is now a disciplinary area of interest.[22] Chemical patterns are unique and yet they share and exhibit similar constraints and behaviors of non-linear systems.[19] Nonlinearities are the main ingredients for pattern formation in chemical systems, as well as in other nonlinear systems, and play important roles only when the system is far from thermodynamic equilibrium. In a closed laboratory chemical experiment, nonequilibrium conditions occur only in the beginning of the reaction. As the reagents are being consumed during the resection, the system evolves irreversibly toward thermodynamic equilibrium. Thus, complex patterns can at most be observed as transients. This is in sharp contrast to living organisms that can sustain themselves by taking in external supplies (food) and excreting unused materials (waste). Another disadvantage of a closed system is the lack of control parameters. These serious restrictions hinder the study of chemical patterns in terms of the well-developed bifurcation theory for complex systems.[11,45,56] Experimental studies in open systems have been common practice in many disciplines, but experiments on chemical patterns were, until recently, conducted in closed systems. In these lectures, I will describe several novel designs for open reactors,[38,52,53,54,55] in which spatiotemporal chemical patterns can be sustained far from thermodynamic equilibrium, and explore phenomena observed in these reactors using oscillating chemical reactions. Due to the limitation of space, I will restrict myself to a few examples and refer interested readers to other references.[14,31,33,40,48]

OSCILLATING CHEMICAL REACTIONS: THE BELOUSOV-ZHABOTINSKII REACTION

Oscillating chemical reactions were discovered as early as 1921,[20,22] but did not attract much attention until the discovery of the now well-known Belousov-Zhabotinskii (BZ) reaction some thirty years later.[10,67,69] The reaction was first discovered by B. P. Belousov[10] in 1958 and later modified by A. M. Zhabotinskii,[69] who brought world attention to the reaction. The reaction has now become the prototype for studies of oscillating chemical reactions even though many different chemical oscillators have been found since then.[22] The reaction involves the oxidation of an organic substrate, e.g., malonic acid, by bromate in a sulfuric acid solution with a metal-ion catalyst, e.g., cerium. Oscillations of the BZ reaction in a homogeneous (well-mixed) system can be observed as periodical concentration variations in certain intermediate species of the reaction, e.g., the bromide ion concentration, and as periodical color changes when a suitable catalyst is used, e.g., from colorless to yellow and back when cerium is used as a catalyst. The visual oscillations are more dramatic when the cerium catalyst is replaced by an iron catalyst, where distinctive red-blue color changes mark the reaction. In addition to temporal oscillations in homogeneous conditions, beautiful spatial patterns of waves, as shown in Figure 1, can be observed in a thin medium of the reagents,[63,64,65,66,68] which will be discussed in another section.

The oscillating phenomena of the BZ reaction can be understood by studying the reaction mechanism[17,69] that was put into its final form by Field, Körös, and Noyes in the early 1970s.[21,23] The mechanism, known as the FKN model,[21] involves 20 different chemical species in a set of 18 reaction steps and has laid the corner stone for the study of the BZ resection. It is not possible, using simplified

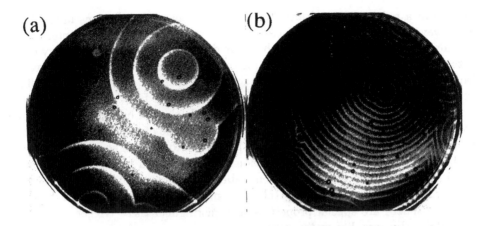

FIGURE 1   Trigger waves in BZ reactions, (a) target waves and (b) spiral waves. Light regions represent the blue wave fronts and dark regions represent the red background.

versions of the FKN model and inexpensive computing power, to obtain qualitative and sometimes quantitative agreements between numerical models and laboratory experiments.[19] Despite the complexities involved in oscillating reactions, simple essential criteria must be met in order to sustain oscillations. They are:

1. The system must be far from thermodynamic equilibrium.
2. Autocatalytic feedback mechanisms exist.

The first condition is essential for the second condition to play a role, and could be met only in the beginning of the reaction in a closed system or in an open reactor called the continuous-flow stirred-tank reactor (CSTR) where continuous feeds of chemicals are used to sustain the oscillations. The second condition provides the nonlinearities needed to produce oscillations and complex behavior. Conditions 1 and 2 are the backbones of the dynamics in oscillating reactions. Oscillations will stop when the conditions are no longer valid. Thus, chemical oscillations in a closed reactor are transients, though they may last as long as several hours in some cases. This limitation can be lifted in a CSTR, which will be discussed later.

CHEMICAL WAVES

The same BZ solution (with iron catalyst) that exhibits oscillations in a homogeneous reactor when left unstirred as a thin layer in a Petri dish will also exhibit spatiotemporal patterns in the form of waves.[1,2,3,12,13,34,35,41,44,46,47,49,62] Beautiful concentric blue wave fronts initiated from some centers will propagate radially outward in a red background after some global transient oscillations in the layer. These waves are called trigger waves because they are initiated from centers called pace makers, which are often gas bubbles or dust particles or the boundary of the container. Figure 1(a) shows an example of the trigger waves. Similar waves can also be observed in excitable BZ media that have chemical receipts slightly different from that of the oscillating BZ media. The blue wave front of the trigger waves has a high concentration of the oxidized iron, while the red background has a high concentration of reduced iron. The concentration profile of the oxidized iron can be mapped out by absorption measurements of monochromatic light at about 488 nm. A schematic diagram of the concentration profile is shown in Figure 2. The wave front has a sharp concentration gradient (about 17 mM/mm of iron concentration) in contrast to the slowly varying (about 1 mM/mm) refractory tail behind the front.[34,35] There are special properties that distinguish these trigger waves from other physical waves, like water waves:

1. There is no attenuation of the wave amplitude; the oxidized iron concentration remains the same at the wave front as the wave propagates.
2. Trigger waves do not interpenetrate each other, but annihilate each other on contact (see the cusp-shaped structures in Figure 1(a)).

The best analogue to trigger waves is fire propagation on grass in a wilderness. The "amplitude" of the fire will be the same as long as there is the same amount of grass ahead of the fire front. The consumed grass behind the front, of course,

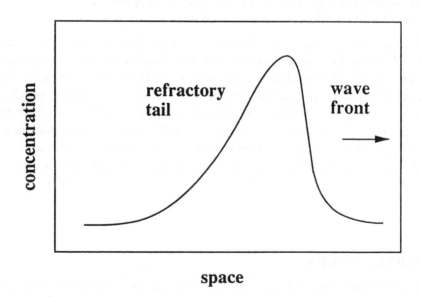

FIGURE 2   Schematic concentration profile of a trigger wave. The wave is traveling to the right as indicated by the arrow.

cannot support another fire front until fresh grass is grown. In the case of the BZ reaction, each wave front consumes only a small percent of the reagents, so that even after a wave front has passed, the medium can support another wave after some refractory time. Thus, successive wave fronts can propagate through the medium many times forming the observed concentric trigger waves. The concentric waves are hence called target waves to label this unique feature, as shown in Figure 1(a).

The origin of the pace makers for the trigger waves has, so far, not yet been settled. Experiments[13] have shown that dust particles, impurities, or bubbles facilitate the formation of pace makers, and that by reducing these heterogeneous centers, waves can be suppressed in an excitable system. But waves can still be initiated in a dust-free oscillating medium. More well-controlled experiments have to be carried out in order to shed light onto the origin of the trigger waves.

Target waves are not the only form of chemical waves. When a wave front is disrupted, by physical, chemical, or optical means, propagation of the wave at the perturbed location can be suppressed. The front will then break into two parts. The ends of the broken waves will evolve into a pair of highly regular, spiral-shaped, counter-rotating waves, called spiral waves (Figure 1(b)).[1,2,3,34,35,44,64,65,68] The tips of these waves turn inward around a rotation center while sending waves in the outward direction. The spiral waves, once produced, are self-sustained and do not have pace makers like target waves. The rotation center (spiral core), about 10 μm in diameter, is a singular region where the chemistry remains quasi-stationary. Beyond the core, variations of chemical concentrations resume the full amplitude of the trigger waves. The wave velocity and length (distance between successive wave fronts) of spiral waves are unique for a given medium, in contrast to that of target waves, where waves of different velocities and wavelengths can be initiated from different pace makers. Detailed measurements[34,35] of the spiral waves have revealed that the shape of the spiral front of isoconcentration level can be approximated by an Archimedean spiral or by an involute of a circle, while numerical results[30] fit closer to the involute. The two curves are asymptotically identical and differ only slightly near the core. Spiral waves are actually quite common in nature, and can be observed in many biological and physical systems. Besides single-armed spirals, "multi-armed" spirals[1] have been reported in BZ reagents. In three-dimensional mediums, fascinating scroll waves have been observed[62,63] and studied in great detail numerically.[66] The wave phenomena in BZ reaction has attracted as much, or even more, interest as in the temporal behavior and deserves further discussion.

FORMATION OF SPATIAL PATTERNS

Spatial patterns in chemical reactions arise mainly from the interaction between reaction kinetics and diffusion of different species. Reactions build up local concentrations to create gradients, while diffusion smooths out spatial variations. It is the competition between the two mechanisms that gives rise to spatial patterns from an initial homogeneous background as first suggested by A. M. Turing in 1950.[58] Turing's work provided the theoretical background for pattern formation in chemical systems, as well as biological systems. The waves in the BZ reactions, though not predicted by Turing, provide a paradigm for the study of spatiotemporal patterns. It has been shown that only two variables are needed to describe the propagation of waves in an excitable medium.[59] The equations that govern the dynamics of chemical waves are

$$\frac{\partial u}{\partial t} = \frac{F(u, v)}{\epsilon} + D_u \nabla^2 u \tag{1}$$

and

$$\frac{\partial u}{\partial t} = G(u, v) + D_\nu \nabla^2 \nu \tag{2}$$

where u, v are the independent variables, F and G account for the kinetics of the reaction and are functions of the u, v, and rate constants. D_u and D_v are scaled diffusion coefficients for u and v, respectively. $\epsilon \ll 1$ is a parameter that is proportional to the ratio of the diffusion rate to the chemical reaction rate and plays a crucial role in determining the dynamics of the system. It is easily seen that, because F is scaled by ϵ, u changes much more rapidly than v. The dynamics of Eqs. (1) and (2) are best represented in phase diagrams as shown in Figure 3. Under homogeneous conditions, the equilibrium fixed point (s) of the system corresponds to the intersection of the nullclines of $F(u, v)$ and $G(u, v)$ as shown in Figure 3. The F nullcline has a characteristic S-shaped dependent, while G has a monotonic dependent. If the G nullcline intersects the left branches of the F nullcline, trajectories that start in the neighborhood of the fixed point will flow toward the fixed point quickly, but trajectories that start on the right-hand side of the middle branch of the F nullcline will exhibit a large amplitude excursion before settling back to the fixed point as indicated by the flow of the vector fields in Figure 3(a). This large excursion is a typical property of excitability, where the fixed point is stable to small perturbations, but perturbations about a certain threshold will lead to large amplitude excursions before returning to the fixed point. This excitability when coupled spatially by diffusion is the main ingredient for wave propagation in chemical, as well as biological, systems. Chemicals diffused from the sharp wave front into the neighborhood ahead of the front can excite the quiescent background in that region into a new wave front; thus propagation of the wave front continues. The amplitude and wave velocity depend on the delicate balance between the chemical kinetics and the diffusion. Model studies of Eqs. (1) and (2) with BZ reaction have yielded semi-quantitative agreements with experiments on the dispersion of wave propagation and on the form of spiral waves.[16,30] A side note for Eqs. (1) and (2) is that, when the G nullcline intersects the middle branch of the F nullcline, the fixed point is no longer stable and relaxational oscillations start, as shown in Figure 3(b).

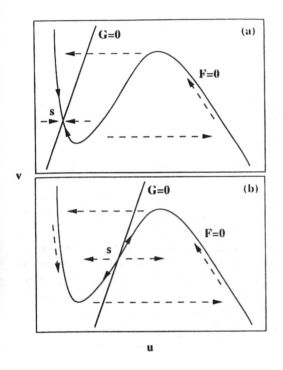

FIGURE 3 Phase diagram of u and v. The solid lines represent the nullclines of $F(u,v)$ and $G(u,v)$ as indicated. s is the fixed point. The arrows indicate the flow fields of the reaction. (a) is excitable and (b) is oscillatory.

NEEDS FOR OPEN REACTORS

The experimental studies of temporal and spatial patterns in chemical reactions discussed so far were conducted in closed systems. In such systems, patterns will evolve irreversibly and uncontrollably toward thermodynamic equilibrium. Any patterns observed are transients and cannot be studied on a long-term basis. Despite the transient nature of patterns in closed systems, detailed studies of these patterns are of fundamental importance and have been studied actively. These studies revealed basic understanding of chemical patterns, but left many questions unanswered, including:

1. Are these patterns stable in an open system?
2. What are the possible patterns?
3. What are the transitions between patterns?

In a closed chemical reactor, one can only set the initial conditions, like the initial concentrations of the reagents. In such systems, there are no tunable parameters that can be varied back and forth; hence, transitions between states cannot be investigated. This puts serious limitations on closed reactors. Thus, it is very desirable to develop open reactors where there are external parameters that one can freely adjust. In the following sections, I will describe some of these open reactors that have been developed over recent years to study temporal and spatial patterns of oscillating chemical reactions.

OPEN TEMPORAL REACTORS
CONTINUOUS-FLOW STIRRED-TANK REACTOR

Chemical processings in open reactors has long been a common practice among the chemical engineering community, but open reactors were first used for the study of oscillating chemical reactions in the mid-1970s by a group of chemists in Bordeaux, France.[20] A schematic diagram of a continuous-flow stirred-tank reactor (CSTR) is shown in Figure 4. Continuous feeding of reagents into the reactor through the inlet is the crucial feature of the CSTR. Peristaltic or precision piston pumps can be used to deliver chemicals into the reactor. A simple overflow mechanism defines the reactor volume and facilitates the removal of chemicals. A rapidly rotating stirrer is used to mix the reagents inside the reactor. Temporal patterns can be monitored or recorded by direct measurement of chemical potentials using ion-sensitive electrodes or optical absorption methods. The temperature of the reactor can be regulated by simply submerging the reactor into a temperature bath. Common control parameters for the CSTR are:

1. initial feed concentrations of reagents,
2. feed rate of chemicals,
3. temperature of the reactor, and
4. rate of mixing in the reactor.

The last control parameter addresses the heterogeneous effect of mixing and has been studied for some mixing sensitive reactions.[36] As for common oscillating reactions, mixing has small effects, and adequate stirring is enough to ensure a homogeneous condition. The second control parameter, the feed rate, is the most commonly used. A variation of this parameter is the residence time τ, which is equal to the volume of the reactor divided by the total feed rate. The residence time represents the duration that reagents stay in the reactor. The third control parameter sets the reaction rates of the system and is often held fixed for convenience.

Owing to the above controllable features, the CSTR is an ideal system with which to study temporal patterns of chemical reactions, where nonequilibrium conditions can be maintained and varied to study stabilities and transitions of different temporal states. This tool, together with the techniques developed for the study of dynamics systems, has made tremendous progress in understanding oscillating reactions.[4,5,8,15,28,32,42,43,50,57] Temporal patterns of oscillating reactions observed in CSTR can be summarized as follows:

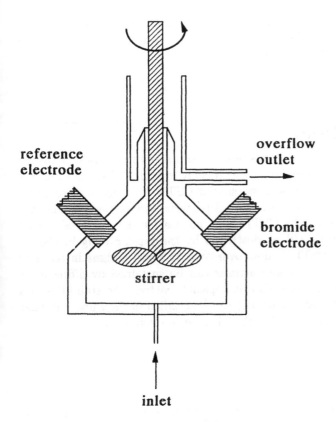

FIGURE 4 Schematic diagram of a continuous-flow stirred-tank reactor.

1. hysteretic transitions between multiple steady states,
2. bifurcations of steady states to limit cycles,
3. birhythmicity-hysteretic behavior between two oscillating states with different frequencies,
4. bifurcations of periodic states to quasi-periodic states with two incommensurate frequencies f_1 and f_2,
5. frequency locking of quasi-periodic states—$f_2 = p/qf_1$, where p, q are integers,
6. period doubling of periodic states—the frequency of the states changes from $f_0 \rightarrow f_0/2 \rightarrow f_2/4 \cdots$,
7. intermittency—irregular bursts of large amplitudes at irregular intervals from a seemingly periodic state, and
8. chaos via several mechanisms—period-doubling cascade, quasi-periodicity, and intermittency.

Model calculations of oscillating reactions in CSTR have reproduced all these findings. The strong evidences from experiments and models have unambiguously demonstrated that complex behavior observed in oscillating chemical reactions in a CSTR is deterministic rather than noisy behavior arising from random driving forces.[4,5,28,42,43,50,57] Moreover, the chaos observed in chemical reactions can be understood in terms of bifurcation theories.

OPEN SPATIAL REACTORS

The study of spatial patterns in open systems, in contrast to temporal patterns, has been lacking. The advancement made with CSTR in temporal studies has not been shared with spatial studies because no similar CSTR reactors, until recently, have been available for spatial patterns. Spatial patterns cannot be retained in a well-mixed environment so that the CSTR is of no use for the study of spatial patterns; any spatial structures will be destroyed by the rapid mixing in the reactor. This drawback of the CSTR is now overcome by novel designs in recently developed open spatial reactors, where patterns can be sustained by diffusion or controllable hydrodynamic flows. In such open spatial reactors, the stabilities and transitions between different spatial patterns can be studied in the same way that temporal patterns are studied in the CSTR's. Three such open spatial reactors will be discussed in this lecture.

RING REACTOR

A schematic diagram of the ring reactor[38] is shown in Figure 5. The main ingredient of the reactor is an annular polyacrylamide gel. The gel suppresses any convective motion generated by concentration gradients or other hydrodynamic instabilities and the formation of bubbles from gases evolved in the BZ reaction,[6] but allows patterns to be formed inside. The patterns formed can be sustained by chemicals diffusing into the gel. The gel is chemically inert to the reaction so that patterns resulting inside the gel are only due to the kinetics of the BZ reaction and the diffusion inside the gel. The gel ring, 1 mm thick, is sandwiched between two fixtures that separate the reactor into two separate compartments such that the gel is in contact with each compartment from either side (see Figure 5). The two compartments serve as reservoirs of chemicals that can be replenished from external feeds.

In the experiment, reagents of the BZ reaction were separated with the sulfuric acid and potassium bromate in the outer compartment, and malonic acid and iron (ferroin) catalyst in the inner compartment. After the start of the experiment, a red front of ferroin could be seen advancing slowly from the inner edge toward the outer edge of the annulus. Then two to three hours later, the circular symmetry of the front started to break, and local irregularities developed. These irregularities would eventually develop into stable states of an equal number of multiple

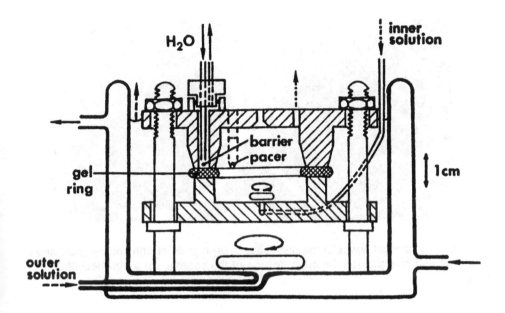

FIGURE 5   Cross section of the ring reactor. See Noszticzius[38] for details.

sources and sinks located around the annular as shown in Figure 6(a) with one source and one sink, where waves are initiated from the source and annihilated at the sink. The source-sink configuration, though it could last as long as the same external feeds are applied, is rather idiosyncratic. A more regular and stable state of rotating waves could be obtained with an appropriate perturbation. This can be done by stimulating waves traveling in two directions, but eliminating waves traveling in one direction, e.g., eliminating the counter-clockwise waves. This will result in uneven waves in the two directions, which will then develop after annihilation of the remaining counter-clockwise waves by the clockwise waves into a state of waves traveling in the clockwise direction. Chemical or optical perturbations can be used to create these one-directional waves. As an example, waves were initiated in the experiment at the pacer (see Figure 5) by adding a solution of higher concentrations of H_2SO_4 and $KBrO_3$ to that location and were inhibited in the counter rotating direction by passing a slow continuous flow of distilled water at the barrier. The perturbations at the pacer and the barrier were then removed after waves traveling in one direction, counter-clockwise, were eliminated. The unevenly spaced waves then relaxed to form a state of regular equally spaced waves rotating in the clockwise direction as shown in Figure 6(b) for a state of seven clockwise waves. These highly regular wave patterns, like pinwheels, rotate faithfully around the annulus as long as the same external controls are applied. (The pattern eventually will disrupt

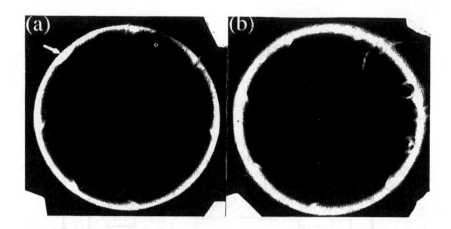

FIGURE 6   Chemical waves in the ring reactor: (a) a state of one source (indicated by an arrow) and one sink (on the opposite side of the source), (b) a pinwheel state of 7 waves.

due to the shrinking and decomposition of the gel in a strong acidic medium, but gels of different compositions could be used to solve this minor problem.) The experiment shows that regular rotating chemical patterns (clockwise as well as counter clockwise) are stable if the number of waves falls within a certain range. This is the first study showing that spatial patterns of chemical reactions can be sustained in an open reactor. In a recent experiment on chemical pinwheels, using an improved ring reactor,[31] symmetry breaking and time dependence of the waves were revealed when the temperature of the reactor was varied. The chemical pinwheel is the first realization of rotating wave structures predicted for nonlinear chemical systems in a circular geometry.[37]

An open reactor using the same principle as in the ring reactor has been developed recently by Castets et al.[14] Instead of ring-shaped gel, they use a narrow rectangular gel. The two opposite long edges are in contact with two chemical reservoirs, which are filled with reagents of a variant of the Chlorite-iodide reaction. The reagents of Chlorite-iodide reaction are separately distributed in such a way that neither solution in the reservoirs is active. Reaction takes place only inside the gel. They observe patterns of stripes parallel to the gel edges in the direction of the concentration gradient. These stripes then break up into lines of periodic spots of characteristic size of about 0.2 mm which seems to be nongeometrically related. The periodic spots are stable and form only for a well-defined range of concentrations. The pattern exhibits the symmetry-breaking phenomenon in the direction transverse to the imposed gradient and is interpreted as the first experimental evidence of a genuine Turing structure[58] as discussed in the previous section. Future studies will certainly verify the extraordinary finding.

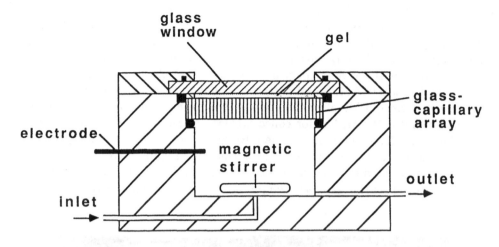

FIGURE 7   Cross-section of the CFUR. See Tam[52] for details.

CONTINUOUSLY FED UNSTIRRED REACTOR (CFUR)

A cross-sectional view of a CFUR designed by Tam et al.[52] is shown in Figure 7. A disk-shaped, diameter 2.54 cm, polyacrylamide gel (same as in the ring reactor) is sandwiched horizontally between a glass window and a 1 mm thick glass capillary array with evenly spaced 10 μm diameter capillaries packed in a hexagonal matrix. (A nitrocellulose membrane, not shown in Figure 7, is placed between the gel and the capillary array to provide a white background for enhancement of pattern visualization.) The gel is 1 mm thick and is sufficiently thin for two-dimensional wave propagation. The gel communicates through the capillary array with a CSTR below the array; hence, patterns formed in the gel can be sustained by chemicals from the CSTR. The glass capillary array is a crucial feature of the reactor; it provides uniform feed that is in a direction perpendicular to the plane in which patterns can form, and ensures that only vertical mass transport occurs between the gel and the CSTR. If horizontal transport were not suppressed, spatial pattern formation could occur already in the capillary array. This effect is obvious when the capillary array is replaced by a fitted glass disc where mass transport can occur in all directions in the disc.

BZ reaction was used in the experiment. Reagents were delivered into the CSTR by high-pressure piston pumps so that bubble formation from gases produced in the reaction was suppressed by holding 10 atm pressure in the reactor. Bubbles are undesirable because they can block the capillaries and cause inhomogeneous feed to the gel. Feeds of fresh reagents were combined into a single line before entering the CSTR. The feed rate and the concentrations of reagents were kept constant, except the concentration of $NaBrO_3$ that was varied over the range 0.01–0.05 M and was used as the control parameter in the experiment. Patterns were digitized using a 512×480 pixel, 8-bit resolution imaging system. The intensity amplitudes of the digitized images are directly related to the concentrations of the oxidized iron.

Homogeneous states of a uniform background are observed for $NaBrO_3$ concentrations below 0.017 M. Spiral patterns similar to those observed in Petri dish experiments form quickly after the start of the experiment for $NaBrO_3$ concentrations around 0.018 M, but for concentrations above about 0.025 M, irregular waves form first (see Figure 8(a)). These irregular waves can last a long time until one or more spiral waves appear. The formation of spiral waves from irregular waves can be initiated by a perturbation. A simple method is to stop the stirrer in the CSTR for a short time to create temporary spatial inhomogeneities in the CSTR

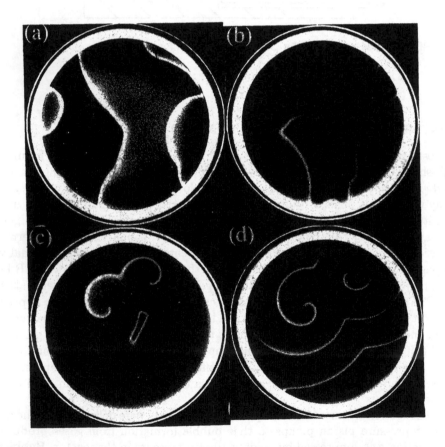

FIGURE 8 States showing the formation of spiral waves in the CFUR by a perturbation as described in the text. (a) irregular waves before the perturbation, and (b) 12 minutes, (c) 24 minutes, and (d) 60 minutes after the perturbation.

and, hence, in the feeds to the gel. This perturbation breaks up a wave into two disconnected waves. The tips of the broken waves will then evolve into a pair of spiral cores. Figures 8(b)–(d) show the different stages of the formation of a pair of spiral waves using such perturbation. Spiral waves formed are stable and can last for weeks. Spiral waves can occasionally drift to the walls where they are annihilated. Optical methods can also be used to break the wave front. The transition from homogeneous state to spiral state is shown as intensity amplitudes in Figure 9 as a function of the NaBrO$_3$ concentration. The insert in Figure 9 clearly shows a small hysteretic loop with location and size reproducible within a few percent in runs with different gels. Figure 9 is the first study of the transition of spatial patterns of oscillating reactions in an open reactor. The hysteretic behavior observed in Figure 9 could simply be due to the excitability of the BZ reaction, but a detailed calculation may shed more light into the nature of the transition.

The CFUR is a very useful tool in the study of two-dimensional chemical waves. An example is to study the dynamics of a single spiral wave. The tip of a spiral wave is known from experiments and numerical simulations to rotate in a circle about a core region. But, meandering tips of spiral waves have also been observed in BZ reagents.[2,7,9,25,29,48]

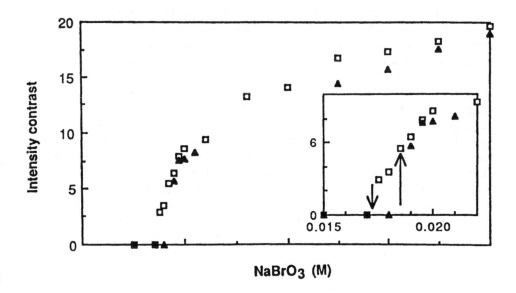

NaBrO$_3$ (M)

FIGURE 9 The intensity contrast, which is the difference between the maximum intensity of a spiral wave front (the oxidized state) and the minimum intensity (the reduced state), is shown as a function of NaBrO$_3$ concentration. (□ decreasing concentration; △, increasing concentration). The appearance (disappearance of the spiral waves with increasing (decreasing) concentration is indicated by ↑ (↓) in the insert.

The meandering effect could have been caused by boundary effects, hydrodynamics, or inhomogeneities. Recent studies of Jahuke, Skaggs, and Winfree[29] on chemical waves in gels loaded with BZ reagents have revealed that the meandering motion is not an artifact, but is inherent in the chemistry used. They observed circular tip orbits for some chemistries, as well as epicycle-like (compound) orbits for others in both experiment and model calculation. Recent detailed studies by Skinner and Swinney[48] on the motion of spiral tip using a CFUR has brought new understanding in the tip motion. They found supercritical transition from simple rotation (one frequency) to compound rotation (two frequencies). (The control parameter used in that experiment was the feed NaBrO$_3$ concentration.) The supercritical transition is consistent with recent model studies using a simple kinetic for the BZ reaction.[9] Skinner and Swinney[48] conjectured that the refractory tail of the wave plays a dominant role in controlling the tip motion. This can be verified by future model studies.

COUETTE REACTOR

The transport of reagents in the ring reactor and the CFUR is through molecular diffusion, which is about 10^{-5} cm^2/s. Spatial structures resulting from reaction-diffusion mechanisms have a characteristic size proportional to the square root of the product of the diffusion coefficient and the averaged reaction time. Thus, spatial structures of BZ reactions in gels have a typical size of above few millimeters. Large spatial structures can be obtained by enhancing the mass transport that can be easily modeled in simulations, but not in experiments. The Couette reactor[27,40,53] is designed to have the unique property of controllable mass transport. A schematic diagram of the Couette reactor[53] is shown in Figure 10. It consists of a fluid filling the gap between two concentric cylinders with the inner cylinder rotating and the outer cylinder at rest. This arrangement is exactly the Couette-Taylor system[26] used in hydrodynamic studies of shear flows; but instead of nonreacting fluid, the fluid used in the Couette reactor is now the BZ reagent. Experiments[18,51] indicate that the transport arising from the hydrodynamic flow at sufficiently high cylinder speed can be modeled as a one-dimensional diffusion process in the axial direction. (The effective diffusion has a simple power dependence on the rotation rate of the inner cylinder and can be orders of magnitude larger than molecular diffusion.) Thus, the reactor is effectively a one-dimensional reaction-diffusion system where the transport, effective diffusion, can be controlled by varying the rotation rate. Patterns in the Couette reactor can be sustained by feeds at the ends as shown in Figure 10.

A dual substrate glucose-acetone BZ reaction was used in the experiment because the reaction produces an insignificant amount of gas compared to the usual bubble-producing, malonic acid BZ system.[39] Bubbles are undesirable in the Couette reactor because they interfere with the flow state and change the effective

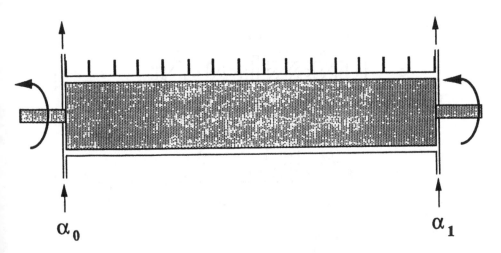

FIGURE 10   The Couette reactor α_0 and α_1 are the feed rates at $Z = 0$ and $Z = 1$, respectively. The vertical lines show the location of the 16 evenly spaced Ag–AgBr electrodes.

diffusion coefficient. Another advantage of the dual substrate system is that the reaction exhibits only steady and simple periodic states in homogeneous systems; thus, any complex spatiotemporal behavior observed in the Couette reactor must arise from an interplay between the simple local dynamics and effective diffusion. In the experiment, the oxidizer (KBrO$_3$) was fed into the cylindrical annulus at one end, $Z = 0$, while the reducers (glucose and acetone) were fed at the other end, $Z = 1$. The rate of removal of chemicals at each end was carefully adjusted to match the feed rate at that end so that there was no net axial flow. The feed rate of the oxidizer was varied from 0–35 ml/h, while the feed rate α_1 of the reducer was fixed at 10 ml/h (H$_2$SO$_4$ was fed at both ends, while MnSO$_4$ (catalyst) was fed only at $Z = 1$). Three rotation rates of 6, 9, and 12 Hz, corresponding to an effective diffusion coefficient of 0.12, 0.16, and 0.22 cm^2/s estimated from Tam et al.,[51] respectively, were studied in the experiment. The concentration of one of the intermediate species, bromide ion, was measured at 16 locations with ion-selective electrodes (Ag–AgBr) spaced equally along the axial extent of the reactor. Time series, power spectra, and phase portraits were used to identify the spatiotemporal patterns for each α_0. Figures 11(a)–(c) show bifurcation sequences obtained as the feed rate α_0 is varied at the rotation rates of 6, 9, and 12 Hz, respectively. Data were obtained with both increasing and decreasing α_0 and no hysteresis was observed.

At low α_0, only time-independent states with spatial bromide concentration variations are observed. The first transition is from a time-independent state to an oscillating state with increasing α_0 for all rotation rates. Figures 12(a) and 13(b) show the power spectrum and the phase portrait of an oscillating state, respectively.

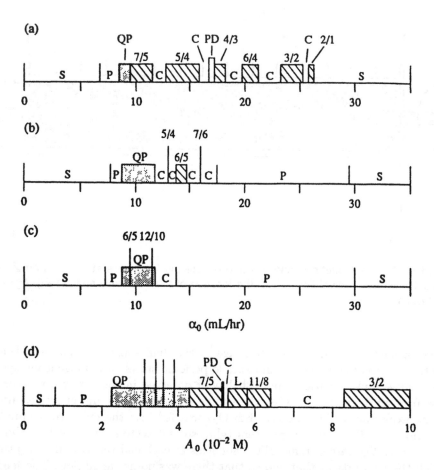

FIGURE 11   (a)–(c) Bifurcation sequences observed in the Couette reactor as a function of α_0 for cylinder rotation rates of 6, 9, and 12 Hz, respectively. (d) Bifurcation sequence for a reaction diffusion model as a function of bromate feed concentration A_0 (for D = 0.078 cm^2/s).[53,61] The labels for the different dynamical regions are: steady (S), periodic (P), quasi-periodic (QP), frequency-locked (p/q is the frequency locking ratio, where p and q are integers), period-doubled (PD), and chaotic (C); the region labeled L in (d) corresponds to many different frequency-locked states, and the four long vehicle lines in (d) are very narrow windows of frequency-locked states.

The periodic oscillations are neither traveling waves nor standing waves—the phase has a non-monotonic spatial dependence. The amplitudes of the oscillations are not the same everywhere along the reactor; they are large near the $Z = 0$ end and small near the $Z = 1$ end. At higher α_0 after the first transition, a second transition to a quasi-periodic state occurs. The frequencies obtained from time series and

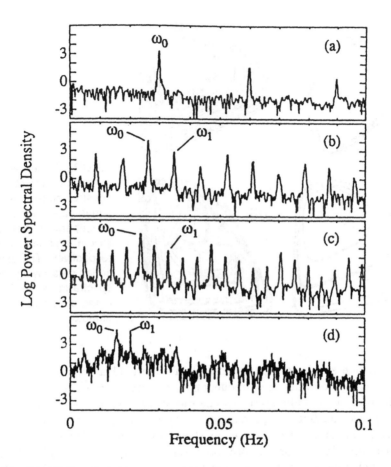

FIGURE 12   (a)–(d) Power spectra determined from the time series data for, respectively, periodic ($\alpha_0 = 8.0$ mL/h), quasi-periodic ($\alpha_0 = 9.0$ mL/h), frequency-locked ($\alpha_0 = 10.0$ mL/h), and chaotic states ($\alpha_0 = 16.0$ mL/h), obtained in each case from a bromide electrode located at $Z = 0.2$ with rotation rate of 6 Hz and $\alpha_0 = 10.0$ mL/h.

power spectra (Figure 12(b)) of this state are incommensurate and localized; α_0 (the frequency of the original periodic state) is dominant near $Z = 0$, while the new frequency ω_1 is dominant near $Z = 1$. A phase portrait of a quasi-periodic state is shown in Figure 13(b). No third frequency is observed with increasing α_0, instead the two frequencies of the quasi-periodic state can become locked together at some integer ratio to form a frequency-locked state. Figure 12(c) and 13(c) show the power spectrum and the phase portrait of a frequency-locked state, respectively. Further increase in α_0 leads to a chaotic state, then (except at 12 Hz) to many frequency-locked and chaotic states as shown in Figures 11(a)–(c). Chaotic states are identified by their power spectra (Figure 13(d)) and phase portraits (Figure

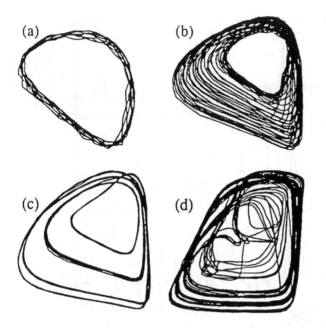

(a)

(b)

(c)

(d)

FIGURE 13   (a)–(d) Phase portraits of attractors constructed from time series data[24] for periodic, quasi-periodic, frequency-locked, and chaotic states, respectively. The same time series data giving the power spectra in Figure 12 are used in the construction of the attractors.

12(d)). In one special case, labeled PD in Figure 11(a), the transition to chaos is via a period-doubling of a frequency-locked state. Presumably other observed transitions from periodic states to chaotic states also proceed through a period-doubling cascade, but the parameter range is too small to detect. At very high α_0, only steady states are observed as shown in Figures 11(a)–(c), The bifurcation sequences are similar for all rotation rates, except that, at high rotation rates, complex behavior occurs for a smaller range in α_0. This behavior is due to the fact that the higher the mass transport rate (effective mixing), the closer the system is toward homogeneous conditions as in a CSTR where only steady and simple periodic states are observed.

Results at lower rotation rates have revealed more complicated dynamics, but at that rotation rate, the effective diffusion model for mass transport is no longer applicable and interpretation of the results is not trivial. The bifurcation sequences in Figures 11(a)–(c), except for a shift of a few percent in the transition α_0 values, are reproducible from run to run. Recent model calculations of a one-dimensional reaction-diffusion system capture the observed phenomena.[53,60,61] Figure 11(d) shows a bifurcation sequence obtained from a model calculation with only two species for the local chemistry and a single effective diffusion of $0.078 cm^2/s$ (close to 6 Hz rotation rate) for both species. The correspondence between the numerical results and the experimental results at 6 Hz is remarkable. Model calculations at higher rotation rates are also similar to the experimental results at high rotation rates.[61] The similarity between the model and the experiment is to some degree fortuitous. The boundary conditions and control parameters are different, having flow rate in the experiment and feed concentration in the model. Furthermore, the two-species model is for the malonic acid BZ system, and the relation to the glucose-acetone system is unknown. Despite this, the qualitative agreements between experiment and model suggest that similar bifurcation sequences can be observed in other chemistries in a one-dimensional reactor with opposing oxidation and reduction gradients.

DISCUSSION

The study of chemical patterns has reached another height since the discovery of the BZ reaction. This progress is possible primarily due to the adoption of open reactors to study chemical reactions and the advance in the studies of dynamical systems. The CSTR has proven to be a critical tool in the study of temporal patterns. The success in the CSTR can now be seen in the open spatial reactors. Sustained spatiotemporal patterns, transitions and bifurcations sequences of spatial patterns are, for the first time, obtained using these reactors. With these newly developed open spatial reactors, it should now be straightforward to study sustained spatial patterns in other geometries and boundary conditions, and new phenomena are expected to be discovered. The study of chemical patterns in open systems may shed light on understanding the complexity of pattern formation in living things.

ACKNOWLEDGMENTS

I thank Professor Dan Stein and Professor Lynn Nadel for the invitation to lecture at SFI. I am indebted to Z. Noszticzius, J. A. Vastano, W. Horsthemke, W. D. McCormick, H. L. Swinney, Q. Ouyang, and P. DeKepper for their collaborations in the works reported in these lectures.

REFERENCES

1. Agladze, K. I., and V. I. Krinsky. "Multi-Armed Vortices in an Active Chemical Medium." *Nature* **296** (1982): 424–426.
2. Agladze, K. I., A. V. Panfilov, and A. N. Rudenko. "Nonstationary Rotation of Spiral Waves: Three-Dimensional Effect." *Physica* **29D** (1988): 409–415.
3. Agladze, K. I., V. I. Krinsky, and A. M. Pertsov. "Chaos in the Non-Stirred Belousov-Zhabotinskii Reaction is Induced by Interaction of Waves and Stationary Dissipative Structures." *Nature* **308** (1984): 834–835.
4. Argoul, F., A. Arneodo, P. Richetti, J. C. Roux, and H. L. Swinney. "Chemical Chaos: From Hints to Confirmation." *Acc. Chem. Res.* **20** (1987): 436–422.
5. Argoul, F., A. Arneodo, P. Richetti, and J. C. Roux. "From Quasiperiodicity to Chaos in the Belousov-Zhabotinskii Reaction I. Experiment." *J. Chem. Phys.* **86** (1987): 3325–3338.
6. Avnir, D., and M. Kagan. "Spatial Structures Generated by Chemical Reactions at Interfaces." *Nature* **307** (1984): 717–721.
7. Barkey, D. "A Coupled-Map Lattice for Simulating Waves in Excitable Media." In *Nonlinear Structures in Physical Systems*, edited by L. Lam and H. C. Morris. New York: Springer, 1990.
8. Barkley, D. E. "Studies of the Complex Dynamics of The Belousov-Zhabotinskii Reaction." Ph.D. thesis, University of Texas at Austin, 1988.
9. Barkley, D., M. Kness, and L. S. Tuckerman. "Spiral-Waves Dynamics in a Simple Model of Excitable Media: The Transition from Simple to Compound Rotation." *Phys. Rev. A* (1990): to appear.
10. Belousov, B. P. "A Periodic Reaction and Its Mechanism" In *Sbornik Referatov po Radiatsionni Meditsine*, 145. Moscow: Medgiz, 1958.
11. Bergé, P., Y. Pomeau, and C. Vidal. *Order Within Chaos.* New York: Wiley, 1984.
12. Bodet, J. M., J. Ross, and C. Vidal. "Experiments on Phase Diffusion Waves." *J. Chem. Phys.* **86** (1987): 4418–4424.
13. Bodet, J. M., C. Vidal, A. Pacault, and F. Argoul. "Experimental Study of Target Patterns Exhibited by the B. Z. Reaction." In *Nonequilibrium Dynamics in Chemical Systems*, edited by C. Vidal and A. Pacault, 102–106. Heidelberg: Springer-Verlag, 1984.
14. Castets, V., E. Dulos, J. Boissonade, and P. DeKepper. "Experimental Evidence of a Sustained Standing Turing-Type Nonequilibrium Chemical Pattern." *Phys. Rev. Lett.* **64** (1990): 2953–2956.
15. Coffman, K. G., W. D. McCormick, Z. Noszticzius, R. H. Simoyi, and H. L. Swinney. "Universality, Multiplicity and the Effect of Iron Impurities in the Belousov-Zhabotinskii Reaction." *J. Chem. Phys.* **86** (1987): 119–129.
16. Dockery, J. D., J. P. Keener, and J. J. Tyson. "Dispersion of Traveling Waves in the Belousov-Zhabotinskii Reaction." *Physica D* **30** (1988): 177–191.

17. Edelson, D., R. J. Field, and R. M. Noyes. "Mechanistic Details of the Belousov-Zhabotinskii Oscillations." *Int. J. Chem. Kinet.* **7** (1975): 417–432.
18. Enokida, Y., K. Nakata, and A. Suzuki. "Axial Turbulent Diffusion in Fluid Between Rotating Coaxial Cylinders." *AIChE J.* **35** (1989): 1211–1214.
19. Epstein, I. R. "Chemical Oscillations and Nonlinear Chemical Dynamics." In *1989 Lectures in Complex Systems*, edited by E. Jen. Santa Fe Institute Studies in the Sciences of Complexity, Lect. Vol. II, 213–269. Redwood City, CA: Addison-Wesley, 1989.
20. Epstein, I. R., K. Kustin, P. DeKepper, and M. Orban. "Oscillating Chemical Reactions." *Sci. Am.* **248** (1983): 112–123.
21. Field, R. J., E. Körös, and R. M. Noyes. "Oscillations in Chemical Systems. Part 2. Thorough Analysis of Temporal Oscillations in the Ce-BrO$_3$-Malonic Acid System." *J. Am. Chem. Soc.* **34** (1972): 8649–8664.
22. Field, R. J., and M. Burger. *Oscillations and Traveling Waves in Chemical Systems*. New York: Wiley, 1985.
23. Field, R. J., and R. M. Noyes. "Oscillations in Chemical Systems. Part 4. Limit Cycle Behavior in a Model of a Real Chemical Reaction." *J. Chem. Phys.* **60** (1974): 1877–1884.
24. Fraser, A. M., and H. L. Swinney. "Independent Coordinates for Strange Attraction from Mutual Information." *Phys. Rev. A* **33** (1986): 1134–1140.
25. Gerhardt, M., H. Schuster, and J. J. Tyson. "A Cellular Automaton Model of Excitable Media Including Curvature and Dispersion." *Science* **247** (1990): 1563–1566.
26. Gorman, M., and H. L. Swinney. "Spatial and Temporal Characteristics of Modulated Waves in the Circular Couette System." *J. Fluid Mech.* **117** (1982): 123–142.
27. Grutzner, J. B., E. A. Patrick, P. J. Pellechia, and M. Vera. "The Continuously Rotated Cellular Reactor." *J. Am. Chem. Soc.* **110** (1988): 726–728.
28. Hudson, J. L., M. Hart, and D. Marinko. "An Experimental Study of Multiple Peak Periodic and Nonperiodic Oscillations in the Belousov-Zhabotinskii Reaction." *J. Chem. Phys.* **71** (1979): 1601–1606.
29. Jahnke, W., W. E. Skaggs, and A. T. Winfree. "Chemical Vortex Dynamics in the Belousov-Zhabotinsky Reaction and in the 2-Variable Originator Model." *J. Phys. Chem.* **93** (1989): 740–749.
30. Keener, J. P., and J. J. Tyson. "Spiral Waves in the Belousov-Zhabotinskii Reaction." *Physica D* **21** (1986): 307–324.
31. Kreisberg, N., W. D. McCormick, and H. L. Swinney. "Symmetry Breaking in a Chemical Pinwheel." *J. Chem. Phys.* **91** (1989): 6532–6533.
32. Maselko, J., and H. L. Swinney. "Complex Periodic Oscillations and Farey Arithmetic in the Belousov-Zhabotinskii Reaction." *J. Chem. Phys.* **85** (1986): 6430–6441.
33. Maselko, J., and K. Showalter. "Chemical Waves on Spherical Surfaces." *Nature* **339** (1989): 609–611.

34. Müller, S. C., T. Plesser, and B. Hess. "Two-Dimensional Spectrophotometry of Spiral Wave Propagation in the Belousov-Zhabotinskii Reaction." *Physica D* **24** (1987): 87–96.

35. Müller, S. C., T. Plesser, and B. Hess. "The Structure of the Core of the Spiral Waves in the Belousov-Zhabotinskii." *Science* **230** (1985): 661–623.

36. Nagypal, I., and I. R. Epstein. "Fluctuations and Stirring Rate Effects in the Cholorite-Thiosulfate Reaction." *J. Phys. Chem.* **90** (1986): 6285–6292.

37. Nicolis, G., and I. Prigogine. *Self-Organization in Nonequilibrium Systems*, 153–156. New York: Wiley, 1977.

38. Noszticzius, Z., W. Horsthemke, W. D. McCormick, H. L. Swinney, and W. Y. Tam. "Sustained Chemical Waves in an Annular Gel Reactor: A Chemical Pinwheel." *Nature* **329** (1987): 619-620.

39. Ouyang, Q., W. Y. Tam, P. DeKepper, W. D. McCormick, Z. Noszticzius, and H. L. Swinney. "Bubble-Free Belousov-Zhabotinskii-Type Reactions." *J. Phys. Chem.* **91** (1987): 2181–2184.

40. Ouyang, Q., J. Boissonade, J. C. Roux, and P. DeKepper. "Sustained Reaction Diffusion Structures in an Open Reactor." *Phys. Lett. A.* **134** (1989): 282–286.

41. Ross, J., S. C. Müller, and C. Vidal. "Chemical Waves." *Science* **240** (1988): 460–464.

42. Roux. J. C., J. S. Turner, W. D. McCormick, and H. L. Swinney. "Experimental Observations of Complex Dynamics in a Chemical Reaction." In *Nonlinear Problems: Present and Future*, edited by A. R. Bishop, D. K. Campbell, and B. Nicolaenko, 409–422. Amsterdam: North-Holland, 1982.

43. Roux, J. C., R. H. Simoyi, and H. L. Swinney. "Observation of a Strange Attractor." *Physica D* **8** (1983): 257–266.

44. Saul, A., and K. Showalter. "Propagating Reaction-Diffusion Fronts." In *Oscillation and Traveling Wave in Chemical System*, edited by R. J. Fields and M. Burger, 419–439. New York: Wiley, 1985.

45. Schuster, H. G. *Deterministic Chaos*. Weinheim: Physik-Verlag, 1984.

46. Showalter, K. J. "Trigger Waves in the Acidic Bromate Oxidation of Ferroin." *J. Phys. Chem.* **85** (1981): 440–450.

47. Showalter, K., R. M. Noyes, and H. Turner. "Detailed Studies of Trigger Waves Initiation and Detection." *J. Am. Chem. Soc.* **101** (1979): 7463–7469.

48. Skinner, G. S., and H. L. Swinney. "Periodic to Quasiperiodic Transition of Chemical Spiral Rotation." *Physica D*, to appear.

49. Smoes, M. L. "Chemical Waves in the Oscillatory Zhabotinskii System. A Transition from Temporal to Spatio-temporal Organization." In *Dynamics of Synergetic Systems*, edited by H. Haken, 80–96. Berlin: Springer, 1980.

50. Swinney, H. L., and J. C. Roux. "Chemical Chaos." In *Nonequilibrium Dynamics in Chemical Systems*, edited by C. Vidal and A. Pacault, 124–140. Berlin: Springer, 1984.

51. Tam, W. Y., and H. L. Swinney. "Mass Transport in Turbulent Couette-Taylor Flow." *Phys. Rev. A* **36** (1987): 1374–1381.

52. Tam, W. Y., W. Horsthemke, Z. Noszticzius, and H. L. Swinney. "Sustained Spiral Waves in a Continuously Fed Unstirred Chemical Reactor." *J. Chem. Phys.* **88** (1988): 3395–3396.
53. Tam, W. Y., J. A. Vastano, H. L. Swinney, and W. Horsthemke. "Regular and Chaotic Chemical Spatiotemporal Patterns." *Phys. Rev. Lett.* **61** (1988): 2163–2166.
54. Tam, W. Y. "Pattern Formation in Chemical Systems." In *Nonlinear Structures in Physical Systems*, edited by L. Lam and H. C. Morris. New York: Springer, 1990. (Figure 1 on p. 89 should read: Continuously Fed Unstirred Reactor (CFUR).)
55. Tam, W. Y., and H. L. Swinney. "Spatiotemporal Patterns in a One-Dimensional Open Reaction-Diffusion System." *Physica. D* **46** (1990): 10–22.
56. Thompson, J. M. T., and H. B. Stewart. *Nonlinear Dynamics and Chaos.* New York: Wiley, 1986.
57. Turner, J. S., J. C. Roux, W. D. McCormick, and H. L. Swinney. "Alternating Periodic and Chaotic Regimes in a Chemical Reaction-Experiment and Theory." *Phys. Lett. A* **85** (1981): 9–12.
58. Turing, A. M. "The Chemical Basis of Morphogenesis." *Phil. Trans. R. Soc.* **B237** (1952): 37–72.
59. Tyson, J. J., and P. C. Fife. "Target Patterns in a Realistic Model of the Belousov-Zhabotinskii Reaction." *J. Chem. Phys.* **73** (1980): 2224–2236.
60. Vastano, J. A. "Bifurcation in Spatiotemporal Systems." Ph.D. thesis, The University of Texas at Austin, 1988.
61. Vastano, J. A., T. Russo, and H. L. Swinney. "Bifurcation to Spatiotemporal Chaos in a Reaction-Diffusion System." *Physica D* **40** (1990): 10–22.
62. Welsh, B. J., J. Gomatam, and A. E. Burgess. "Three-Dimensional Chemical Waves in the Belousov-Zhabotinskii Reaction." *Nature* **304** (1983): 611–614.
63. Winfree, A. T. "Organizing Centers for Chemical Waves in Two and Three Dimensions." In *Oscillation and Traveling Wave in Chemical Systems*, edited by R. J. Field and M. Burger, 441–472. New York: Wiley, 1985.
64. Winfree, A. T. "Spiral Waves of Chemical Activity." *Science* **175** (1972): 634–636.
65. Winfree, A. T. "Rotating Chemical Reactions." *Sci. Am.* **230** (1974): 82–95.
66. Winfree, A. T., and S. H. Strogatz. "Organizing Centers for Three-Dimensional Chemical Waves." *Nature* **311** (1984): 611–615.
67. Winfree, A. T. "The Pre-history of the Belousov-Zhabotinskii Reaction." *J. Chem. Educ.* **61** (1984): 661–663.
68. Zaikin, A. N., and A. M. Zhabotinskii. "Concentration Wave Propagation in Two-Dimensional Liquid-Phase Self-Oscillating System." *Nature* **225** (1970): 535–537.
69. Zhabotinskii, A. M. "Periodic Processes of the Oxidation of Malonic Acid in Solution (Study of the Kinetics of Belousov's Reaction)." *Biofizika* **9** (1964): 306–311.

Neil A. Gershenfeld† and Andreas S. Weigend‡
†MIT Media Laboratory, 20 Ames Street, Cambridge, MA 02139;
e-mail: neilg@media.mit.edu.
‡ Department of Information Systems, Stern School of Business, New York University, 44 W 4th Street, New York, NY 10012;
e-mail: aweigend@stern.nyu.edu.

The Future of Time Series: Learning and Understanding

This chapter originally appeared in *Time Series Prediction: Forecasting the Future and Understanding the Past*, edited by A. S. Weigend and N. A. Gershenfeld, 1–70. Santa Fe Institute Studies in the Sciences of Complexity, Proc. Vol. XV. Reading, MA: Addison-Wesley, 1994. Reprinted by permission.

Throughout scientific research, measured time series are the basis for characterizing an observed system and for predicting its future behavior. A number of new techniques (such as state-space reconstruction and neural networks) promise insights that traditional approaches to these very old problems cannot provide. In practice, however, the application of such new techniques has been hampered by the unreliability of their results and by the difficulty of relating their performance to those of mature algorithms. This chapter reports on a competition run through the Santa Fe Institute in which participants from a range of relevant disciplines applied a variety of time series analysis tools to a small group of common data sets in order to help make meaningful comparisons among their approaches. The design and the results of this competition are described, and the historical

and theoretical backgrounds necessary to understand the successful entries
are reviewed.

1. INTRODUCTION

The desire to predict the future and understand the past drives the search for laws
that explain the behavior of observed phenomena; examples range from the irregu-
larity in a heartbeat to the volatility of a currency exchange rate. If there are known
underlying deterministic equations, in principle they can be solved to forecast the
outcome of an experiment based on knowledge of the initial conditions. To make
a forecast if the equations are not known, one must find both the rules governing
system evolution and the actual state of the system. In this chapter we will focus on
phenomena for which underlying equations are not given; the rules that govern the
evolution must be inferred from regularities in the past. For example, the motion
of a pendulum or the rhythm of the seasons carry within them the potential for
predicting their future behavior from knowledge of their oscillations without requiring insight into the underlying mechanism. We will use the terms "understanding"
and "learning" to refer to two complementary approaches taken to analyze an unfamiliar time series. *Understanding* is based on explicit mathematical insight into
how systems behave, and *learning* is based on algorithms that can emulate the
structure in a time series. In both cases, the goal is to explain observations; we will
not consider the important related problem of using knowledge about a system for
controlling it in order to produce some desired behavior.

Time series analysis has three goals: forecasting, modeling, and characterization. The aim of *forecasting* (also called *predicting*) is to accurately predict the
short-term evolution of the system; the goal of *modeling* is to find a description that
accurately captures features of the long-term behavior of the system. These are not
necessarily identical: finding governing equations with proper long-term properties
may not be the most reliable way to determine parameters for good short-term
forecasts, and a model that is useful for short-term forecasts may have incorrect
long-term properties. The third goal, system *characterization*, attempts with little
or no *a priori* knowledge to determine fundamental properties, such as the number
of degrees of freedom of a system or the amount of randomness. This overlaps with
forecasting but can differ: the complexity of a model useful for forecasting may not
be related to the actual complexity of the system.

Before the 1920s, forecasting was done by simply extrapolating the series
through a global fit in the time domain. The beginning of "modern" time series
prediction might be set at 1927 when Yule[176] invented the autoregressive technique
in order to predict the annual number of sunspots. His model predicted the next
value as a weighted sum of previous observations of the series. In order to obtain
"interesting" behavior from such a linear system, outside intervention in the form of
external shocks must be assumed. For the half-century following Yule, the reigning
paradigm remained that of linear models driven by noise.

However, there are simple cases for which this paradigm is inadequate. For example, a simple iterated map, such as the logistic equation (Eq. (11), in Section 3.2), can generate a broadband power spectrum that cannot be obtained by a linear approximation. The realization that apparently complicated time series can be generated by very simple equations pointed to the need for a more general theoretical framework for time series analysis and prediction.

Two crucial developments occurred around 1980; both were enabled by the general availability of powerful computers that permitted much longer time series to be recorded, more complex algorithms to be applied to them, and the data and the results of these algorithms to be interactively visualized. The first development, state-space reconstruction by time-delay embedding, drew on ideas from differential topology and dynamical systems to provide a technique for recognizing when a time series has been generated by deterministic governing equations and, if so, for understanding the geometrical structure underlying the observed behavior. The second development was the emergence of the field of machine learning, typified by neural networks, that can adaptively explore a large space of potential models. With the shift in artificial intelligence from rule-based methods toward data-driven methods,[1] the field was ready to apply itself to time series, and time series, now recorded with orders of magnitude more data points than were available previously, were ready to be analyzed with machine-learning techniques requiring relatively large data sets.

The realization of the promise of these two approaches has been hampered by the lack of a general framework for the evaluation of progress. Because time series problems arise in so many disciplines, and because it is much easier to describe an algorithm than to evaluate its accuracy and its relationship to mature techniques, the literature in these areas has become fragmented and somewhat anecdotal. The breadth (and the range in reliability) of relevant material makes it difficult for new research to build on the accumulated insight of past experience (researchers standing on each other's toes rather than shoulders).

Global computer networks now offer a mechanism for the disjoint communities to attack common problems through the widespread exchange of data and information. In order to foster this process and to help clarify the current state of time series analysis, we organized the Santa Fe Time Series Prediction and Analysis Competition under the auspices of the Santa Fe Institute during the fall of 1991. The goal was not to pick "winners" and "losers," but rather to provide a structure for researchers from the many relevant disciplines to compare quantitatively the results of their analyses of a group of data sets selected to span the range of studied problems. To explore the results of the competition, a NATO Advanced Research Workshop was held in the spring of 1992; workshop participants included members of the competition advisory board, representatives of the groups that had collected the data, participants in the competition, and interested observers. Although the

---

[1]Data sets of hundreds of megabytes are routinely analyzed with massively parallel supercomputers, using parallel algorithms to find near neighbors in multidimensional spaces.[7,154]

participants came from a broad range of disciplines, the discussions were framed by the analysis of common data sets and it was (usually) possible to find a meaningful common ground. In this overview chapter we describe the structure and the results of this competition and review the theoretical material required to understand the successful entries; much more detail is available in the articles by the participants in this volume.

2. THE COMPETITION

The planning for the competition emerged from informal discussions at the Complex Systems Summer School at the Santa Fe Institute in the summer of 1990; the first step was to assemble an advisory board to represent the interests of many of the relevant fields.[2] With the help of this group we gathered roughly 200 megabytes of experimental time series for possible use in the competition. This volume of data reflects the growth of techniques that use enormous data sets (where automatic collection and processing is essential) over traditional time series (such as quarterly economic indicators, where it is possible to develop an intimate relationship with each data point).

In order to be widely accessible, the data needed to be distributed by ftp over the Internet, by electronic mail, and by floppy disks for people without network access. The latter distribution channels limited the size of the competition data to a few megabytes; the final data sets were chosen to span as many of a desired group of attributes as possible given this size limitation (the attributes are shown in Figure 2). The final selection was:

A. **A clean physics laboratory experiment.** 1,000 points of the fluctuations in a far-infrared laser, approximately described by three coupled nonlinear ordinary differential equations (Hübner et al.[73])

B. **Physiological data from a patient with sleep apnea.** 34,000 points of the heart rate, chest volume, blood oxygen concentration, and EEG state of a sleeping patient. These observables interact, but the underlying regulatory mechanism is not well understood (Rigney et al.[123]).

C. **High-frequency currency exchange rate data.** Ten segments of 3,000 points each of the exchange rate between the Swiss franc and the U.S. dollar. The average time between two quotes is between one and two minutes (Lequarré[89]). If the market was efficient, such data should be a random walk.

---

[2]The advisors were Leon Glass (biology), Clive Granger (economics), Bill Press (astrophysics and numerical analysis), Maurice Priestley (statistics), Itamar Procaccia (dynamical systems), T. Subba Rao (statistics), and Harry Swinney (experimental physics).

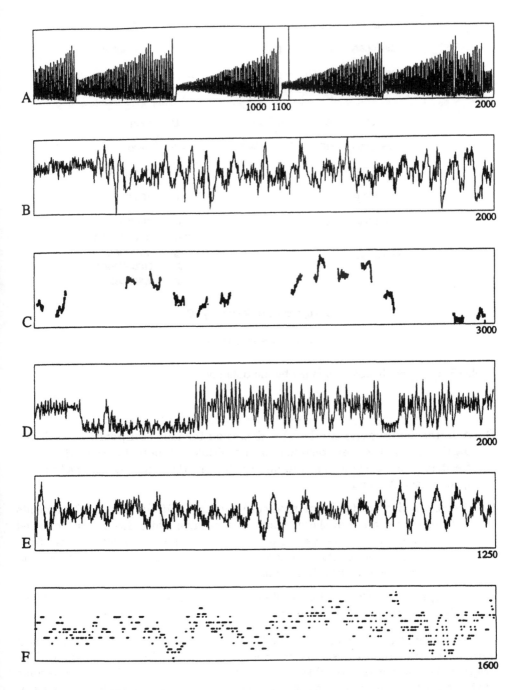

FIGURE 1   Sections of the competition data sets.

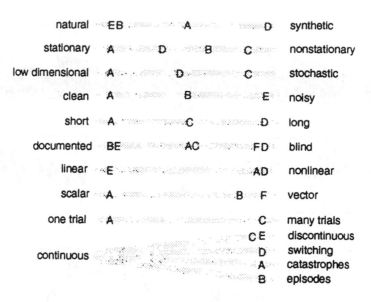

FIGURE 2 Some attributes spanned by the data sets.

D. **A numerically generated series designed for this competition.** A driven particle in a four-dimensional nonlinear multiple-well potential (nine degrees of freedom) with a small nonstationarity drift in the well depths. (Details are given in the Appendix.)

E. **Astrophysical data from a variable star.** 27,704 points in 17 segments of the time variation of the intensity of a variable white dwarf star, collected by the *Whole Earth Telescope* (Clemens[25]). The intensity variation arises from a superposition of relatively independent spherical harmonic multiplets, and there is significant observational noise.

F. **A fugue.** J. S. Bach's final (unfinished) fugue from *The Art of the Fugue*, added after the close of the formal competition (Dirst & Weigend[34]).

The amount of information available to the entrants about the origin of each data set varied from extensive (Data Sets B and E) to blind (Data Set D). The original files will remain available. The data sets are graphed in Figure 1, and some of the characteristics are summarized in Figure 2. The appropriate level of description for models of these data ranges from low-dimensional stationary dynamics to stochastic processes.

After selecting the data sets, we next chose **competition tasks** appropriate to the data sets and research interests. The participants were asked to:

- predict the (withheld) continuations of the data sets with respect to given error measures,
- characterize the systems (including aspects such as the number of degrees of freedom, predictability, noise characteristics, and the nonlinearity of the system),
- infer a model of the governing equations, and
- describe the algorithms employed.

The data sets and competition tasks were made publicly available on August 1, 1991, and competition entries were accepted until January 15, 1992. Participants were required to describe their algorithms. (Insight in some previous competitions was hampered by the acceptance of proprietary techniques.) One interesting trend in the entries was the focus on prediction, for which three motivations were given: (i) because predictions are falsifiable, insight into a model used for prediction is verifiable; (ii) there are a variety of financial incentives to study prediction; and (iii) the growth of interest in machine learning brings with it the hope that there can be universally and easily applicable algorithms that can be used to generate forecasts. Another trend was the general failure of simplistic "black-box" approaches—in all successful entries, exploratory data analysis preceded the algorithm application.[3]

It is interesting to compare this time series competition to the previous state of the art as reflected in two earlier competitions.[95],[94] In these, a very large number of time series was provided (111 and 1001, respectively), taken from business (forecasting sales), economics (predicting recovery from the recession), finance, and the social sciences. However, all of the series used were very short, generally less than 100 values long. Most of the algorithms entered were fully automated, and most of the discussion centered on linear models.[4] In the Santa Fe Competition all of the successful entries were fundamentally nonlinear and, even though significantly more computer power was used to analyze the larger data sets with more complex models, the application of the algorithms required more careful manual control than in the past.

---

[3]The data, analysis programs, and summaries of the results are available by anonymous ftp from ftp.santafe.edu, as described in the Appendix to this volume. In the competition period, on average 5 to 10 people retrieved the data per day, and 30 groups submitted final entries by the deadline. Entries came from the U.S., Europe (including former communist countries), and Asia, ranging from junior graduate students to senior researchers.

[4]These discussions focused on issues such as the order of the linear model. Chatfield[23] summarizes previous competitions.

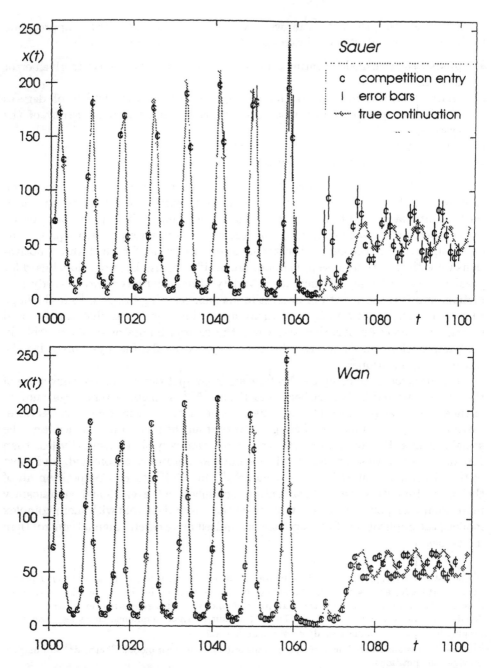

FIGURE 3 The two best predicted continuations for Data Set A, by Sauer and by Wan. Predicted values are indicated by "c," predicted error bars by vertical lines. The true continuation (not available at the time when the predictions were received) is shown in grey (the points are connected to guide the eye).

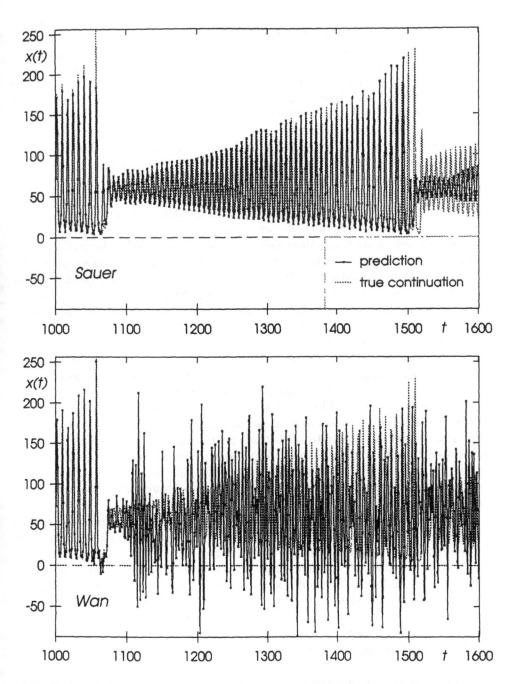

FIGURE 4 Predictions obtained by the same two models as in the previous figure, but continued 500 points further into the future. The solid line connects the predicted points; the grey line indicates the true continuation.

As an example of the results, consider the intensity of the laser (Data Set A; see Figure 1). On the one hand, the laser can be described by a relatively simple "correct" model of three nonlinear differential equations, the same equations that Lorenz[92] used to approximate weather phenomena. On the other hand, since the 1,000-point training set showed only three of four collapses, it is difficult to predict the next collapse based on so few instances.

For this data set we asked for predictions of the next 100 points as well as estimates of the error bars associated with these predictions. We used two measures to evaluate the submissions. The first measure (normalized mean squared error) was based on the predicted values only; the second measure used the submitted error predictions to compute the likelihood of the observed data given the predictions. The Appendix to this chapter gives the definitions and explanations of the error measures as well as a table of all entries received. We would like to point out a few interesting features. Although this single trial does not permit fine distinctions to be made between techniques with comparable performance, two techniques clearly did much better than the others for Data Set A; one used state-space reconstruction to build an explicit model for the dynamics and the other used a connectionist network (also called a neural network). Incidentally, a prediction based solely on visually examining and extrapolating the training data did much worse than the best techniques, but also much better than the worst.

Figure 3 shows the two best predictions. Sauer[135] attempts to understand and develop a *representation* for the geometry in the system's state space, which is the best that can be done without knowing something about the system's governing equations, while Wan[163] addresses the issue of *function approximation* by using a connectionist network to learn to emulate the input-output behavior. Both methods generated remarkably accurate predictions for the specified task. In terms of the measures defined for the competition, Wan's squared errors are one-third as as large as Sauer's, and—taking the predicted uncertainty into account—Wan's model is four times more likely than Sauer's.[5] According to the competition scores for Data Set A, this puts Wan's network in the first place.

A different picture, which cautions the hurried researcher against declaring one method to be universally superior to another, emerges when one examines the evolution of these two prediction methods further into the future. Figure 4 shows the same two predictors, but now the continuations extend 500 points beyond the 100 points submitted for the competition entry (no error estimates are shown).[6] The neural network's class of potential behavior is much broader than what can be generated from a small set of coupled ordinary differential equations, but the state-space model is able to reliably forecast the data much further because its explicit description can correctly capture the character of the long-term dynamics.

[5] The likelihood ratio can be obtained from Table 2 in the Appendix[135] as $\exp(-3.5)/\exp(-4.8)$.

[6] Furthermore, we invite the reader to compare Figure 5 by Sauer[135] (p. 191) with Figure 13 by Wan[163] (p. 213). Both entrants start the competition model at the same four (new) different points. The squared errors are compared in Table 2 (Sauer[135]).

In order to understand the details of these approaches, we will detour to review the framework for (and then the failure of) linear time series analysis.

---

3. LINEAR TIME SERIES MODELS

Linear time series models have two particularly desirable features: they can be understood in great detail and they are straightforward to implement. The penalty for this convenience is that they may be entirely inappropriate for even moderately complicated systems. In this section we will review their basic features and then consider why and how such models fail. The literature on linear time series analysis is vast; a good introduction is the very readable book by Chatfield,[24] many derivations can be found (and understood) in the comprehensive text by Priestley,[120] and a classic reference is Box and Jenkins' book.[8] Historically, the general theory of linear predictors can be traced back to Kolmogorov[78] and to Wiener.[174]

Two crucial assumptions will be made in this section: the system is assumed to be linear and stationary. In the rest of this chapter we will say a great deal about relaxing the assumption of linearity; much less is known about models that have coefficients that vary with time. To be precise, unless explicitly stated (such as for Data Set D), we assume that the underlying equations do not change in time, i.e., *time invariance* of the system.

3.1 ARMA, FIR, AND ALL THAT

There are two complementary tasks that need to be discussed: understanding how a given model behaves and finding a particular model that is appropriate for a given time series. We start with the former task. It is simplest to discuss separately the role of external inputs (moving average models) and internal memory (autoregressive models).

3.1.1 PROPERTIES OF A GIVEN LINEAR MODEL

Moving average (MA) models.   Assume we are given an external input series $\{e_t\}$ and want to modify it to produce another series $\{x_t\}$. Assuming linearity of the system and causality (the present value of x is influenced by the present and N past values of the input series e), the relationship between the input and output is

$$x_t = \sum_{n=0}^{N} b_n e_{t-n} = b_0 e_t + b_1 e_{t-1} + \cdots + b_N e_{t-N} \, . \tag{1}$$

This equation describes a convolution filter: the new series x is generated by an Nth-order filter with coefficients $b_0, \cdots, b_n$ from the series e. Statisticians and econometricians call this an Nth-order *moving average* model, MA(N). The origin of this (sometimes confusing) terminology can be seen if one pictures a simple smoothing filter which averages the last few values of series e. Engineers call this a *finite impulse response* (FIR) filter, because the output is guaranteed to go to zero at N time steps after the input becomes zero.

Properties of the output series x clearly depend on the input series e. The question is whether there are characteristic features independent of a specific input sequence. For a linear system, the response of the filter is independent of the input. A characterization focuses on properties of the system, rather than on properties of the time series. (For example, it does not make sense to attribute linearity to a time series itself, only to a system.)

We will give three equivalent characterizations of an MA model: in the time domain (the impulse response of the filter), in the frequency domain (its spectrum), and in terms of its autocorrelation coefficients. In the first case, we assume that the input is nonzero only at a single time step t_0 and that it vanishes for all other times. The response (in the time domain) to this "impulse" is simply given by the b's in Eq. (1): at each time step the impulse moves up to the next coefficient until, after N steps, the output disappears. The series $b_N, b_{N-1}, \cdots, b_0$ is thus the impulse response of the system. The response to an arbitrary input can be computed by superimposing the responses at appropriate delays, weighted by the respective input values ("convolution"). The transfer function thus completely describes a linear system, i.e., a system where the superposition principle holds: the output is determined by impulse response and input.

Sometimes it is more convenient to describe the filter in the frequency domain. This is useful (and simple) because a convolution in the time domain becomes a product in the frequency domain. If the input to a MA model is an impulse (which has a flat power spectrum), the discrete Fourier transform of the output is given by $\sum_{n=0}^{N} b_n \exp(-i2\pi nf)$ (see, for example, Box & Jenkins,[8] p.69). The power spectrum is given by the squared magnitude of this:

$$\left| 1 + b_1 e^{-i2\pi 1 f} + b_2 e^{-i2\pi 2 f} + \cdots + b_N e^{-i2\pi N f} \right|^2 . \tag{2}$$

The third way of representing yet again the same information is, in terms of the autocorrelation coefficients, defined in terms of the mean $\mu = \langle x_t \rangle$ and the variance $\sigma^2 = \langle (x_t - \mu)^2 \rangle$ by

$$\rho_\tau \equiv \frac{1}{\sigma^2} \langle (x_t - \mu)(x_{t-\tau} - \mu) \rangle . \tag{3}$$

The angular brackets $\langle \cdot \rangle$ denote expectation values, in the statistics literature often indicated by $E\{\cdot\}$. The autocorrelation coefficients describe how much, on average, two values of a series that are τ time steps apart co-vary with each other. (We will later replace this linear measure with mutual information, suited also to describe nonlinear relations.) If the input to the system is a stochastic process with input values at different times uncorrelated, $\langle e_i e_j \rangle = 0$ for $i \neq j$, then all of the cross terms will disappear from the expectation value in Eq. (3), and the resulting autocorrelation coefficients are

$$
\rho_\tau = \begin{cases} \dfrac{1}{\sum_{n=0}^{N} b_n^2} \displaystyle\sum_{n=\tau}^{N} b_n b_{n-|\tau|} & |\tau| \le N \ , \\ 0 & |\tau| > N \ . \end{cases} \tag{4}
$$

Autoregressive (AR) models. MA (or FIR) filters operate in an open loop without feedback; they can only transform an input that is applied to them. If we do not want to drive the series externally, we need to provide some feedback (or memory) in order to generate internal dynamics:

$$
x_t = \sum_{m=1}^{M} a_m x_{t-m} + e_t \ . \tag{5}
$$

This is called an Mth-order *autoregressive model* (AR(M)) or an *infinite impulse response* (IIR) filter (because the output can continue after the input ceases). Depending on the application, e_t can represent either a controlled input to the system or noise. As before, if e is white noise, the autocorrelations of the output series x can be expressed in terms of the model coefficients. Here, however—due to the feedback coupling of previous steps—we obtain a set of linear equations rather than just a single equation for each autocorrelation coefficient. By multiplying Eq. (5) by $x_{t-\tau}$, taking expectation values, and normalizing (see Box & Jenkins,[8] p.54), the autocorrelation coefficients of an AR model are found by solving this set of linear equations, traditionally called the *Yule-Walker equations*,

$$
\rho_\tau = \sum_{m=1}^{M} a_m \rho_{\tau-m}, \qquad \tau > 0 . \tag{6}
$$

Unlike the MA case, the autocorrelation coefficient need not vanish after M steps. Taking the Fourier transform of both sides of Eq. (5) and rearranging terms shows that the output equals the input times $(1 - \sum_{m=1}^{M} a_m \exp(-i2\pi m f))^{-1}$. The power spectrum of output is thus that of the input times

$$
\frac{1}{|1 - a_1 e^{-i2\pi 1 f} - a_2 e^{-i2\pi 2 f} - \cdots - a_M e^{-i2\pi M f}|^2} \ . \tag{7}
$$

To generate a specific realization of the series, we must specify the initial conditions, usually by the first M values of series x. Beyond that, the input term e_t is crucial for the life of an AR model. If there was no input, we might be disappointed by the series we get: depending on the amount of feedback, after iterating it for a while, the output produced can only decay to zero, diverge, or oscillate periodically.[7]

Clearly, the next step in complexity is to allow both AR and MA parts in the model; this is called an ARMA(M, N) model:

$$x_t = \sum_{m=1}^{M} a_m x_{t-m} + \sum_{n=0}^{N} b_n e_{t-n} \,. \tag{8}$$

Its output is most easily understood in terms of the *z-transform* (Oppenheim & Schafer[107]), which generalizes the discrete Fourier transform to the complex plane:

$$X(z) \equiv \sum_{t=-\infty}^{\infty} x_t z^t \,. \tag{9}$$

On the unit circle, $z = \exp(-i2\pi f)$, the z-transform reduces to the discrete Fourier transform. Off the unit circle, the z-transform measures the rate of divergence or convergence of a series. Since the convolution of two series in the time domain corresponds to the multiplication of their z-transforms, the z-transform of the output of an ARMA model is

$$\begin{aligned} X(z) &= A(z)X(z) + B(z)E(z) \\ &= \frac{B(z)}{1 - A(z)}\, E(z) \end{aligned} \tag{10}$$

(ignoring a term that depends on the initial conditions). The input z-transform $E(z)$ is multiplied by a transfer function that is unrelated to it; the transfer function will vanish at zeros of the MA term ($B(z) = 0$) and diverge at poles ($A(z) = 1$) due to the AR term (unless cancelled by a zero in the numerator). As $A(z)$ is an Mth-order complex polynomial, and $B(z)$ is Nth-order, there will be M poles and N zeros. Therefore, the z-transform of a time series produced by Eq. (8) can be decomposed into a rational function and a remaining (possibly continuous) part due to the input. The number of poles and zeros determines the number of *degrees of freedom* of the system (the number of previous states that the dynamics retains). Note that since only the ratio enters, there is no unique ARMA model. In the extreme cases, a finite-order AR model can always be expressed by an infinite-order MA model, and vice versa.

ARMA models have dominated all areas of time series analysis and discrete-time signal processing for more than half a century. For example, in speech recognition and synthesis, Linear Predictive Coding (Press et al.,[119] p.571) compresses

---

[7] In the case of a first-order AR model, this can easily be seen: if the absolute value of the coefficient is less than unity, the value of x exponentially decays to zero; if it is larger than unity, it exponentially explodes. For higher-order AR models, the long-term behavior is determined by the locations of the zeroes of the polynomial with coefficients a_i.

speech by transmitting the slowly varying coefficients for a linear model (and possibly the remaining error between the linear forecast and the desired signal) rather than the original signal. If the model is good, it transforms the signal into a small number of coefficients plus residual white noise (of one kind or another).

3.1.2 FITTING A LINEAR MODEL TO A GIVEN TIME SERIES

Fitting the coefficients. The Yule-Walker set of linear equations (Eq. (6)) allowed us to express the autocorrelation coefficients of a time series in terms of the AR coefficients that generated it. But there is a second reading of the same equations: they also allow us to estimate the coefficients of an $AR(M)$ model from the observed correlational structure of an observed signal.[8] An alternative approach views the estimation of the coefficients as a regression problem: expressing the next value as a function of M previous values, i.e., linearly regress x_{t+1} onto $\{x_t, x_{t-1}, \ldots, x_{t-(M-1)}\}$. This can be done by minimizing squared errors: the parameters are determined such that the squared difference between the model output and the observed value, summed over all time steps in the fitting region, is as small as possible. There is no comparable conceptually simple expression for finding MA and full ARMA coefficients from observed data. For all cases, however, standard techniques exist, often expressed as efficient recursive procedures (Box & Jenkins[8]; Press et al.[119]).

Although there is no reason to expect that an arbitrary signal was produced by a system that can be written in the form of Eq. (8), it is reasonable to attempt to approximate a linear system's true transfer function (z-transform) by a ratio of polynomials, i.e., an ARMA model. This is a problem in function approximation, and it is well known that a suitable sequence of ratios of polynomials (called Padé approximants; see Press et al.,[119] p.200) converges faster than a power series for arbitrary functions.

Selecting the (order of the) model. So far we have dealt with the question of how to estimate the coefficients from data for an ARMA model of order (M, N), but have not addressed the choice for the order of the model. There is not a unique best choice for the values or even for the number of coefficients to model a data set—as the order of the model is increased, the fitting error decreases, but the test error of the forecasts beyond the training set will usually start to increase at some point because the model will be fitting extraneous noise in the system. There are several heuristics to find the "right" order (such as the Akaike Information Criterion (AIC), Akaike[3]; Sakomoto et al.[132])—but these heuristics rely heavily on the linearity of the model and on assumptions about the distribution from which the errors are drawn. When it is not clear whether these assumptions hold, a simple approach

---

[8] In statistics, it is common to emphasize the difference between a given model and an estimated model by using different symbols, such as $\hat{a}$ for the estimated coefficients of an AR model. In this paper, we avoid introducing another set of symbols; we hope that it is clear from the context whether values are theoretical or estimated.

(but wasteful in terms of the data) is to hold back some of the training data and use these to evaluate the performance of competing models. Model selection is a general problem that will reappear even more forcefully in the context of nonlinear models, because they are more flexible and, hence, more capable of modeling irrelevant noise.

3.2 THE BREAKDOWN OF LINEAR MODELS

We have seen that ARMA coefficients, power spectra, and autocorrelation coefficients contain the same information about a linear system that is driven by uncorrelated white noise. Thus, *if and only if* the power spectrum is a useful characterization of the relevant features of a time series, an ARMA model will be a good choice for describing it. This appealing simplicity can fail entirely for even simple nonlinearities if they lead to complicated power spectra (as they can). Two time series can have very similar broadband spectra but can be generated from systems with very different properties, such as a linear system that is driven stochastically by external noise, and a deterministic (noise-free) nonlinear system with a small number of degrees of freedom. One of the key problems addressed in this chapter is how these cases can be distinguished—linear operators definitely will not be able to do the job.

Let us consider two nonlinear examples of discrete-time maps (like an AR model, but now nonlinear):

■ The first example can be traced back to Ulam[160]: the next value of a series is derived from the present one by a simple parabola

$$x_{t+1} = \lambda\, x_t\, (1 - x_t)\,. \tag{11}$$

Popularized in the context of population dynamics as an example of a "simple mathematical model with very complicated dynamics,"[97] it has been found to describe a number of controlled laboratory systems such as hydrodynamic flows and chemical reactions, because of the universality of smooth unimodal maps.[26]) In this context, this parabola is called the *logistic map* or *quadratic map*. The value x_t deterministically depends on the previous value x_{t-1}; λ is a parameter that controls the qualitative behavior, ranging from a fixed point (for small values of λ) to deterministic chaos. For example, for $\lambda = 4$, each iteration destroys one bit of information. Consider that, by plotting x_t against x_{t-1}, each value of x_t has two equally likely predecessors or, equally well, the average slope (its absolute value) is two: if we know the location within ϵ before the iteration, we will on average know it within 2ϵ afterward. This exponential increase in uncertainty is the hallmark of deterministic chaos ("divergence of nearby trajectories").

■ The second example is equally simple: consider the time series generated by the map

$$x_t = 2x_{t-1} \ (\text{mod } 1). \qquad (12)$$

The action of this map is easily understood by considering the position x_t written in a binary fractional expansion (i.e., $x_t = 0.d_1 d_2 \ldots = (d_1 \times 2^{-1}) + (d_2 \times 2^{-2}) + \ldots$): each iteration shifts every digit one place to the left ($d_i \leftarrow d_{i+1}$). This means that the most significant digit d_1 is discarded and one more digit of the binary expansion of the initial condition is revealed. This map can be implemented in a simple physical system consisting of a classical billiard ball and reflecting surfaces, where the x_t are the successive positions at which the ball crosses a given line.[103])

Both systems are completely deterministic (their evolutions are entirely determined by the initial condition x_0), yet they can easily generate time series with broadband power spectra. In the context of an ARMA model a broadband component in a power spectrum of the output must come from external noise input to the system, but here it arises in two one-dimensional systems as simple as a parabola and two straight lines. Nonlinearities are essential for the production of "interesting" behavior in a deterministic system, the point here is that even simple nonlinearities suffice.

Historically, an important step beyond linear models for prediction was taken in 1980 by Tong and Lim[155] (see also Tong[156]). After more than five decades of approximating a system with *one* globally linear function, they suggested the use of *two* functions. This *threshold autoregressive model* (TAR) is globally nonlinear: it consists of choosing one of two local linear autoregressive models based on the value of the system's state. From here, the next step is to use many local linear models; however, the number of such regions that must be chosen may be very large if the system has even quadratic nonlinearities (such as the logistic map). A natural extension of Eq. (8) for handling this is to include quadratic and higher order powers in the model; this is called a Volterra series.[161]

TAR models, Volterra models, and their extensions significantly expand the scope of possible functional relationships for modeling time series, but these come at the expense of the simplicity with which linear models can be understood and fit to data. For nonlinear models to be useful, there must be a process that exploits features of the data to guide (and restrict) the construction of the model; lack of insight into this problem has limited the use of nonlinear time series models. In the next sections we will look at two complementary solutions to this problem: building explicit models with state-space reconstruction, and developing implicit models in a connectionist framework. To understand why both of these approaches exist and why they are useful, let us consider the nature of scientific modeling.

4. UNDERSTANDING AND LEARNING

Strong models have strong assumptions. They are usually expressed in a few equations with a few parameters, and can often explain a plethora of phenomena. In weak models, on the other hand, there are only a few domain-specific assumptions. To compensate for the lack of explicit knowledge, weak models usually contain many more parameters (which can make a clear interpretation difficult). It can be helpful to conceptualize models in the two-dimensional space spanned by the axes data-poor↔data-rich and theory-poor↔theory-rich. Due to the dramatic expansion of the capability for automatic data acquisition and processing, it is increasingly feasible to venture into the theory-poor and data-rich domain.

Strong models are clearly preferable, but they often originate in weak models. (However, if the behavior of an observed system does not arise from simple rules, they may not be appropriate.) Consider planetary motion.[54] Tycho Brahe's (1546–1601) experimental observations of planetary motion were accurately described by Johannes Kepler's (1571–1630) phenomenological laws; this success helped lead to Isaac Newton's (1642–1727) simpler but much more general theory of gravity which could derive these laws; Henri Poincaré's (1854–1912) inability to solve the resulting three-body gravitational problem helped lead to the modern theory of dynamical systems and, ultimately, to the identification of chaotic planetary motion.[145,146]

As in the previous section on linear systems, there are two complementary tasks: discovering the properties of a time series generated from a given model, and inferring a model from observed data. We focus here on the latter, but there has been comparable progress for the former. Exploring the behavior of a model has become feasible in interactive computer environments, such as Cornell's `dstool`,[9] and the combination of traditional numerical algorithms with algebraic, geometric, symbolic, and artificial intelligence techniques is leading to automated platforms for exploring dynamics.[1,9,175] For a nonlinear system, it is no longer possible to decompose an output into an input signal and an independent transfer function (and thereby find the correct input signal to produce a desired output), but there are adaptive techniques for controlling nonlinear systems[72,108] that make use of techniques similar to the modeling methods that we will describe.

The idea of weak modeling (data-rich and theory-poor) is by no means new—an ARMA model is a good example. What is new is the emergence of weak models (such as neural networks) that combine broad generality with insight into how to manage their complexity. For such models with broad approximation abilities and few specific assumptions, the distinction between memorization and generalization becomes important. Whereas the signal-processing community sometimes uses the term *learning* for any adaptation of parameters, we need to contrast learning without generalization to learning with generalization. Let us consider the widely

[9]Available by anonymous ftp from `macomb.tn.cornell.edu` in `pub/dstool`.

and wildly celebrated fact that neural networks can learn to implement the exclusive OR (XOR). But—what kind of learning is this? When four out of four cases are specified, no generalization exists! Learning a truth table is nothing but rote memorization: learning XOR is as interesting as memorizing the phone book. More interesting—and more realistic—are real-world problems, such as the prediction of financial data. In forecasting, nobody cares how well a model fits the training data— only the quality of future predictions counts, i.e., the performance on novel data or the *generalization* ability. Learning means extracting regularities from training examples that do transfer to new examples.

Learning procedures are, in essence, statistical devices for performing inductive inference. There is a tension between two goals. The immediate goal is to fit the training examples, suggesting devices as general as possible so that they can learn a broad range of problems. In connectionism, this suggests large and flexible networks, since networks that are too small might not have the complexity needed to model the data. The ultimate goal of an inductive device is, however, its performance on cases it has not yet seen, i.e., the quality of its predictions outside the training set. This suggests—at least for noisy training data—networks that are not too large since networks with too many high-precision weights will pick out idiosyncrasies of the training set and will not generalize well.

An instructive example is polynomial curve fitting in the presence of noise. On the one hand, a polynomial of too low an order cannot capture the structure present in the data. On the other hand, a polynomial of too high an order, going through all of the training points and merely interpolating between them, captures the noise as well as the signal and is likely to be a very poor predictor for new cases. This problem of fitting the noise in addition to the signal is called *overfitting*. By employing a regularizer (i.e., a term that penalizes the complexity of the model) it is often possible to fit the parameters and to select the relevant variables at the same time. Neural networks, for example, can be cast in such a Bayesian framework.[14]

To clearly separate memorization from generalization, the true continuation of the competition data was kept secret until the deadline, ensuring that the continuation data could not be used by the participants for tasks such as parameter estimation or model selection.[10] Successful forecasts of the withheld *test set* (also called *out-of-sample predictions*) from the provided *training set* (also called *fitting set*) were produced by two general classes of techniques: those based on state-space reconstruction (which make use of explicit understanding of the relationship between the internal degrees of freedom of a deterministic system and an observable of the system's state in order to build a model of the rules governing the measured behavior of the system), and connectionist modeling (which uses potentially rich models along with learning algorithms to develop an implicit model of the system). We will see that neither is uniquely preferable. The domains of applicability are

---

[10]After all, predictions are hard, particularly those concerning the future.

not the same, and the choice of which to use depends on the goals of the analysis (such as an understandable description vs. accurate short-term forecasts).

4.1 UNDERSTANDING: STATE-SPACE RECONSTRUCTION

Yule's original idea for forecasting was that future predictions can be improved by using immediately preceding values. An ARMA model, Eq. (8), can be rewritten as a dot product between vectors of the time-lagged variables and coefficients:

$$x_t = \mathbf{a} \cdot \mathbf{x}_{t-1} + \mathbf{b} \cdot \mathbf{e}_t, \tag{13}$$

where $\mathbf{x}_t = (x_t, x_{t-1}, \ldots, x_{t-(d-1)})$, and $\mathbf{a} = (a_1, a_2, \ldots, a_d)$. (We slightly change notation here: what was M (the order of the AR model) is now called d (for dimension).) Such lag vectors, also called tapped delay lines, are used routinely in the context of signal processing and time series analysis, suggesting that they are more than just a typographical convenience.[11]

In fact, there is a deep connection between time-lagged vectors and underlying dynamics. This connection was was proposed in 1980 by Ruelle,[127] Packard et al.,[109] and Takens[149] (he published the first proof), and later strengthened by Sauer et al.[134] Delay vectors of sufficient length are not just a representation of the state of a linear system—it turns out that delay vectors can recover the full geometrical structure of a nonlinear system. These results address the general problem of inferring the behavior of the intrinsic degrees of freedom of a system when a function of the state of the system is measured. If the governing equations and the functional form of the observable are known in advance, then a Kalman filter is the optimal linear estimator of the state of the system.[20,24] We, however, focus on the case where there is little or no *a priori* information available about the origin of the time series.

There are four relevant (and easily confused) spaces and dimensions for this discussion[12]:

1. The *configuration space* of a system is the space "where the equations live." It specifies the values of all of the potentially accessible physical degrees of freedom of the system. For example, for a fluid governed by the Navier-Stokes partial differential equations, these are the infinite-dimensional degrees of freedom associated with the continuous velocity, pressure, and temperature fields.

---

[11] For example, the spectral test for random number generators is based on looking for structure in the space of lagged vectors of the output of the source; these will lie on hyperplanes for a linear congruential generator $x_{t+1} = ax_t + b \pmod{c}$ (Knuth,[77] p.90).

[12] The first point (configuration space and potentially accessible degrees of freedom) will not be used again in this chapter. On the other hand, the dimension of the solution manifold (the actual degrees of freedom) will be important both for characterization and for prediction.

2. The *solution manifold* is where "the solution lives," i.e., the part of the configuration space that the system actually explores as its dynamics unfolds (such as the support of an attractor or an integral surface). Due to unexcited or correlated degrees of freedom, this can be much smaller than the configuration space; the dimension of the solution manifold is the number of parameters that are needed to uniquely specify a distinguishable state of the overall system. For example, in some regimes the infinite physical degrees of freedom of a convecting fluid reduce to a small set of coupled ordinary differential equations for a mode expansion.[92] Dimensionality reduction from the configuration space to the solution manifold is a common feature of dissipative systems: dissipation in a system will reduce its dynamics onto a lower dimensional subspace.[150]

3. The *observable* is a (usually) one-dimensional function of the variables of configuration, an example is Eq. (51) in the Appendix. In an experiment, this might be the temperature or a velocity component at a point in the fluid.

4. The *reconstructed state space* is obtained from that (scalar) observable by combining past values of it to form a lag vector (which for the convection case would aim to recover the evolution of the components of the mode expansion).

Given a time series measured from such a system—and no other information about the origin of the time series—the question is: What can be deduced about the underlying dynamics?

Let $\mathbf{y}$ be the state vector on the solution manifold (in the convection example the components of $\mathbf{y}$ are the magnitude of each of the relevant modes), let $d\mathbf{y}/dt = \mathbf{f}(\mathbf{y})$ be the governing equations, and let the measured quantity be $x_t = x(\mathbf{y}(t))$ (e.g., the temperature at a point). The results to be cited here also apply to systems that are described by iterated maps. Given a delay time τ and a dimension d, a lag vector $\mathbf{x}$ can be defined,

$$\text{lag vector}: \ \mathbf{x}_t = (x_t, x_{t-\tau}, \ldots, x_{t-(d-1)\tau}). \tag{14}$$

The central result is that the behavior of $\mathbf{x}$ and $\mathbf{y}$ will differ only by a smooth local invertible change of coordinates (i.e., the mapping between $\mathbf{x}$ and $\mathbf{y}$ is an embedding, which requires that it be diffeomorphic) for almost every possible choice of $\mathbf{f}(\mathbf{y}), x(\mathbf{y})$, and τ, as long as d is large enough (in a way that we will make precise), x depends on at least some of the components of $\mathbf{y}$, and the remaining components of $\mathbf{y}$ are coupled by the governing equations to the ones that influence $\mathbf{x}$. The proof of this result has two parts: a local piece, showing that the linearization of the embedding map is almost always nondegenerate, and a global part, showing

1000 points

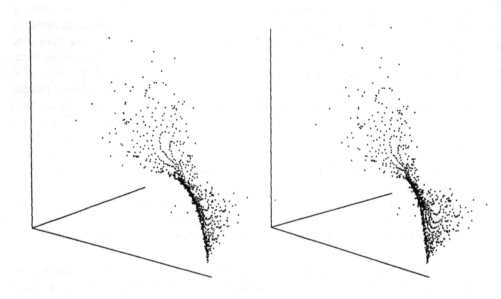

25000 points

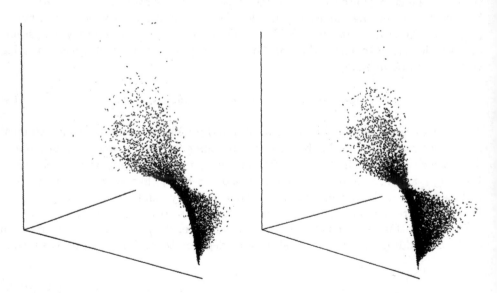

FIGURE 5 Stereo pairs for the three-dimensional embedding of Data Set A. The shape of the surface is apparent with just the 1,000 points that were given.

that this holds everywhere. If τ tends to zero, the embedding will tend to lie on the diagonal of the embedding space and, as τ is increased, it sets a length scale for the reconstructed dynamics. There can be degenerate choices for τ for which the embedding fails (such as choosing it to be exactly equal to the period of a periodic system), but these degeneracies almost always will be removed by an arbitrary perturbation of τ. The intrinsic noise in physical systems guarantees that these results hold in all known nontrivial examples, although in practice, if the coupling between degrees of freedom is sufficiently weak, then the available experimental resolution will not be large enough to detect them (see Casdagli et al.[16] for further discussion of how noise constrains embedding).[13]

Data Set A appears complicated when plotted as a time series (Figure 1). The simple structure of the system becomes visible in a figure of its three-dimensional embedding (Figure 5). In contrast, high-dimensional dynamics would show up as a structureless cloud in such a stereo plot. Simply plotting the data in a stereo plot allows one to guess a value of the dimension of the manifold of around two, not far from computed values of 2.0–2.2. In Section 6, we will discuss in detail the practical issues associated with choosing and understanding the embedding parameters.

Time-delay embedding differs from traditional experimental measurements in three fundamental respects:

1. It provides detailed information about the behavior of degrees of freedom other than the one that is directly measured.
2. It rests on probabilistic assumptions and—although it has been routinely and reliably used in practice—it is not guaranteed to be valid for any system.
3. It allows precise questions only about quantities that are invariant under such a transformation, since the reconstructed dynamics have been modified by an unknown smooth change of coordinates.

This last restriction may be unfamiliar, but it is surprisingly unimportant: we will show how embedded data can be used for forecasting a time series and for characterizing the essential features of the dynamics that produced it. We close this section by presenting two extensions of the simple embedding considered so far.

Filtered embedding generalizes simple time-delay embedding by presenting a linearly transformed version of the lag vector to the next processing stage. The lag vector $\mathbf{x}$ is trivially equal to itself times an identity matrix. Rather than using the identity matrix, the lag vector can be multiplied by any (not necessarily square) matrix. The resulting vector is an embedding if the rank of the matrix is equal to or larger than the desired embedding dimension. (The window of lags can be larger than the final embedding dimension, which allows the embedding procedure

---

[13]The Whitney embedding theorem from the 1930s (see Guillemin & Pollack,[63] p. 48) guarantees that the number of independent observations d required to embed an arbitrary manifold (in the absence of noise) into a Euclidean embedding space will be no more than twice the dimension of the manifold. For example, a two-dimensional Möbius strip can be embedded in a three-dimensional Euclidean space, but a two-dimensional Klein bottle requires a four-dimensional space.

to include additional signal processing.) A specific example, used by Sauer,[135] is embedding with a matrix produced by multiplying a discrete Fourier transform, a low-pass filter, and an inverse Fourier transform; as long as the filter cut-off is chosen high enough to keep the rank of the overall transformation greater than or equal to the required embedding dimension, this will remove noise but will preserve the embedding. There are a number of more sophisticated linear filters that can be used for embedding,[107] and we will also see that connectionist networks can be interpreted as sets of nonlinear filters.

A final modification of time-delay embedding that can be useful in practice is **embedding by expectation values.** Often the goal of an analysis is to recover not the detailed trajectory of $\mathbf{x}(t)$, but rather to estimate the probability distribution $p(\mathbf{x})$ for finding the system in the neighborhood of a point $\mathbf{x}$. This probability is defined over a measurement of duration T in terms of an arbitrary test function $g(\mathbf{x})$ by

$$\frac{1}{T} \int_0^T g(\mathbf{x}(t))dt = \langle g(\mathbf{x}(t)) \rangle_t$$
$$= \int g(\mathbf{x})p(\mathbf{x}) \, d\mathbf{x}. \tag{15}$$

Note that this is an empirical definition of the probability distribution for the observed trajectory; it is not equivalent to assuming the existence of an invariant measure or of ergodicity so that the distribution is valid for all possible trajectories.[112] If a complex exponential is chosen for the test function

$$\langle e^{i\mathbf{k}\cdot\mathbf{x}(t)} \rangle = \langle e^{i\mathbf{k}\cdot(x_t, x_{t-\tau}, \ldots, x_{t-(d-1)\tau})} \rangle$$
$$= \int e^{i\mathbf{k}\cdot\mathbf{x}} p(\mathbf{x}) \, d\mathbf{x}, \tag{16}$$

we see that the time average of this is equal to the Fourier transform of the desired probability distribution (this is just a characteristic function of the lag vector). This means that, if it is not possible to measure a time series directly (such as for very fast dynamics), it can still be possible to do time-delay embedding by measuring a set of time-average expectation values and then taking the inverse Fourier transform to find $p(\mathbf{x})$.[52] We will return to this point in Section 6.2 and show how embedding by expectation values can also provide a useful framework for distinguishing measurement noise from underlying dynamics.

We have seen that time-delay embedding, while appearing similar to traditional state-space models with lagged vectors, makes a crucial link between behavior in the reconstructed state space and the internal degrees of freedom. We will apply this insight to forecasting and characterizing deterministic systems later (in Sections 5.1 and 6.2). Now, we address the problem of what can be done if we are unable to understand the system in such explicit terms. The main idea will be to learn to emulate the behavior of the system.

4.2 LEARNING: NEURAL NETWORKS

In the competition, the majority of contributions, and also the best predictions for each set used connectionist methods. They provide a convenient language game for nonlinear modeling. Connectionist networks are also known as neural networks, parallel distributed processing, or even as "brain-style computation"; we use these terms interchangeably. Their practical application (such as by large financial institutions for forecasting) has been marked by (and marketed with) great hope and hype.[65,138]

Neural networks are typically used in pattern recognition, where a collection of features (such as an image) is presented to the network, and the task is to assign the input feature to one or more classes. Another typical use for neural networks is (nonlinear) regression, where the task is to find a smooth interpolation between points. In both these cases, all the relevant information is presented simultaneously. In contrast, time series prediction involves processing of patterns that evolve over time—the appropriate response at a particular point in time depends not only on the current value of the observable but also on the past. Time series prediction has had an appeal for neural networkers from the very beginning of the field. In 1964, Hu applied Widrow's adaptive linear network to weather forecasting. In the post-backpropagation era, Lapedes and Farber[85] trained their (nonlinear) network to emulate the relationship between output (the next point in the series) and inputs (its predecessors) for computer-generated time series, and Weigend, Huberman, and Rumelhart[165,168] addressed the issue of finding networks of appropriate complexity for predicting observed (real-world) time series. In all these cases, temporal information is presented spatially to the network by a time-lagged vector (also called tapped delay line).

A number of ingredients are needed to specify a neural network:

- its interconnection architecture,
- its activation functions (that relate the output value of a node to its inputs),
- the cost function that evaluates the network's output (such as squared error),
- a training algorithm that changes the interconnection parameters (called weights) in order to minimize the cost function.

The simplest case is a network **without hidden units**: it consists of one output unit that computes a weighted linear superposition of d inputs, $\text{out}^{(t)} = \sum_{i=1}^{d} w_i x_i^{(t)}$. The superscript $^{(t)}$ denotes a specific "pattern"; $x_i^{(t)}$ is the value of the ith input of that pattern.[14] w_i is the weight between input i and the output. The network output can also be interpreted as a dot-product $\mathbf{w} \cdot \mathbf{x}^{(t)}$ between the weight vector $\mathbf{w} = (w_1, \cdots, w_d)$ and an input pattern $\mathbf{x}^{(t)} = (x_1^{(t)}, \ldots, x_d^{(t)})$.

---

[14]In the context of time series prediction, $x_i^{(t)}$ can be the ith component of the delay vector, $x_i^{(t)} = x_{t-i}$.

Given such an input-output relationship, the central task in learning is to find a way to change the weights such that the actual output $\text{out}^{(t)}$ gets closer to the desired output or $\text{target}^{(t)}$. The closeness is expressed by a cost function, for example, the squared error $E^{(t)} = (\text{out}^{(t)} - \text{target}^{(t)})^2$. A learning algorithm iteratively updates the weights by taking a small step (parametrized by the learning rate η) in the direction that decreases the error the most, i.e., following the negative of the local gradient.[15] The "new" weight $\widetilde{w}_i$, after the update, is expressed in terms of the "old" weight w_i as

$$\widetilde{w}_i = w_i - \eta \frac{\partial E^{(t)}}{\partial w_i} = w_i + 2\eta \underbrace{x_i}_{\text{activation}} \underbrace{(\text{out}^{(t)} - \text{target}^{(t)})}_{\text{error}}. \tag{17}$$

The weight-change $(\widetilde{w}_i - w_i)$ is proportional to the product of the activation going into the weight and the size of error, here the deviation $(\text{out}^{(t)} - \text{target}^{(t)})$. This rule for adapting weights (for linear output units with squared errors) goes back to Widrow and Hoff.[173]

If the input values are the lagged values of a time series and the output is the prediction for the next value, this simple network is equivalent to determining an $AR(d)$ model through least squares regression: the weights at the end of training equal the coefficients of the AR model.

Linear networks are very limited—exactly as limited as linear AR models. The key idea responsible for the power, potential, and popularity of connectionism is the insertion of one of more layers of **nonlinear hidden units** (between the inputs and output). These nonlinearities allow for interactions between the inputs (such as products between input variables) and thereby allow the network to fit more complicated functions. (This is discussed further in the subsection on neural networks and statistics below.)

The simplest such nonlinear network contains only one hidden layer and is defined by the following components:

- There are d *inputs*.
- The inputs are fully connected to a layer of nonlinear *hidden units*.
- The hidden units are connected to the one linear *output unit*.
- The output and hidden units have adjustable offsets or *biases b*.
- The *weights w* can be positive, negative, or zero.

The response of a unit is called its activation value or, in short, *activation*. A common choice for the nonlinear *activation function* of the hidden units is a

---

[15] Eq. (17) assumes updates after each pattern. This is a stochastic approximation (also called "on-line updates" or "pattern mode") to first average the errors over the entire training set, $E = 1/N \sum_t E^{(t)}$ and then update (also called "batch updates" or "epoch mode"). If there is repetition in the training set, learning with pattern updates is faster.

composition of two operators: an affine mapping followed by a sigmoidal function. First, the inputs into a hidden unit h are linearly combined and a bias b_h is added:

$$\xi_h^{(t)} = \sum_{i=1}^{d} w_{hi} x_i^{(t)} + b_h . \tag{18}$$

Then, the output of the unit is determined by passing $\xi_h^{(t)}$ through a sigmoidal function ("squashing function") such as

$$S(\xi_h^{(t)}) = \frac{1}{1 + e^{-a\xi_h^{(t)}}} = \frac{1}{2}\left(1 + \tanh\frac{a}{2}\xi_h^{(t)}\right), \tag{19}$$

where the slope a determines the steepness of the response.

In the introductory example of a linear network, we have seen how to change the weights when the activations are known at both ends of the weight. How do we update the weights to the hidden units that do not have a target value? The revolutionary (but in hindsight obvious) idea that solved this problem is the chain rule of differentiation. This idea of **error backpropagation** can be traced back to Werbos,[170] but only found widespread use after it was independently invented by Rumelhart et al.[128,129] at a time when computers had become sufficiently powerful to permit easy exploration and successful application of the backpropagation rule.

As in the linear case, weights are adjusted by taking small steps in the direction of the negative gradient, $-\partial E/\partial w$. The weight-change rule is still of the same form (activation into the weight) × (error signal from above). The activation that goes into the weight remains unmodified (it is the same as in the linear case). The difference lies in the error signal. For weights between hidden unit h and the output, the error signal for a given pattern is now $(\text{out}^{(t)} - \text{target}^{(t)}) \times S'(\xi_h^{(t)})$; i.e., the previous difference between prediction and target is now multiplied with the derivative of the hidden unit activation function taken at $\xi_h^{(t)}$. For weights that do not connect directly to the output, the error signal is computed recursively in terms of the error signals of the units to which it directly connects, and the weights of those connections. The weight change is computed locally from incoming activation, the derivative, and error terms from above multiplied with the corresponding weights.[16] Clear derivations of backpropagation can be found in Rumelhart, Hinton, and Williams[128] and in the textbooks by Hertz, Krogh, and Palmer[68] (p.117) and Kung[81] (p.154). The theoretical foundations of backpropagation are laid out clearly by Rumelhart et al.[130]

---

[16] All update rules are local. The flip side of the locality of the update rules discussed above is that learning becomes an iterative process. Statisticians usually do not focus on the emergence of iterative local solutions.

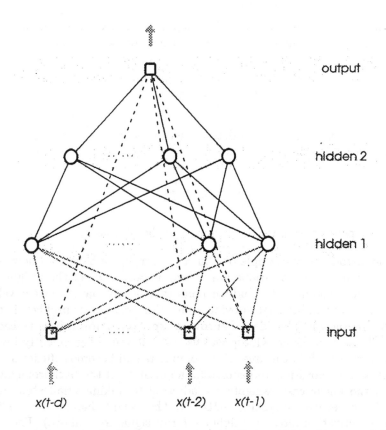

output

hidden 2

hidden 1

input

x(t-d)   x(t-2)   x(t-1)

FIGURE 6 Architecture of a feedforward network with two hidden layers and direct connections from the input to the output. The lines correspond to weight values. The dashed lines represent direct connections from the inputs to the (linear) output unit. Biases are not shown.

Figure 6 shows a typical network; activations flow from the bottom up. In addition to a second layer of (nonlinear) hidden units, we also include direct (linear) connections between each input and the output. Although not used by the competition entrants, this architecture can extract the linearly predictable part early in the learning process and free up the nonlinear resources to be employed where they are really needed. It can be advantageous to choose different learning rates for different parts of the architecture, and thus not follow the gradient exactly.[164] In Section 6.3.2 we will describe yet another modification, i.e., sandwiching a bottleneck hidden layer between two additional (larger) layers.

NEURAL NETWORKS AND STATISTICS. Given that feedforward networks with hidden units implement a nonlinear regression of the output onto the inputs, what features do they have that might give them an advantage over more traditional methods? Consider polynomial regression. Here the components of the input vector $(x_1, x_2, \ldots, x_d)$ can be combined in pairs $(x_1x_2, x_1x_3, \ldots)$, in triples $(x_1x_2x_3, x_1x_2x_4, \ldots)$, etc., as well as in combinations of higher powers. This vast number of possible terms can approximate any desired output surface. One might be tempted to conjecture that feedforward networks are able to represent a larger function space with fewer parameters. This, however, is not true: Cover[27] and Mitchison and Durbin[100] showed that the "capacity" of both polynomial expansions and networks is proportional to the number of parameters. The real difference between the two representations is in the kinds of constraints they impose. For the polynomial case, the number of possible terms grows rapidly with the input dimension, making it sometimes impossible to use even all of the second-order terms. Thus, the necessary selection of which terms to include implies a decision to permit only specific pairwise or perhaps three-way interactions between components of the input vector. A layered network, rather than limiting the *order* of the interactions, limits only the total *number* of interactions and learns to select an appropriate combination of inputs. Finding a simple representation for a complex signal might require looking for such simultaneous relationships among many input variables. A small network is already potentially fully nonlinear. Units are added to increase the number of features that can be represented (rather than to increase the model order in the example of polynomial regression).

NEURAL NETWORKS AND MACHINE LEARNING. Theoretical work in connectionism ranges from reassuring proofs that neural networks with sigmoid hidden units can essentially fit any well-behaved function and its derivative[5,32,47,171] to results on the ability to generalize.[66][17] Neural networks have found their place in (and helped develop) the broader field of machine learning which studies algorithms for learning from examples. For time series prediction, this has included genetic algorithms,[80,99,109] for a recent monograph on genetic programming, Boltzmann machines,[69] and conventional AI techniques.[82] The increasing number of such techniques that arrive with strong claims about their performance is forcing the machine learning community to pay greater attention to methodological issues. In this sense, the comparative evaluations of the Santa Fe Competition can also be viewed as a small stone in the mosaic of machine learning.

[17]This paper by Haussler (on PAC [probably approximately correct] learning) will be published in the collection edited by Smolensky, Mozer, and Rumelhart.[142] That collection also contains theoretical connectionist results by Vapnik (on induction principles), Judd (on complexity of learning), and Rissanen (on information theory and neural networks).

5. FORECASTING

In the previous section we have seen that analysis of the geometry of the embedded data and machine learning techniques provide alternative approaches to discovering the relationship between past and future points in a time series. Such insight can be used to forecast the unknown continuation of a given time series. In this section we will consider the details of how prediction is implemented, and in the following section we will step back to look at the related problem of characterizing the essential properties of a system.

5.1 STATE-SPACE FORECASTING

If an experimentally observed quantity arises from deterministic governing equations, it is possible to use time-delay embedding to recover a representation of the relevant internal degrees of freedom of the system from the observable. Although the precise values of these reconstructed variables are not meaningful (because of the unknown change of coordinates), they can be used to make precise forecasts because the embedding map preserves their geometrical structure. In this section we explain how this is done for a time series that has been generated by a deterministic system; in Section 6.2 we will consider how to determine whether or not this is the case (and, if so, what the embedding parameters should be) and, in Section 5.2, how to forecast systems that are not simply deterministic.

Figure 5 is an example of the structure that an embedding can reveal. Notice that the surface appears to be single-valued; this, in fact, must be the case if the system is deterministic and if the number of time lags used is sufficient for an embedding. Differential equations and maps have unique solutions forward in time; this property is preserved under a diffeomorphic transformation and so the first component of an embedded vector must be a unique function of the preceding values $x_{t-\tau}, ..., x_{t-(d-1)\tau}$ once d is large enough. Therefore, the points must lie on a single-valued hypersurface. Future values of the observable can be read off from this surface if it can be adequately estimated from the given data set (which may contain noise and is limited in length).

Using embedding for forecasting appears—at first sight—to be very similar to Yule's original AR model: a prediction function is sought based on time-lagged vectors. The crucial difference is that understanding embedding reduces forecasting to recognizing and then representing the underlying geometrical structure, and once the number of lags exceeds the minimum embedding dimension, this geometry will not change. A global linear model (AR) must do this with a single hyperplane. Since this may be a very poor approximation, there is no fundamental insight into how to choose the number of delays and related parameters. Instead, heuristic rules such as the AIC[18] are used (and vigorously debated). The ease of producing and exploring pictures, such as Figure 5, with modern computers has helped clarify the importance of this point.

Early efforts to improve global linear AR models included systematically increasing the order of interaction ("bilinear" models[19]; Granger & Anderson[57]). splitting the input space across one variable and allowing for two AR models (threshold autoregressive models, Tong & Lim[155] and using the nonlinearities of a Volterra expansion. A more recent example of the evolutionary improvements are adaptive fits with local splines (Multivariate Adaptive Regression Splines, MARS; Friedman,[46] Lewis et al.[90]). The "insight" gained from a model such as MARS describes what parameter values are used for particular regions of state space, but it does not help with deeper questions about the nature of a system (such as how many degrees of freedom there are, or how trajectories evolve). Compare this to forecasting based on state-space embedding, which starts by testing for the presence of identifiable geometrical structure and then proceeds to model the geometry, rather than starting with (often inadequate) assumptions about the geometry. This characterization step (to be discussed in detail in Section 6.2) is crucial: simple state-space forecasting becomes problematic if there is a large amount of noise in the system, or if there are nonstationarities on the time scale of the sampling time. Assuming that sufficiently low-dimensional dynamics has been detected, the next step is to build a model of the geometry of the hypersurface in the embedding space that can interpolate between measured points and can distinguish between measurement noise and intrinsic dynamics. This can be done by both local and global representations (as well as by intermediate hybrids).

Farmer and Sidorowich[37] introduced *local linear models* for state-space forecasting. The simple idea is to recognize that any manifold is locally linear (i.e., locally a hyperplane). Furthermore, the constraint that the surface must be single-valued allows noise transverse to the surface (the generic case) to be recognized and eliminated. Broomhead and King[11] use *Singular Value Decomposition* (SVD) for this projection; the distinction between local and global SVD is crucial. (Fraser[41] points out some problems with the use of SVD for nonlinear systems.) Smith[141] discusses the relationship between local linear and nonlinear models, as well as the relationship between local and global approaches. Finally, Casdagli and Weigend[18]

---

[18]For linear regression, it is sometimes possible to "correct" for the usually over-optimistic estimate. An example is to multiply the fitting error with $(N + k)/(N - k)$, where N is the number of data points and k is the number of parameters of the model.[3,102,132] Moody[102] extended this for nonlinear regression and used a notion of effective number of parameters for a network that has converged. Weigend and Rumelhart[166,167] focused on the increase of the effective network size (expressed as the effective number of hidden units) as a function on training time.

[19]A bilinear model contains second-order interactions between the inputs, i.e., $x_i x_j$. The term "bilinear" comes from the fact that two inputs enter linearly into such products.

specifically explore the continuum between local and global models by varying the size of the local neighborhood used in the local linear fit. Before returning to this issue in Section 6.3 where we will show how this variation relates to (and characterizes) a system's properties, we now summarize the method used by Tim Sauer in his successful entry. (Details are given by Sauer.[135])

In his competition entry, shown in Figure 3, Sauer used a careful implementation of local-linear fitting that had five steps:

1. Low-pass embed the data to help remove measurement and quantization noise. This low-pass filtering produces a smoothed version of the original series. (We explained such filtered embedding at the end of Section 4.1.)
2. Generate more points in embedding space by (Fourier) interpolating between the points obtained from Step 1. This is to increase the coverage in embedding space.
3. Find the k nearest neighbors to the point of prediction (the choice of k tries to balance the increasing bias and decreasing variance that come from using a larger neighborhood).
4. Use a local SVD to project (possibly very noisy) points onto the local surface. (Even if a point is very far away from the surface, this step forces the dynamics back on the reconstructed solution manifold.)
5. Regress a linear model for the neighborhood and use it to generate the forecast.

Because Data Set A was generated by low-dimensional smooth dynamics, such a local linear model is able to capture the geometry remarkably well based on the relatively small sample size. The great advantage of local models is their ability to adhere to the local shape of an arbitrary surface; the corresponding disadvantage is that they do not lead to a compact description of the system. Global expansions of the surface reverse this trade-off by providing a more manageable representation at the risk of larger local errors. Giona et al.[55] give a particularly nice approach to global modeling that builds an orthogonal set of basis functions with respect to the natural measure of the attractor rather than picking a fixed set independent of the data. If $x_{t+1} = f(\mathbf{x}_t)$, $\rho(\mathbf{x})$ is the probability distribution for the state vector $\mathbf{x}$, and $\{p_i\}$ denotes a set of polynomials that are orthogonal with respect to this distribution:

$$\langle p_i(\mathbf{x})p_j(\mathbf{x})\rangle = \int p_i(\mathbf{x})p_j(\mathbf{x})\rho(\mathbf{x})\,d\mathbf{x} = \delta_{ij}\,, \tag{20}$$

then the expansion coefficients

$$f(\mathbf{x}) = \sum_i a_i p_i(\mathbf{x}) \tag{21}$$

can be found from the time average by the orthogonality condition:

$$a_i = \langle f(\mathbf{x}_t)p_i(\mathbf{x}_t)\rangle = \langle x_{t+1}p_i(\mathbf{x}_t)\rangle = \lim_{N\to\infty}\frac{1}{N}\sum_{t=1}^{N} x_{t+1}p_i(\mathbf{x}_t)\,. \tag{22}$$

The orthogonal polynomials can be found from Gram-Schmidt orthogonalization on the moments of the time series. This expansion is similar in spirit to embedding by expectation values presented in Eq. (16).

In between global and local models lie descriptions such as **radial basis functions**.[12,15,116,117,141] A typical choice is a mixture of (spherically symmetric) Gaussians, defined by

$$f(\mathbf{x}) = \sum_i w_i e^{-(\mathbf{x} - \mathbf{c}_i)^2/(2\sigma_i^2)} . \qquad (23)$$

For each basis function, three quantities have to be determined: its center, $\mathbf{c}_i$; its width, σ_i; and the weight, w_i. In the simplest case, all the widths and centers are fixed. (The centers are, for example, placed on the observed data points.) In a weaker model, these assumptions are relaxed: the widths can be made adaptive, the constraint of spherical Gaussians can be removed by allowing for a general covariance matrix, and the centers can be allowed to adapt freely.

An important issue in function approximation is whether the adjustable parameters are all "after" the nonlinearities (for radial basis functions this corresponds to fixed centers and widths), or whether some of them are also "before" the nonlinearities (i.e., the centers and/or widths can be adjusted). The advantage of the former case is that the only remaining free parameters, the weights, can be estimated by matrix inversion. Its disadvantage is an exponential "curse of dimensionality."[20] In the latter case of adaptive nonlinearities, parameter estimation is harder, but can always be cast in an error backpropagation framework, i.e., solved with gradient descent. The surprising—and promising—result is that in this case when the adaptive parameters are "before" the nonlinearity, the curse of dimensionality is only linear with the dimension of the input space.[5]

A goal beyond modeling the geometry of the manifold is to find a set of differential equations that might have produced the time series. This is often a more compact (and meaningful) description, as shown for simple examples by Cremers and Hübler,[29] and by Crutchfield and McNamara.[30] An alternative goal, trying to characterize symbol sequences (which might be obtained by a coarse quantization of real-valued data where each bin has a symbol assigned to it), is suggested by Crutchfield and Young,[31] who try to extract the rules of an automaton that could have generated the observed symbol sequence. Such approaches are interesting but are not yet routinely applicable.

---

[20] The higher the dimension of a data set of a given size, the more sparse the data set appears. If the average distance ϵ to the nearest point is to remain constant, the number of points N needed to cover the space increases exponentially with the dimension of the space d, $N \propto (1/\epsilon)^d$.

5.2 CONNECTIONIST FORECASTING

State-space embedding "solves" the forecasting problem for a low-dimensional deterministic system. If there is understandable structure in the embedding space, it can be detected and modeled; the open questions have to do with finding good representations of the surface and with estimating the reliability of the forecast. This approach of reconstructing the geometry of the manifold will fail if the system is high-dimensional, has stochastic inputs, or is nonstationary, because in these cases, there is no longer a simple surface to model.

Neural networks do not build an explicit description of a surface. On the one hand, this makes it harder to interpret them even for simple forecasting tasks. On the other hand, they promise to be applicable (and mis-applicable) to situations where simpler, more explicit approaches fail: much of their promise comes from the hope that they can learn to emulate unanticipated regularities in a complex signal. This broad attraction leads to results like those seen in Table 2 in the Appendix: the best as well as many of the worst forecasts of Data Set A were obtained with neural networks. The purpose of this section is to examine how neural networks can be used to forecast relatively simple time series (Data Set A, laser) as well as more difficult time series (Data Set D, high-dimensional chaos; Data Set E, currency exchange rates). We will start with *point predictions*, i.e., predictions of a single value at each time step (such as the most likely one), and then turn to the additional estimation of error bars, and eventually to the prediction of the full probability distribution.

A network that is to predict the future must know about the past. The simplest approach is to provide time-delayed samples to its input layer. In Section 4.2 we discussed that, on the one hand, a network without (nonlinear) hidden units is equivalent to an AR model (one linear filter). On the other hand, we showed that with nonliner hidden units, the network combines a number of "squashed" filters.

Eric Wan (at the time a graduate student at Stanford) used a somewhat more difficult architecture for his competition entry. The key modification to the network displayed in Figure 6 is that each connection now becomes an AR filter (tapped delay line). Rather than displaying the explicit buffer of the input units, it suffices then to draw only a single input unit and conceptually move the weight vectors into the tapped delay lines from the input to each hidden unit. This architecture is also known as "time-delay neural network" by Lang, Waibel, and Hinton[84] or (in the spatial domain) as a network with "linked weights," as suggested by le Cun.[87]

We would like to emphasize that these architectures are all examples of *feed-forward* networks: they are trained in "open loop," and there is no feedback of activations in training. (For iterated predictions, however, predictions must be used for the input: the network is run in "closed loop" mode. A more consistent scheme is to train the network in the mode eventually used in prediction; we return to this point at the end of this section.)

Wan's network had 1 input unit, two layers of 12 hidden units each, and 1 output unit. The "generalized weights" of the first layer were tapped delay lines with 25 taps; the second and third layers had 5 taps each. These values are not the result of a simple quantitative analysis, but rather the result of evaluating the performance of a variety of architectures on some part of the available data that Wan had set aside for this purpose. Such careful exploration is important for the successful use of neural networks.

At first sight, selecting an architecture with 1,105 parameters to fit 1,000 data points seems absurd. How is it possible not to overfit if there are more parameters than data points? The key is knowing when to stop. At the outset of training, the parameters have random values, and so changing any one coefficient has little impact on the quality of the predictions. As training progresses and the fit improves, the *effective number of parameters* grows.[166,167] The overall error in predicting points out of the training set will initially decrease as the network learns to do something, but then will begin to increase once the network learns to do too much; the location of the minimum of the "cross validation" error determines when the effective network complexity is right.[21]

The competition entry was obtained by a network that was stopped early. It did very well over the prediction interval (Figure 3), but notice how it fails dramatically after 100 time steps (Figure 4) while the local linear forecast continues on. The reason for this difference is that the local linear model with local SVD (described in Section 5.1) is constrained to stay on the reconstructed surface in the embedding space (any point gets projected onto the reconstructed manifold before the prediction is made), whereas the network does not have a comparable constraint.

Short-term predictors optimize the parameters for forecasting the next step. The same architecture (e.g., Figure 6) can also be used to follow the trajectory more closely in the longer run, at the expense of possibly worse single-step predictions, by using the *predicted* value at the input in training, rather than the *true* value.[22] There is a continuum between these two extremes: in training, the errors from both sources can be combined in a weighted way, e.g., $\lambda \times$ (short-term error) $+ (1 - \lambda) \times$ (long-term error). This mixed error is then used in backpropagation. Since the network produces very poor predictions at the beginning of the training process, it can be advantageous to begin with $\lambda = 1$ (full teacher forcing) and then anneal to the desired value.

[21]Other ideas, besides early stopping, for arriving at networks of an appropriate size include (i) penalizing network complexity by adding a term to the error function, (ii) removing unimportant weights with the help of the Hessian (the second derivative of the error with respect to the weights), and (iii) expressing the problem in a Bayesian framework.

[22]There are a number of terms associated with this distinction. Engineers use the expression "open loop" when the inputs are set to true values, and "closed loop" for the case when the input are given the predicted values. In the connectionist community, the term "teacher forcing" is used when the inputs are set to the true values, and the term "trajectory learning"[118] when the predicted output is fed back to the inputs.

We close the section on point-predictions with some remarks on the question whether predictions for several steps in the future should be made by iterating the single-step prediction T times or by making a noniterated, direct prediction. Farmer and Sidorowich[38] argue that for deterministic chaotic systems, iterated forecasts lead to better predictions than direct forecasts.[23] However, for noisy series the question "iterated vs. direct forecasts?" remains open. The answer depends on issues such as the sampling time (if the sampling is faster than the highest frequency in the system, one-step predictions will focus on the noise), and the complexity of the input-output map (even the simple parabola of the logistic map becomes a polynomial of order 2^T when direct T step predictions are attempted). As Sauer[135] points out, the dichotomy between iterated and direct forecasts is not necessary: combining both approaches leads to better forecasts than using either individually.

5.3 BEYOND POINT-PREDICTIONS

So far we have discussed how to predict the continuation of a time series. It is often desirable and important to also know the confidence associated with a prediction. Before giving some suggestions, we make explicit the two assumptions that minimizing sum squared errors implies in a maximum likelihood framework (which we need not accept):

- The errors of different data points are independent of each other: statistical independence is present if the joint probability between two events is precisely the product between the two individual probabilities. After taking the logarithm, the product becomes a sum. Summing errors thus assumes statistical independence of the measurement errors.

- The errors are Gaussian distributed: i.e., the likelihood of a data point given the prediction is a Gaussian. Taking the logarithm of a Gaussian transforms it into a squared difference. This squared difference can be interpreted as a squared error. Summing squared errors with the same weight for each data point assumes that the uncertainty is the same for each data point (i.e., that the size of the error bar is independent of the location in state space).

The second assumption is clearly violated in the laser data: the errors made by the predictors on Data Set A depend strongly on the location in state space (see, for example, Smith[141]). This is not surprising, since the local properties of the attractor (such as the rate of divergence of nearby trajectories) vary with the location. Furthermore, different regions are sampled unequally since the training set is finite.

The second assumption can be relaxed: in our own research, we have used maximum likelihood networks that successfully estimate the error bars as a function of the network input. These networks had two output units: the first one predicts the value, the second the error bar. This model of a single Gaussian (where mean

---

[23]Fraser[44] points out that this argument can be traced back to Rissanen and Langdon.[124]

and width are estimated depending on the input) is easily generalized to a mixture of Gaussians. It is also possible to explore more general models with the activation distributed over a set of output units, allowing one to predict the probability density.[24] All these models fall within the elegant and powerful probabilistic framework laid out for connectionist modeling by Rumelhart et al.[130] Another approach to estimating the output errors is to use "noisy weights": rather than characterizing each weight only by its value, Hinton and van Camp[70] also encode the precision of the weight. In addition to facilitating the application of the minimum description length principle, this formulation yields a probability distribution of the output for a given input.

To force competition entrants to address the issue of the reliability of their estimates, we required them to submit, for Data Set A, both the predicted values and their estimated accuracies. We set as the error measure the likelihood of the true (withheld) data, computed from the submitted predictions and confidence intervals under a Gaussian model. In the Appendix we motivate this measure and list its values for the entries received. Analyzing the competition entries, we were quite surprised by how little effort was put into estimating the error bars. In all cases, the techniques used to generate the submitted error estimates were much less sophisticated than the models used for the point-predictions.

Motivated by this lack of sophistication, following the close of the competition, Fraser and Dimitriadis[39] used a hidden Markov model (HMM; see, e.g., Rabiner & Juang[121]) to predict the evolution of the entire probability density for the computer-generated Data Set D, with 100,000 points the longest of the data sets. Nonhidden Markov models fall in the same class as feedforward networks: their only notion of state is what is presented to them in the moment—there is no implicit memory or stack. Hidden Markov models introduce "hidden states" that are not directly observable but are built up from the past. In this sense, they resemble recurrent networks which also form a representation of the past in internal memory that can influence decisions beyond what is seen by the immediate inputs.

The states of a hidden Markov model can be discrete or continuous; Fraser and Dimitriadis use a mixture of both: they generate their predictions from a 20-state model using eighth-order linear autoregressive filters. The model parametrizes multivariate normal distributions.

---

[24]On the one hand, a localist representation is suited for the prediction of symbols where we want to avoid imposing a metric. For example, in composing music, it is undesirable to inherit a metric obtained from the pitch value (see Dirst & Weigend[34]). On the other hand, if neighborhood is a sensible distance measure, then higher accuracy can be obtained by smearing out the activation over several units (e.g., Saund[136]).

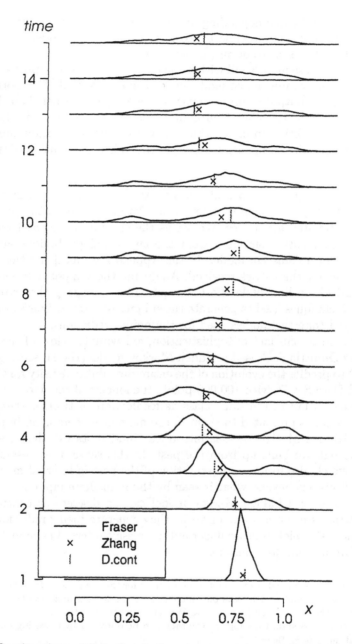

FIGURE 7   Continuations of Data Set D, the computer-generated data. The curves are the predictions for the probability density function from a hidden Markov model.[39] The ×'s indicate the point predictions from a neural network.[177] The "true" continuation is indicated by vertical lines.

Figure 7 shows the evolution of the probability density function according to Fraser and Dimitriadis.[39] They first estimated the roughly 6,000 parameters of their model from the 100,000 points of Data Set D, and then generated several million continuations from their model. We plot the histograms of these continuations (normalized to have the same areas) along with the values obtained by continuing the generation of the data from the differential equations (as described in the Appendix). Unfortunately, the predicted probability density functions by Fraser and Dimitriadis are a lot wider than the uncertainty due to the dynamics of the system and the stochasticity used in generating the series requires: on the time scale of Figure 7, an ensemble of continuations of Data Set D spreads only about 1%, a lot less than the uncertainty predicted by the mixed-state HMM.

In the competition, we received several sets of point-predictions for Data Set D. They are listed and plotted in the Appendix. In Figure 7 we have included the best of these, obtained by Zhang and Hutchinson[177]). They trained 108 simple feedforward networks in a brute force approach: the final run alone—after all initial experimentation and architecture selection—used 100 hours on a Connection Machine CM-2 with 8,192 floating point processors. Each network had between 20 and 30 inputs, either one hidden layer with 100 units or two layers with 30 units each, and up to 5 outputs. The complex, unstructured architecture unfortunately does not allow a satisfying interpretation of the resulting networks.

Beyond physical systems: Financial data. We close this section with some remarks about the forecasts for the exchange rate time series (Data Set C). (No prediction tasks were specified for the biological, astronomical, and musical sets.) The market is quite large: in 1989 the daily volume of the currency markets was estimated to be U.S. $650 billion, and in 1993 the market exceeded U.S. $1 trillion ($= 10^{12}$) on busy days. Lequarré[89] reports that 97% of this is speculation—i.e., only 3% of the trades are "actual" transactions.

The simplest model of a market indicator is the efficient market hypothesis: "you can't do better than predicting that tomorrow's rate is the same as today's." Diebold and Nason,[33] for example, review attempts to forecast foreign exchange rates and conclude that none of the methods succeeded in beating the random walk hypothesis out-of-sample. However, all these academic findings about the unpredictability of the vicissitudes of financial markets are based on daily or weekly rates, the only data available until a few years ago. The great deal of activity in foreign exchange trading suggests that it must be possible to make money with better data.

Recently, high-frequency data have become available, and we were fortunate enough to be able to provide a set of such "tick-by-tick" data for the competition. Data Set C consists of quotes on the time scale of one to two minutes for the exchange rate between the Swiss franc and the U.S. dollar. The market is based on bids and asks to buy and sell. (There is no central market for currency transactions.) Prices are given for a trade of U.S. $10 million, and an offer is good for five seconds(!). In addition to the quote, we included the day of the week and the time after the opening that day (to allow for modeling of intra- and inter-day effects).

In order to balance the desire for statistically significant results and the need to keep the competition prediction task manageable, we asked for forecasts for 10 episodes of 3,000 points each, taken from the period between August 7, 1990 to April 18, 1991.[25] This assessment provided a basic sanity test of the submitted predictors (some of them were worse than chance by a factor of 16), and afterward the best two groups were invited to analyze a much larger sample. The quality of the predictions is expressed in terms of the following ratio of squared errors:

$$\frac{\sum_t \left(\text{observation}_t - \text{prediction}_t\right)^2}{\sum_t \left(\text{observation}_t - \text{observation}_{t-1}\right)^2}. \qquad (24)$$

The denominator simply predicts the last observed value—which is the best that can be done for a random walk. A ratio above 1.0 thus corresponds to a prediction that is worse than chance; a ratio below 1.0 is an improvement over a random walk.[26]

In this second round of evaluation, predictions were made for all the points in the gaps between the training segments (still using the competition data for training). The out-of-sample performance on this extended test set, as reported by Mozer[104] and by Zhang and Hutchinson,[177] is collected in Table 1. Please refer to their articles for more details and further evaluation.

[25] For each trial we asked for six forecasts: 1 minute after the last tick, 15 minutes after the last tick, 60 minutes after the last tick, the closing value of the day of the last tick, the opening value of the next trading day, and the closing value of the fifth trading day (usually one week) after the day of the last tick. One example evaluation is given in the following table. The numbers are the ratio of the sum of squared errors of the submitted predictions by Zhang and Hutchinson[177] (p. 235) divided by the sum of the squared errors obtained by using the last observation as the prediction.

	1 minute	15 minutes	1 hour
training set ("in-sample")	0.889	0.891	0.885
test set ("out-of-sample")	0.697	1.04	0.988

This table shows the crucial difference between training set and test set performances, and suggests that the uncertainty in the numbers is large. Hence, we proposed the evaluation on larger data sets, as described in the main text.

[26] The random walk model used here for comparison is a weak null hypothesis. LeBaron[86] reports statistically significant autocorrelations on Data Set C (see Table 2 in his paper[86] on p. 462). To the degree that the in-sample autocorrelations generalize out-of-sample, this justifies a low-order AR model as a stronger null hypothesis. However, to avoid additional assumptions in our comparison here (such as the order of the model), we decided simply to compare to a random walk.

TABLE 1 Performance on exchange rate predictions expressed as the squared error of the predictor divided by the squared error from predicting no change, as defined in Eq. (24). The numbers are as reported by Mozer[104] (p. 261) for his recurrent networks, and Zhang and Hutchinson[177] (p. 236) for their feedforward networks.

	1 minute ($N = 18,465$)	15 minutes ($N = 7,246$)	1 hour ($N = 3,334$)
Mozer	0.9976	0.9989	0.9965
Zhang & Hutchinson	1.090	1.103	1.098

These results—in particular the last row in Table 1 where all out-of-sample predictions are on average worse than chance—make clear that a naive application of the techniques that worked so well for Data Set A (the laser had less than 1% measurement noise added to deterministic dynamics) and to some degree for Data Set D fails for a data set so close to pure randomness as this financial data set. Future directions include[164]

- using additional information from other sources (such as other currencies, interest rates, financial indicators, as well as information automatically extracted from incoming newswires, "topic spotting"),
- splitting the problem of predicting returns to the two separate tasks of predicting the squared change (volatility) and its direction,
- employing architectures that allow for stretching and compressing of the time series, as well as enhancing the input with features that collapse time in ways typically done by traders,
- implementing trading strategies (i.e., converting predictions to recommendations for actions), and subsequently improving them by backpropagating the actual loss or profit through a pay-off matrix (taking transaction costs into account).

Financial predictions can also serve as good vehicles to stimulate research in areas such as subset selection (finding relevant variables) and capacity control (avoiding overfitting). The full record of 329,112 quotes (bid and ask) from May 20, 1985 to April 12, 1991 is available as a benchmark data set via anonymous ftp.[27] We encourage the reader to experiment (and inform the authors of positive results).

[27] It corresponds to 11.5 MB. Like the data sets used in the competition, it is available via anonymous ftp to `ftp.santafe.edu`.

6. CHARACTERIZATION

Simple systems can produce time series that appear to be complicated; complex systems can produce time series that are complicated. These two extremes have different goals and require different techniques; what constitutes a successful forecast depends on where the underlying system falls on this continuum. In this section we will look at characterization methods that can be used to extract some of the essential properties that lie behind an observed time series, both as an end in itself and as a guide to further analysis and modeling.

Characterizing time series through their frequency content goes back to Schuster's "periodogram."[137] For a simple linear system the traditional spectral analysis is very useful (peaks = modes = degrees of freedom), but different nonlinear systems can have similar featureless broadband power spectra. Therefore, a broadly useful characterization of a nonlinear system cannot be based on its frequency content.

We will describe two approaches that parallel the earlier discussion of forecasting: (1) an explicit analysis of the structure in an embedding space (in Section 6.2 we will introduce the information-theoretic measure of redundancy as an example of understanding by "opening up the box"), and (2) an implicit approach based on analyzing properties of an emulation of the system arrived at through learning (in Section 6.3.1 we will show how local linear methods can be used for characterization, and in Section 6.3.2 we will discuss how neural networks can be used to estimate dimensions, the amount of nonlinearity, and Lyapunov coefficients). Before turning to these more sophisticated analyses, we discuss some simple tests.

6.1 SIMPLE TESTS

This section begins with suggestions for exploring data when no additional information is available. We then turn to time series whose frequency spectra follow a power law where low-frequency structure can lead to artifacts. We then show how surrogate data can be generated with identical linear but different nonlinear structure. Finally, we suggest some ways of analyzing the residual errors.

EXPLORATORY DATA ANALYSIS. The importance of exploring a given data set with a broad range of methods cannot be overemphasized. Besides the many traditional techniques,[158] modern methods of interactive exploration range from interactive graphics with linked plots and virtual movements in a visual space[28] to examination in an auditory space (data sonification). The latter method uses the temporal abilities of our auditory systems—after all, analyzing temporal sequences as static plots has not been a prime goal in human evolution.

---

[28] A good example is the visualization package xgobi. To obtain information about it, send the one line message send index to statlib@lib.stat.cmu.edu.

LINEAR CORRELATIONS. We have seen in Section 3.2 that nonlinear structure can be missed by linear analysis. But that is not the only problem: linear structure that is present in the data can confuse nonlinear analysis. Important (and notorious) examples are processes with power spectra proportional to $|\omega|^{-\alpha}$, which arise routinely in fluctuating transport processes. At the extreme, white noise is defined by a flat spectrum; i.e., its spectral coefficient is $\alpha = 0$. When white noise is integrated (summed up over time), the result is a random walk process. After squaring the amplitude (in order to arrive at the power spectrum), a random walk yields a spectral exponent $\alpha = 2$. The intermediate value of $\alpha = 1$ is seen in everything from electrical resistors to traffic on a highway to music.[36] This intermediate spectral coefficient of $\alpha = 1$ implies that all time-scales (over which the $1/\omega$ behavior holds) are equally important.

Consider the cloud of points obtained by plotting x_t against $x_{t-\tau}$ for such a series with a lag time τ. (This plot was introduced in Section 3.2 in the context of the logistic map whose phase portrait was a parabola.) In general, a distribution can be described by its moments. The first moments (means) give the coordinates of the center of the point cloud; the second moments (covariance matrix) contain the information about how elongated the point cloud is, and how it is rotated with respect to the axes.[29] We are interested here in this elongation of the point cloud; it can be described in terms of the eigenvalues of its correlation matrix. The larger eigenvalue, λ_+, characterizes the extension in the direction of the larger principal axis along the diagonal, and the smaller eigenvalue, λ_-, measures the extension transverse to it. The ratio of these two eigenvalues can be expressed in terms of the autocorrelation function ρ at the lag τ as

$$\frac{\lambda_-}{\lambda_+} = \frac{1 - \rho(\tau)}{1 + \rho(\tau)}. \tag{25}$$

For a power-law spectrum, the autocorrelation function can be evaluated analytically in terms of the spectral exponent α and the exponential integral Ei.[51] For example, for a measurement bandwidth of 10^{-3} to 10^3 Hz and a lag time τ of 1 sec, this ratio is 0.51 for $\alpha = 1$, 0.005 for $\alpha = 2$, and 0.0001 for $\alpha = 3$. As the spectrum drops off more quickly, i.e., as the spectral exponent gets larger, the autocorrelation function decays more slowly. (In the extreme of a delta function in frequency space, the signal is constant in time.) Large spectral exponents thus imply that the shape of the point cloud (an estimate of the probability distribution in the embedding space) will be increasingly long and skinny, regardless of the detailed dynamics and of the value of τ. If the width becomes small compared to the available experimental resolution, the system will erroneously appear to be one-dimensional. There is

---

[29] We here consider only first- and second-order moments. They completely characterize a Gaussian distribution, and it is easy to relate them to the conventional (linear) correlation coefficient; see Duda and Hart.[35] A nonlinear relationship between x_t and $x_{t-\tau}$ is missed by an analysis in terms of the first- and second-order moments. This is why a restriction to up to second-order terms is sometimes called linear.

a simple test for this artifact: whether or not the quantities of interest change as the lag time τ is varied. This effect is discussed in more detail by Theiler.[152]

SURROGATE DATA. Since the autocorrelation function of a signal is equal to the inverse Fourier transform of the power spectrum, any transformation of the signal that does not change the power spectrum will not change the autocorrelation function. It is therefore possible to take the Fourier transform of a time series, randomize the phases (symmetrically, so that the inverse transform remains real), and then take the inverse transform to produce a series that by construction has the same autocorrelation function but will have removed any nonlinear ordering of the points. This creation of sets of surrogate data provides an important test for whether an algorithm is detecting nonlinear structure or is fooled by linear properties (such as we saw for low frequency signals): if the result is the same for the surrogate data, then the result cannot have anything to do with deterministic rules that depend on the specific sequence of the points. This technique has become popular in recent years; Fraser,[43] for example, compares characteristics of time series from the Lorenz equation with surrogate versions. Kaplan[75] and Theiler et al.[153] apply the method of surrogate data to Data Sets A, B, D, and E; the idea is also the basis of Paluš's comparison[110] between "linear redundancy" and "redundancy" (we will introduce the concept of redundancy in the next section).

SANITY CHECKS AND SMOKE ALARMS. Once a predictor has been fitted, a number of sanity checks should be applied to the resulting predictions and their residual errors. It can be useful to look at a distribution of the errors sorted by their size (see, e.g., Smith,[141] Figures 5 and 8 on p. 331 and 336). Such plots distinguish between forecasts that have the same mean squared error but very different distributions (uniformly medium-sized errors versus very small errors along with a few large outliers). Other basic tests plot the prediction errors against the true (or against the predicted) value. This distribution should be flat if the errors are Gaussian distributed, and proportional to the mean for a Poissonian error distribution. The time ordering of the errors can also contain information: a good model should turn the time series into structureless noise for the residual errors; any remaining structure indicates that the predictor missed some features.

Failure to use common sense was readily apparent in many of the entries in the competition. This ranged from forecasts for Data Set A that included large negative values (recall that the training set was strictly positive) to elaborate "proofs" that Data Set D was very low dimensional (following a noise reduction step that had the effect of removing the high-dimensional dynamics). Distinguishing fact, fiction and fallacies in forecasts is often hard: carrying out simple tests is crucial particularly in the light of readily available sophisticated analysis and prediction algorithms that can swiftly and silently produce nonsense.

6.2 DIRECT CHARACTERIZATION VIA STATE SPACE

It is always possible to define a time-delayed vector from a time series, but this certainly does not mean that it is always possible to identify meaningful structure in the embedded data. Because the mapping between a delay vector and the system's underlying state is not known, the precise value of an embedded data point is not significant. However, because an embedding is diffeomorphic (smooth and invertible), a number of important properties of the system will be preserved by the mapping. These include local features such as the number of degrees of freedom, and global topological features such as the linking of trajectories.[98] The literature on characterizing embedded data in terms of such invariants is vast, motivated by the promise of obtaining deep insight into observations, but plagued by the problem that plausible algorithms will always produce a result—whether or not the result is significant. General reviews of this area may be found in Ruelle and Eckmann,[127] Gershenfeld,[50] and Theiler.[151][30]

Just as state-space forecasting leaps over the systematic increase in complexity from linear models to bilinear models, etc., these characterization ideas bypass the traditional progression from ordinary spectra to higher order spectra.[31] They are predated by similar efforts to analyze signals in terms of dimensionality[157] and near-neighbor scaling,[113] but they could not succeed until the relationship between observed and unobserved degrees of freedom was made explicit by time-delay embedding. This was done to estimate degrees of freedom by Russel et al.,[131] and implemented efficiently by Grassberger and Procaccia.[58] Brock, Dechert, and Scheinkman later developed this algorithm into a statistical test, the BDS test,[10] with respect to the null hypothesis of an iid sequence (which was further refined by Green & Savit[61]).

We summarize here an information-based approach due to Fraser[42] that was successfully used in the competition by Paluš[110] (participating in the competition over the network from Czechoslovakia). Although the connection between information theory and ergodic theory has long been appreciated (see, e.g., Petersen[112]), Shaw[140] helped point out the connection between dissipative dynamics and information theory, and Fraser and Swinney[40] first used information-theoretic measures to find optimal embedding lags. This example of the physical meaning of information[83] can be viewed as an application of information theory back to its roots in dynamics: Shannon[139] built his theory of information on the analysis of the single-molecule Maxwell Demon by Szilard in 1929, which in turn was motivated by Maxwell and Boltzmann's effort to understand the microscopic dynamics of the origin of thermodynamic irreversibility (circa 1870).

---

[30]The term "embedding" is used in the literature in two senses. In its wider sense, the term denotes any lag-space representation, whether there is a unique surface or not. In its narrower (mathematical) sense used here, the term applies if and only if the resulting surface is unique, i.e., if a diffeomorphism exists between the solution manifold in configuration space and the manifold in lag space.

[31]Chaotic processes are analyzed in terms of bispectra by Subba Rao.[144]

Assume that a time series $x(t)$ has been digitized to integer values lying between 1 and N. If a total of n_T points have been observed, and a particular value of x is recorded n_x times, then the probability of seeing this value is estimated to be $p_1(x) = n_x/n_T$.[32] (The subscript of the probability indicates that we are at present considering one-dimensional distributions (histograms). It will soon be generalized to d-dimensional distributions.) In terms of this probability, the **entropy** of this distribution is given by

$$H_1(N) = - \sum_{x=1}^{N} p_1(x) \log_2 p_1(x) \,. \qquad (26)$$

This is the average number of bits required to describe an isolated observation, and can range from 0 (if there is only one possible value for x) to $\log_2 N$ (if all values of x are equally likely and hence the full resolution of x is required).

In the limit $N \to 1$, there is only one possible value and so the probability of seeing it is unity, thus $H_1(1) = 0$. As N is increased, the entropy grows as $\log N$ if all values are equally probable; it will reach an asymptotic value of $\log M$ independent of N if there are M equally probable states in the time series; and if the probability distribution is more complicated, it can grow as $D_1 \log N$ where D_1 is a constant ≤ 1 (the meaning of D_1 will be explained shortly). Therefore, the dependence of H_1 on N provides information about the resolution of the observable.

The probability of seeing a specific lag vector $\mathbf{x}_t = (x_t, x_{t-\tau}, \ldots, x_{t-(d-1)\tau})$ (see Eq. (14)) in d-dimensional lag space is similarly estimated by counting the relative population of the corresponding cell in the d-dimensional array: $p_d(\mathbf{x}) = n_\mathbf{x}/n_T$. The probability of seeing a particular *sequence* of D embedded vectors $(\mathbf{x}_t, \ldots, \mathbf{x}_{t-(D-1)\tau})$ is just $p_{d+D}(x_t, \ldots, x_{t-(d+D-1)\tau})$ because each successive vector is equal to the preceding one with the coordinates shifted over one place and a new observation added at the end. This means that the joint probability of d delayed observations, p_d, is equivalent to the probability of seeing a single point in the d-dimensional embedding space (or the probability of seeing a sequence of $1 + d - n$ points in a smaller n-dimensional space). In terms of p_d, the **joint entropy** or **block entropy** is

$$H_d(\tau, N) =$$
$$- \sum_{x_t=1}^{N} \cdots \sum_{x_{t-(d-1)\tau}=1}^{N} p_d(x_t, x_{t-\tau}, \ldots, x_{t-(d-1)\tau}) \log_2 p_d(x_t, x_{t-\tau}, \ldots, x_{t-(d-1)\tau}) \,.$$
$$(27)$$

---

[32]Note that there can be corrections to such estimates if one is interested in the expectation value of functions of the probability.[59]

This is the average number of bits needed to describe a sequence. (The range of the sum might seem strange at first sight, but keep in mind that we are assuming that x_t is quantized to integers between 1 and N.) In the limit of small lags, we obtain

$$\lim_{\tau \to 0} p_d(x_t, x_{t-\tau}, \ldots, x_{t-(d-1)\tau}) = p_1(x)$$
$$\Rightarrow H_d(0, N) = H_1(N). \tag{28}$$

In the opposite limit, if successive measurements become uncorrelated at large times, the probability distribution will factor:

$$\lim_{\tau \to \infty} p_d(x_t, x_{t-\tau}, \ldots, x_{t-(d-1)\tau}) = p_1(x_t)p_1(x_{t-\tau}), \ldots, p_1(x_{t-(d-1)\tau})$$
$$\Rightarrow \lim_{\tau \to \infty} H_d(\tau, N) = dH_1(N). \tag{29}$$

We will return to the τ dependence later; for now assume that the delay time τ is small but nonzero.

We have already seen that $\lim_{N \to 1} H_d(\tau, N) = 0$. The limit of large N is best understood in the context of the **generalized dimensions** D_q.[67] These are defined by the scaling of the moments of the d-dimensional probability distribution $p_d(\mathbf{x})$ as the number of bins N tends to infinity (i.e., the bin sizes become very small, corresponding to an increasing resolution):

$$D_q = \lim_{N \to \infty} \frac{1}{q-1} \frac{\log_2 \sum_{\mathbf{x_i}} p_d(\mathbf{x_i})^q}{-\log_2 N}. \tag{30}$$

For simple geometrical objects such as lines or surfaces, the D_q's are all equal to the integer topological dimension (1 for a line, 2 for a surface,...). For fractal distributions they need not be an integer, and the q dependence is related to how singular the distribution is. The values of the D_q's are typically similar, and there are strong bounds on how much they can differ (Beck[6]). D_2 measures the scaling of pairs of points (it is the Grassberger-Procaccia[58] correlation dimension; see also Kantz[74] and Pineda & Sommerer[115]), and D_1 provides the connection with entropy:

$$\lim_{q \to 1} D_q = \lim_{N \to \infty} \frac{\sum_{\mathbf{x_i}} p_d(\mathbf{x_i}) \log_2 p_d(\mathbf{x_i})}{-\log_2 N}$$
$$= \lim_{N \to \infty} \frac{H_d(\tau, N)}{\log_2 N}. \tag{31}$$

As N is increased, the prefactor to the logarithmic growth of the entropy is the generalized dimension D_1 of the probability distribution. If the system is deterministic, so that its solutions lies on a low-dimensional attractor, the measured dimension D_1 will equal the dimension of the attractor if the number of time delays used is large enough. If the number of lags is too small, or if the successive observations in the time series are uncorrelated, then the measured dimension will equal the number of lags. The dimension of an attractor measures the number of local directions available to the system and so it (or the smallest integer above it if the dimension is fractal) provides an estimate of the number of degrees of freedom needed to describe a state of the system. If the underlying system has n degrees of freedom, the minimum embedding dimension to recover these dynamics can be anywhere between n and $2n$, depending on the geometry.

The d-dependence of the entropy can be understood in terms of the concept of **mutual information**. The mutual information between two samples is the difference between their joint entropy and the sum of their scalar entropies:

$$
\begin{aligned}
I_2(\tau, N) = & - \sum_{x_t=1}^{N} p_1(x_t) \log_2 p_1(x_t) - \sum_{x_{t-\tau}=1}^{N} p_1(x_{t-\tau}) \log_2 p_1(x_{t-\tau}) \\
& + \sum_{x_t=1}^{N} \sum_{x_{t-\tau}=1}^{N} p_2(x_t, x_{t-\tau}) \log_2 p_2(x_t, x_{t-\tau}) \\
= & \ 2H_1(\tau, N) - H_2(\tau, N) \,.
\end{aligned}
\tag{32}
$$

If the samples are statistically independent (this means by definition that the probability distribution factors, i.e., $p_2(x_t, x_{t-\tau}) \equiv p_1(x_t)p_1(x_{t-\tau})$), then the mutual information will vanish: no knowledge can be gained for the second sample by knowing the first. On the other hand, if the first sample uniquely determines the second sample ($H_2 = H_1$), the mutual information will equal the scalar entropy $I_2 = H_1$. In between these two cases, the mutual information measures in bits the degree to which knowledge of one variable specifies the other.

The mutual information can be generalized to higher dimensions either by the **joint mutual information**

$$
I_d(\tau, N) = dH_1(\tau, N) - H_d(\tau, N)
\tag{33}
$$

or by the **incremental mutual information** or **redundancy** of one sample

$$
R_d(\tau, N) = H_1(\tau, N) + H_{d-1}(\tau, N) - H_d(\tau, N) \,.
\tag{34}
$$

The redundancy measures the average number of bits about an observation that can be determined by knowing $d - 1$ preceding observations. Joint mutual information and redundancy are related by $R_d = I_d - I_{d-1}$.

For systems governed by differential equations or maps, given enough points and enough resolution, the past must uniquely determine the future[33] (to a certain horizon, depending on the finite resolution).

If d is much less than the minimum embedding dimension, then the $d-1$ previous observations do not determine the next one, and so the value of the redundancy will approach zero:

$$p_d(x_t, x_{t-\tau}, \ldots, x_{t-(d-1)\tau}) = p_1(x_t)p_{d-1}(x_{t-\tau}, \ldots, x_{t-(d-1)\tau})$$
$$\Rightarrow H_d = H_1 + H_{d-1} \Rightarrow R_d = 0. \tag{35}$$

On the other hand, if d is much larger than the required embedding dimension, then the new observation will be entirely redundant:[34]

$$p_d(x_t, x_{t-\tau}, \ldots, x_{t-(d-1)\tau}) = p_{d-1}(x_{t-\tau}, \ldots, x_{t-(d-1)\tau})$$
$$\Rightarrow H_d = H_{d-1} \Rightarrow R_d = H_1. \tag{36}$$

The minimum value of d for which the redundancy converges (if there is one) is equal to the minimum embedding dimension at that resolution and delay time, i.e., the size of the smallest Euclidean space that can contain the dynamics without trajectories crossing. Before giving the redundancy for Data Set A of the competition (in Figure 8), we provide relations to other quantities, such as the Lyapunov exponents and the prediction horizon.

The **source entropy** or **Kolmogorov-Sinai entropy**, $h(\tau, N)$, is defined to be the asymptotic rate of increase of the information with each additional measurement given unlimited resolution:

$$h(\tau, N) = \lim_{N \to \infty} \lim_{d \to \infty} H_d(\tau, N) - H_{d-1}(\tau, N). \tag{37}$$

The limit of infinite resolution is usually not needed in practice: the source entropy reaches its maximum asymptotic value once the resolution is sufficiently fine to produce a generating partition (Petersen,[112] p.243). The source entropy is important because the *Pesin identity* relates it to the sum of the positive Lyapunov exponents[127]:

$$h(\tau) = \tau h(1) = \tau \sum_i \lambda_i^+. \tag{38}$$

---

[33] Note that the converse remains true for differential equations but need not be true for maps: for example, neither of the two maps introduced in Section 3.2 has a unique preimage in time—they cannot be inverted. Given that most computer simulations approximate the continuous dynamics of differential equations by discrete dynamics, this is a potentially rich source for artifacts. Lorenz[93] shows how discretization in time can create dynamical features that are impossible in the continuous-time system; see also Grebogi et al.[60] Rico-Martinez, Kevrekidis, and Adomaitis[122] discuss noninvertibility in the context of neural networks.

[34] To be precise, $H_d = H_{d-1}$ is only valid for (1) the limit of short times τ, (2) discrete measurements, and (3) the noise-free case.

The **Lyapunov exponents** λ_i are the eigenvalues of the local linearization of the dynamics (i.e., a local linear model), measuring the average rate of divergence of the principal axes of an ensemble of nearby trajectories. They can be found from the Jacobian (if it is known) or by following trajectories.[13] Diverging trajectories reveal information about the system which is initially hidden by the measurement quantization. The amount of this information is proportional to the expansion rate of the volume, which is given by the sum of the positive exponents.

If the sequence of a time series generated by differential equations is reversed, the positive exponents will become negative ones and vice versa, and so the sum of negative exponents can be measured on a reversed series.[111] If the time series was produced by a discrete map rather than a continuous flow (from differential equations), then the governing equations need not be reversible; for such a system the time-reversed series will no longer be predictable. Therefore, examining a series backward as well as forward is useful for determining whether a dynamical system is invertible and, if it is, the rate at which volumes contract. Note that reversing a time series is very different from actually running the dynamics backward; there may not be a natural measure for the backward dynamics and, if there is one, it will usually not be the same as that of the forward dynamics. Since Lyapunov exponents are defined by time averages (and hence with respect to the natural measure), they will also change.

If the embedding dimension d is large enough, the redundancy is just the difference between the scalar entropy and an estimate of the source entropy:

$$R_d(\tau, N) \approx H_1(\tau, N) - h(\tau, N). \tag{39}$$

In the limit of small lags,

$$H_{d-1}(0, N) = H_d(0, N) \Rightarrow R_d(0, N) = H_1(N), \tag{40}$$

and for long lags

$$\lim_{\tau \to \infty} H_d(\tau, N) = d H_1(\tau, N) \Rightarrow R_d(\infty, N) = 0. \tag{41}$$

The value of τ where the redundancy vanishes provides an estimate of the limit of predictability of the system at that resolution (prediction horizon) and will be very short if d is less than the minimum embedding dimension. Once d is larger than the embedding dimension (if there is one), then the redundancy will decay much more slowly, and the slope for small τ will be the source entropy:

$$R_d(\tau, N) = H_1(N) - \tau h(1). \tag{42}$$

We have seen that it is possible from the block entropy (Eq. (27)) and the redundancy (Eq. (34)) to estimate the resolution, minimum embedding dimension, information dimension D_1, source entropy, and the prediction horizon of the data, and hence learn about the number of degrees of freedom underlying the time series and the rate at which it loses memory of its initial conditions. It is also possible to test for nonlinearity by comparing the redundancy with a linear analog defined in terms of the correlation matrix.[110] The redundancy can be efficiently computed with an $O(N)$ algorithm by sorting the measured values on a simple fixed-resolution binary tree.[53] This tree sort is related to the frequently rediscovered fact that box-counting algorithms (such as are needed for estimating dimensions and entropies) can be implemented in high-dimensional space with an $O(N \log N)$ algorithm requiring no auxiliary storage by sorting the appended indices of lagged vectors.[115] Equal-probability data structures can be used (at the expense of computational complexity) to generate more reliable unbiased entropy estimates.[40]

Figure 8 shows the results of our redundancy calculation for Data Set A. This figure was computed using 10,000 data points sorted on the three most significant bits. Note that three lags are sufficient to retrieve most of the predictable structure. This is in agreement with the exploratory stereo pairs (Figure 5), where the three dimensions plotted appear to be sufficient for an embedding. Furthermore, we can read off from Figure 8 that the system has lost memory of its initial condition after about 100 steps.

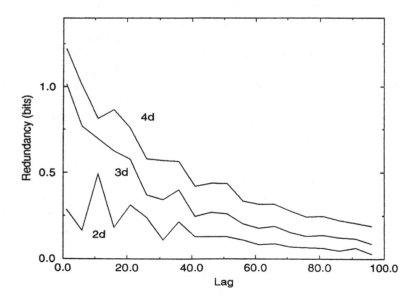

FIGURE 8   The redundancy (incremental mutual information) of Data Set A as a function of the number of time steps. The figure indicates that three past values are sufficient to retrieve most of the predictable structure and that the system has lost the memory of its initial conditions after roughly 100 steps.

There are many other approaches to characterizing embedded data (and choosing embedding parameters): Liebert and Schuster[91] comment on a good choice for the delay time by relating the first minimum of mutual information plotted as a function of the lag time (suggested by Fraser & Swinney[40]) to the generalized correlation integral. Alesić[4] plots the distances between images of close points as a function of the number of lags: at the minimum embedding dimension the distance suddenly becomes small. Savit and Green,[61] and Pi and Peterson[114] exploit the notion of continuity of a function (of the conditional probabilities in the lag space as the number of lags increases). Kennel, Brown, and Abarbanel[76] look for "false neighbors" (once the space is large enough for the attractor to unfold, they disappear). Further methods are presented by Kaplan, and by Pineda and Sommerer.[115] Gershenfeld[51] discusses how high-dimensional probability distributions limit the most complicated deterministic system that can be distinguished from a stochastic one. (The expected fraction of "typical" points in the interior of a distribution is increased; this is true because the ratio of the volume of a thin shell near the surface to the overall volume tends to 1 with increasing dimension.) Work needs to be done to understand the relations, strengths, and weaknesses of these algorithms. However, regardless of the algorithm employed, it is crucial to understand the nature of the errors in the results (both statistical uncertainty and possible artifacts), and to remember that there is no "right" answer for the time delay τ and the number of lags d—the choices will always depend on the goal.

The entries for the analysis part of the competition showed good agreement in the submitted values of the correlation dimension for Data Set A (2.02, 2.05, 2.06, 2.07, 2.2), but the estimates of the the positive Lyapunov exponent (either directly or from the source entropy) were more scattered (.024, .037, .07, .087, .089 bits/step). There were fewer estimates of these quantities for the other data sets (because they were more complicated) and they were more scattered; estimates of the degrees of freedom of Data Set D ranged from 4 to 8.

Insight into embedding can be used for more than characterization; it can also be used to distinguish between measurement noise and intrinsic dynamics. If the system is known, this can be done with a Wiener filter, a two-sided linear filter that estimates the most likely current value (rather than a future value). This method, however, requires advance knowledge of the power spectra of both the desired signal and of the noise, and the recovery will be imperfect if these spectra overlap (Priestley[120], p. 775; Press,[119] p. 574). State-space reconstruction through time-average expectations (Eq. (16)) provides a method for signal separation that requires information only about the noise. If the true observable $x(t)$ is corrupted by additive measurement noise $n(t)$ that is uncorrelated with the system, then the expectation will factor:

$$\langle e^{i\mathbf{k}\cdot(\mathbf{x}(t)+\mathbf{n}(t))}\rangle = \langle e^{i\mathbf{k}\cdot\mathbf{x}(t)}\rangle\langle e^{i\mathbf{k}\cdot\mathbf{n}(t)}\rangle\,; \tag{43}$$

$\mathbf{k}$ is the wave vector indexing the Fourier transform of the state-space probability density. The noise produces a $\mathbf{k}$-dependent correction in the embedding space; if the noise is uncorrelated with itself on the time scale of the lag time τ (as for additive white noise), then this correction can be estimated and removed solely from knowledge of the probability distribution for the noise.[52] This algorithm requires sufficient data to accurately estimate the expectation values; Marteau and Abarbanel,[96] Sauer,[133] and Kantz[74] describe more data-efficient alternatives based on distinguishing between the low-dimensional system's trajectory and the high-dimensional behavior associated with the noise. Much less is known about the much harder problem of separating noise that enters into the dynamics.[62]

6.3 INDIRECT CHARACTERIZATION: UNDERSTANDING THROUGH LEARNING

In Section 3.1 we showed that a linear time system is fully characterized by its Fourier spectrum (or equivalently by its ARMA coefficients or its autocorrelation function). We then showed how we have to go beyond that in the case of nonlinear systems and focused in Section 6.2 on the properties of the observed points in embedding space. As with forecasting, we move from the direct approach to the case where the attempt to understand the system directly does not succeed: we now show examples of how time series can be characterized through forecasting. The price for this appealing generality will be less insight into the meaning of the results. Both classes of algorithms that were successful in the competition will be put to work for characterization; in Section 6.3.1 we use local linear models to obtain DVS plots ("deterministic vs. stochastic") and, in Section 6.3.2, we explain how properties of the system that are not directly accessible can be extracted from connectionist networks that were trained to emulate the system.

6.3.1 CHARACTERIZATION VIA LOCAL LINEAR MODELS: DVS PLOTS. Forecasting models often possess some "knobs" that can be tuned. The dependence of the prediction error on the settings of these knobs can sometimes reveal—indirectly— some information about properties of the system. In local linear modeling, examples of such knobs are the number of delay values d and the number of neighbors k used to construct the local linear model. Casdagli[16] introduces the term "deterministic vs. stochastic modeling" (DVS) for the turning of these knobs, and Casdagli and Weigend[19] apply the idea to the competition data.

In Figure 9 we show the out-of-sample performance for a local linear model on three of the Santa Fe data sets (laser data A.con, computer-generated data D1.dat, and the heart data B2.dat) as a function of the number of neighbors k used for the linear fit. The left side of each plot corresponds to a simple look-up of the neighbor closest in lag space; the right corresponds to a global linear model that fits a hyperplane through all points. In all three panels the scale of the y-axis (absolute errors) is the same. (Before applying the algorithm, all series were normalized to unit variance.)

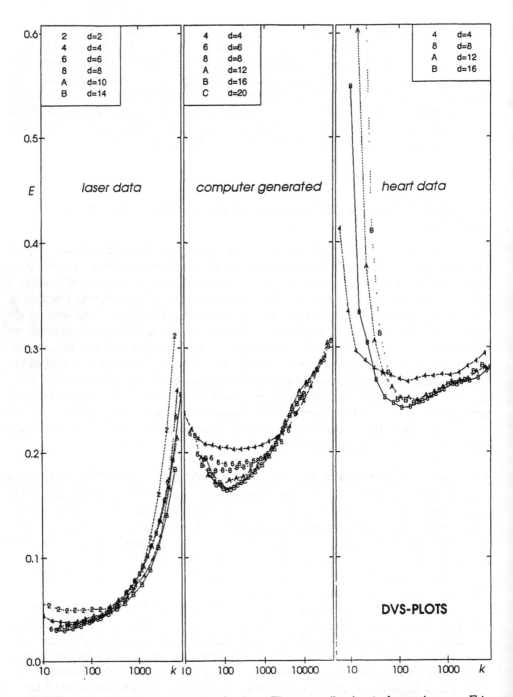

FIGURE 9   Deterministic vs. stochastic plots: The normalized out-of-sample error E is shown as function of the number of neighbors k used to construct a local linear model of order d.

The first observation is the overall size of the out-of-sample errors:[35] The laser data are much more predictable than the computer-generated data, which in turn are more predictable than the heart data. The second observation concerns the shape of the three curves. The location of the minimum shifts from the left extreme for the laser (next-neighbor look-up), through a clear minimum for the computer-generated data (for about one hundred neighbors), to a rather flat behavior beyond a sharp drop for the heart data.

So far, we have framed this discussion along the axis (local linear)↔(global linear). We now stretch the interpretation of the fit in order to infer properties of the generating system. Casdagli[16] uses the term "deterministic" for the left extreme of local linear models, and "stochastic" for the right extreme of global linear models. He motivates this description with computer-generated examples as well as with experimental data from cellular flames, EEG, fully developed turbulence, and the sunspot series. Deterministic nonlinear systems are often successfully modeled by local linear models: the error is small for small neighborhoods, but increases as more and more neighbors are included in the fit. When the neighborhood size is too large, the hyperplane does not accurately approximate the (nonlinear) manifold of the data, and the out-of-sample error increases. This is indeed the case for the laser. The local linear forecasts are 5 to 10 times more accurate than global linear ones, suggesting the interpretation of the laser as a nonlinear deterministic chaotic system. On the other hand, if a system is indeed linear and stochastic, then using smaller neighborhoods makes the out-of-sample predictions worse, due to overfitting of the noise. This is apparent for the heart data, which clearly shows overfitting for small neighborhood sizes and therefore rules out simple deterministic chaos. The DVS plot alone is not powerful enough to decide whether nonlinearities are present in the system or not.[36] Finally, the middle example of computer-generated data falls between these two cases. Nonlinearity, but not low-dimensional chaos, is suggested here since the short-term forecasts at the minimum are between 50% to 100% more accurate than global linear models.

Apart from the number of neighbors, the other knob to turn is the order of the linear model, i.e., the number of time delays d. The order of the AR model that makes successful predictions provides an upper bound on the minimum embedding

[35] We give average absolute errors since they are more robust than squared errors, but the qualitative behavior is the same for squared errors. The numerical values are obtained by dividing the sum of these "linear" errors by the size of the test example (500 points in all three series). Furthermore, the delay between each past sample ("lag time") is chosen to be one time step in all three series, but the prediction time ("lead time") is chosen to be one time step for the laser, two time steps for the computer-generated data, and four time steps for the heart rate predictions to reflect the different sampling rates with respect to the timescale of the dynamics. The predictions for the heart data were obtained from a bivariate model, using both the heart rate and the chest volume as input. The details are given in the article by Casdagli and Weigend.[19]

[36] A comparison to DVS plots for financial data (not shown here) suggests that there are more nonlinearities in the heart than in financial data.

dimension. For the laser, $d = 2$ is clearly worse than $d = 4$, and the lowest out-of-sample errors are reached for $d = 6$. This indeed is an upper estimate for $d = 3$ (compare to Figures 5 and 8). For the computer-generated data, the quality of the predictions continually increases from $d = 4$ to $d = 12$ and saturates at $d = 16$. For the heart data—since the DVS plots give no indications of low-dimensional chaos— it does not make sense to give an embedding dimension into which the geometry can be disambiguated.

We close this section with a statistical perspective on DVS plots: although their interpretation is necessarily somewhat qualitative, they nicely reflect the trade-off between bias and variance. A weak local linear model has a low bias (it is very flexible), but the parameters have a high variance (since there are only a few data points for estimating each parameter). A strong global linear model has a large model bias, but the parameters can be estimated with a small variance since many data points are available to determine the fit (see, e.g., Geman, Bienenstock, & Doursat[48]). The fact that the out-of-sample error is a combination of "true" noise and model mismatch is not limited to DVS plots but should be kept in mind as a necessary limitation of any error-based analysis.

6.3.2 CHARACTERIZATION VIA CONNECTIONIST MODELS. Connectionist models are more flexible than local linear models. We first show how it is possible to extract characteristic properties such as the minimal embedding dimension or the manifold dimension from the network, and then indicate how the network's emulation of the system can be used to estimate Lyapunov coefficients.

For simple feedforward networks (i.e., no direct connections between input and output, and no feedback loops), it is relatively easy to see how the hidden units can be used to discover hidden dimensions:

■ **Vary network size.** In the early days of backpropagation, networks were trained with varying numbers of hidden units and the "final" test error (when the training "had converged") was plotted as a function of the number of hidden units: it usually first drops and then reaches a minimum; the number of hidden units when the minimum is reached can be viewed as a kind of measure of the degrees of freedom of the system. A similar procedure can determine a kind of embedding dimension by systematically varying the number of input units.

Problems with this approach are that these numbers can be strongly influenced by the *choice of the activation function* and the search algorithm employed. Ignoring the search issue for the moment: if, for example, sigmoids are chosen as the activation function, we obtain the manifold dimension *as expressed by sigmoids,* which is an upper limit to the true manifold dimension. Saund[136] suggested, in the context of nonlinear dimensionality reduction, to sandwich the hidden layer (let us now call it the "central" hidden layer) between two (large) additional hidden layers. An interpretation of this architecture is that the time-delay input representation is transformed nonlinearly by the first "encoding" layer of hidden units; if there are more hidden units than inputs, it is an expansion into a higher dimensional space. The goal is to find a representation that makes it easy for the network subsequently to parametrize the manifold with as few parameters as possible (done by the central hidden layer). The prediction is obtained by linearly combining the activations of the final "decoding" hidden layer that follows the central hidden layer.[37]

Although an expansion with the additional sandwich layers reduces the dependence on the specific choice of the activation function, a small size of the bottleneck layer can make the *search* (via gradient descent in backpropagation) hard: overfitting even occurs for small networks, before they have reached their full potential.[169] There are two approaches to this problem: to penalize network complexity, or to use an oversized network and analyze it.

■ **Penalize network complexity or prune.** Most of the algorithms that try to produce small networks have been applied to time series prediction, e.g., "weight elimination,"[165] "soft weight sharing,"[105] and "optimal brain

damage" (developed by le Cun, Denker, & Solla,[88] and applied to time series by Svarer, Hansen, & Larsen.[147] All of these researchers apply their algorithms to the sunspot series and end up with networks of three hidden units. Finding appropriate parameters for the regularizer can be tricky; we now give a method for dimension estimation that does not have such parameters but does require post-training analysis.

■ **Analyze oversized networks.** The idea here is to use a large network that easily reaches the training goal (and also easily overfits).[38] The spectrum of the eigenvalues of the *covariance matrix of the (central) hidden unit activations* is computed as a function of training time. The covariance $C_{ij} = \langle (S_i - \overline{S}_i)(S_j - \overline{S}_j) \rangle$ describes the two-point interaction between the activations of the two hidden units i and j. ($\overline{S}_i = \langle S_i \rangle$ is the average activation of hidden unit i.) The number of significantly sized eigenvalues of the covariance matrix (its effective rank) serves as a measure of the effective dimension of

[37] This approach can also be used for cleaning and for compressing time series. In cleaning, we use a filter network that tries to remove noise in the series: the output hopefully corresponds to a noise-reduced version of the signal corresponding to a time at the center of the input time window, rather than a prediction beyond the window. In compression, the network is to reproduce the entire input vector at the output after piping it through a bottleneck layer with a small number of hidden units. The signal at the hidden units is a compressed version of the larger input vector. The three cases of prediction, cleaning, and compression are just different parametrizations of the same manifold, allocating more resources to the areas appropriate for the specific task.

[38] The network is initialized with very small weights—large enough to break the symmetry but small enough to keep the hidden sigmoids in their linear range. The weights grow as the network learns. In this sense, training time can be viewed in a regularization framework as a complexity term that penalizes weights according their size, strongly at first, and later relaxes.

the hidden unit space.[166,167] It expresses the number of parameters needed to parametrize the solution manifold of the dynamical system in terms of the primitives. Using the sandwich-expansion idea (described on the previous page), the effect of the specific primitives can be reduced. We have also used mutual information to capture dependencies between hidden units; this measure is better suited than linear correlation or covariance if the hidden units have nonlinear functional relations.

All of these approaches have to be used with caution as estimates of the true dimension of the manifold. We have pointed out above that the estimate can be too large (for example, if the sigmoid basis functions are not suitable for the manifold, or if the network is overfitting). But it can also be too small (for example, if the network has essentially learned nothing), as often is the case for financial data (either because there is nothing to be emulated or because the training procedure or the architecture was not suited to the data).

In addition to dimension, networks can be used to extract other properties of the generating system. Here we point to a few possibilities that help locate a series within the space of attributes outlined in Figure 2.

Nonlinearity. DVS plots analyze the error as a function of the nonlinearity of the model (smaller neighborhoods $\Rightarrow$ more nonlinear). Rather than basing the analysis on the errors, we can use a property of the network to characterize the amount of nonlinearity. Weigend et al.[165] analyze the distribution of the activations S of sigmoidal hidden units: they show that the ratio of the quadratic part of the Taylor expansion of a sigmoid with respect to the linear part, i.e., $|f''(\xi)|/|f'(\xi)|$, can be expressed in terms of network parameters (the activation S, the net-input ξ, the activation function f, and the slope a are defined in Section 4.2) as $(a|1 - 2S|)$. The distribution of this statistic (averaged over patterns and hidden units) can be used in addition to the simple comparison of the out-of-sample error of the network to the out-of-sample error of a linear model.

Lyapunov exponents. It is notoriously difficult to estimate Lyapunov exponents from short time records of noisy systems. The hope is that if a network has reliably learned how to emulate such a system, the exponents can be found through the network. This can be done by looking at the out-of-sample errors as a function of prediction time,[165] by using the Jacobian of the map implemented by the network,[49,106] or by using the trained network to generate time series of any length needed for the application of standard techniques.[13]

This section on characterization started with important simple tests that apply to any data set and algorithm, continued with redundancy as an example of the detailed information than can be found by analyzing embedded data, and closed with learning algorithms that are more generally applicable but less explicitly understandable. Connectionist approaches to characterization, which throw a broad model (and a lot of computer time) at the data, should be contrasted with the traditional statistical approach of building up nonlinearities by systematically adding terms to a narrow model (which can be estimated much faster than a neural network can be trained) and hoping that the system can be captured by such extensions. This is another example of the central theme of the trade-off that must be made between model flexibility and specificity. The blind (mis)application of these techniques can easily produce meaningless results, but taken together and used thoughtfully, they can yield deep insights into the behavior of a system of which a time series has been observed.

These themes have recurred throughout our survey of new techniques for time series forecasting and characterization. We have seen results that go far beyond what is possible within the canon of linear systems analysis, but we have also seen unprecedented opportunities for the analysis to go astray. We have shown that it can be possible to anticipate, detect, and prevent such errors, and to relate new algorithms to traditional practice, but these steps, as necessary as they are, require significantly more effort. The possibilities that we have presented will hopefully help motivate such an effort. We close this chapter with some thoughts about future directions.

7. THE FUTURE

We have surveyed the results of what appears to be a steady progress of insight over ignorance in analyzing time series. Is there a limit to this development? Can we hope for the discovery of a universal forecasting algorithm that will predict everything about all time series? The answer is emphatically "no!" Even for completely deterministic systems, there are strong bounds on what can be known. The search for a universal time series algorithm is related to Hilbert's vision of reducing all of mathematics to a set of axioms and a decision procedure to test the truth of assertions based on the axioms (*Entscheidungsproblem*); this culminating dream of mathematical research was dramatically dashed by Gödel[56] [39] and then by Turing. The most familiar result of Turing is the undecidability of the halting problem: it is not possible to decide in advance whether a given computer program will eventually halt.[159] But since Turing machines can be implemented with dynamical systems,[45,103] a universal algorithm that can directly forecast the value of a time series at any future time would need to contain a solution to the halting problem, because it would be able to predict whether a program will eventually halt by examining the program's output. Therefore, there cannot be a universal forecasting algorithm.

[39] Hofstadter paraphrases Gödel's Theorem as: "All consistent axiomatic formulations of number theory include undecidable propositions" (Hofstadter,[71] p. 17).

The connection between computability and time series goes deeper than this. The invention of Turing machines and the undecidability of the halting problem were side results of Turing's proof of the existence of uncomputable real numbers. Unlike a number such as π, for which there is a rule to calculate successive digits, he showed that there are numbers for which there cannot be a rule to generate their digits. If one was unlucky enough to encounter a deterministic time series generated by a chaotic system with an initial condition that was an uncomputable real number, then the chaotic dynamics would continuously reveal more and more digits of this number. Correctly forecasting the time series would require calculating unseen digits from the observed ones, which is an impossible task.

Perhaps we can be more modest in our aspirations. Instead of seeking complete future knowledge from present observations, a more realistic goal is to find the best model for the data, and a natural definition of "best" is the model that requires the least amount of information to describe it. This is exactly the aim of *Algorithmic Information Theory*, independently developed by Chaitin,[21] Kolmogorov,[79] and Solomonoff.[143] Classical information theory, described in Section 6.2, measures information with respect to a probability distribution of an ensemble of observations of a string of symbols. In algorithmic information theory, information is measured within a single string of symbols by the number of bits needed to specify the shortest algorithm that can generate them. This has led to significant extensions of Gödel and Turing's results (Chaitin[22]) and, through the *Minimum Description Length* principle, it has been used as the basis for a general theory of statistical inference (Wallace & Boulton[162]; Rissanen[125,126]). Unfortunately, here again we run afoul of the halting problem. There can be no universal algorithm to find the shortest program to generate an observed sequence because we cannot determine whether an arbitrary candidate program will continue to produce symbols or will halt (e.g., see Cover & Thomas,[28] p.162).

Although there are deep theoretical limitations on time series analysis, the constraints associated with specific domains of application can nevertheless permit strong useful results (such as those algorithms that performed well in the competition), and can leave room for significant future development. In this chapter, we have ignored many of the most important time series problems that will need to be resolved before the theory can find widespread application, including:

- *Building parametrized models for systems with varying inputs.* For example, we have shown how the stationary dynamics of the laser that produced Data Set A can be correctly inferred from an observed time series, but how can a family of models be built as the laser pump energy is varied in order to gain insight into how this parameter enters into the governing equations? The problem is that for a linear system it is possible to identify an internal transfer function that is independent of the external inputs, but such a separation is not possible for nonlinear systems.

- *Controlling nonlinear systems.* How can observations of a nonlinear system and access to some of its inputs be used to build a model that can then be used to guide the manipulation of the system into a desired state? The control of nonlinear systems has been an area of active research; approaches to this problem include both explicit embedding models[9,72,108] and implicit connectionist strategies.[101,172]

■ *The analysis of systems that have spatial as well as temporal structure.* The transition to turbulence in a large aspect-ratio convection cell is an experimental example of a spatiotemporal structure,[148] and cellular automata and coupled map lattices have been extensively investigated to explore the theoretical relationship between temporal and spatial ordering.[64] A promising approach is to extend time series embedding (in which the task is to find a rule to map past observations to future values) to spatial embedding (which aims to find a map between one spatial, or spatiotemporal, region and another). Unfortunately, the mathematical framework underlying time-delay embedding (such as the uniqueness of state-space trajectories) does not simply carry over to spatial structures. Afraimovich et al.[2] explore the prospects for developing such a theory of spatial embedding. A related problem is the combination of data from different sources, ranging from financial problems (using multiple market indicators) to medical data (such as Data Set B).

■ *The analysis of nonlinear stochastic processes.* Is it possible to extend embedding to extract a model for the governing equations for a signal generated by a stationary nonlinear stochastic differential equation? Probabilistic approaches to embedding such as the use of expectation values (Eq. (16)) and hidden Markov models[39] may point toward an approach to this problem.

■ *Understanding versus learning.* How does understanding (explicitly extracting the geometrical structure of a low-dimensional system) relate to learning (adaptively building models that emulate a complex system)? When a neural network correctly forecasts a low-dimensional system, it has to have formed a representation of the system. What is this representation? Can it be separated from the network's implementation? Can a connection be found between the entrails of the internal structure in a possibly recurrent network, the accessible structure in the state-space reconstruction, the structure in the time series, and ultimately the structure of the underlying system?

The progress in the last decade in analyzing time series has been remarkable and is well witnessed by the contributions to this volume. Where once time series analysis was shaped by linear systems theory, it is now possible to recognize when an apparently complicated time series has been produced by a low-dimensional nonlinear system, to characterize its essential properties, and to build a model that can be used for forecasting. At the opposite extreme, there is now a much richer framework for designing algorithms that can learn the regularities of time series that do not have a simple origin. This progress has been inextricably tied to the arrival of the routine availability of significant computational resources (making it possible to collect large time series, apply complex algorithms, and interactively visualize the results), and it may be expected to continue as the hardware improves.

General access to computer networks has enabled the widespread distribution and collection of common material, a necessary part of the logistics of the competition, thereby identifying interesting results where they once might have been overlooked (such as from students, and from researchers in countries that have only recently been connected to international computer networks). This synchronous style of research is complementary to the more common asynchronous mode of relatively independent publication, and may be expected to become a familiar mechanism for large-scale scientific progress. We hope that, in the short term, the data and results from this competition can continue to serve as basic reference benchmarks for new techniques. We also hope that, in the long term, they will be replaced by more worthy successors and by future comparative studies building on this experience.

In summary, in this overview chapter we started with linear systems theory and saw how ideas from fields such as differential topology, dynamical systems, information theory, and machine learning have helped solve what had appeared to be very difficult problems, leading to both fundamental insights and practical applications. Subject to some ultimate limits, there are grounds to expect significant further extensions of these approaches for handling a broader range of tasks. We predict that a robust theory of nonlinear time series prediction and analysis (and nonlinear signal processing in general) will emerge that will join spectrum analysis and linear filters in any scientist's working toolkit.

ACKNOWLEDGMENTS

An international interdisciplinary project such as this requires the intellectual and financial support of a range of institutions; we would particularly like to thank the Santa Fe Institute, the NATO Science and Technology Directorate, the Harvard Society of Fellows, the MIT Media Laboratory, the Xerox Palo Alto Research Center, and, most importantly, all of the researchers who contributed to the Santa Fe Time Series Competition.

APPENDIX

We describe the prediction tasks and competition results in more detail. We have included only those entries that were received by the close of the competition; many people have since refined their analyses. This appendix is not intended to serve as an exhaustive catalog of what is possible; it is just a basic guide to what has been done.

DATA SET A: LASER

Submissions were evaluated in two ways. First, using the predicted values $\widehat{x}_k$ only (in addition to the observed values x_k), we compute the *normalized mean squared error*:

$$\text{NMSE}(N) = \frac{\sum_{k \in T} (\text{observation}_k - \text{prediction}_k)^2}{\sum_{k \in T} (\text{observation}_k - \text{mean}_T)^2} \approx \frac{1}{\widehat{\sigma}_T^2} \frac{1}{N} \sum_{k \in T} (x_k - \widehat{x}_k)^2 \,,$$

(44)

where $k = 1 \cdots N$ enumerates the points in the withheld test set T, and mean_T and $\widehat{\sigma}_T^2$ denote the sample average and sample variance of the observed values (targets) in T. A value of $\text{NMSE} = 1$ corresponds to simply predicting the average.

Second, the submitted error bars $\widehat{\sigma}_k$ were used to compute the likelihood of the observed data, given the predicted values and the predicted error bars, based on an assumption of independent Gaussian errors. Although this assumption may not be justified, it provides a simple form that captures some desirable features of error weighting. Since the original real-valued data is quantized to integer values, the probability of seeing a given point x_k is found by integrating the Gaussian distribution over a unit interval (corresponding to the rounding error of 1 bit of the analog-to-digital converter) centered on x_k:

$$p(x_k | \widehat{x}_k, \widehat{\sigma}_k) = \frac{1}{\sqrt{2\pi\widehat{\sigma}_k^2}} \int_{x_k - 0.5}^{x_k + 0.5} \exp\left(\frac{-(\xi - \widehat{x}_k)^2}{2\widehat{\sigma}_k^2}\right) d\xi \,.$$

(45)

If the predicted error is large, then the computed probability will be relatively small, independent of the value of the predicted point. If the predicted error is small and the predicted point is close to the observed value, then the probability will be large, but if the predicted error is small and the prediction is not close to the observed value, then the probability will be very small. The potential reward, as well as the risk, is greater for a confident prediction (small error bars). Under the assumption of independent errors, the likelihood of the whole test for the the observed data given the submitted model is then

$$p(\text{D}|\text{M}) = \prod_{k=1}^{N} p(x_k | \widehat{x}_k, \widehat{\sigma}_k) \,.$$

(46)

TABLE 2 Entries received before the deadline for the prediction of Data Set A (laser). We give the normalized mean squared error (NMSE), and the negative logarithm of the likelihood of the data given the predicted values and predicted errors. Both scores are averaged over the prediction set of 100 points.

code	method	type	computer	time	NMSE(100)	-log(lik.)
W	conn	1-12-12-1; lag 25,5,5	SPARC 2	12 hrs	0.028	3.5
Sa	loc lin	low-pass embd, 8 dim, 4nn	DEC 3100	20 min	0.080	4.8
McL	conn	feedforward, 200-100-1	CRAY Y-MP	3 hrs	0.77	5.5
N	conn	feedforward, 50-20-1	SPARC 1	3 wks.	1.0	6.1
K	visual	look for similar stretches	SG Iris	10 sec	1.5	6.2
L	visual	look for similar stretches			0.45	6.2
M	conn	feedforward, 50-350-50-50	386 PC	5 days	0.38	6.4
Can	conn	recurrent, 4-4c-1	VAX 8530	1 hr	1.4	7.2
U	tree	k-d tree; AIC	VAX 6420	20 min	0.62	7.3
A	loc lin	21 dim, 30 nn	SPARC 2	1 min	0.71	10.
P	loc lin	3 dim time delay	Sun	10 min	1.3	–
Sw	conn	feedforward	SPARC 2	20 hrs	1.5	–
Y	conn	feedforward, weight-decay	SPARC 1	30 min	1.5	–
Car	linear	Wiener filter, width 100	MIPS 3230	30 min	1.9	–

Finally, we take the logarithm of this probability of the data given the model (this turns products into sums and also avoids numerical problems), and then scale the result by the size of the data set N. This defines the *negative average log likelihood*:

$$-\frac{1}{N}\sum_{k=1}^{N}\log p(x_k|\widehat{x}_k,\widehat{\sigma}_k)\,. \tag{47}$$

We give these two statistics for the submitted entries for Data Set A in Table 2 along with a brief summary of the entries. The best two entries (W=Wan, Sa=Sauer) are shown in Figures 3 and 4 in the main text. Figure 10 displays all predictions received for Data Set A. The symbols (×) correspond to the predicted values, the vertical bars to the submitted error bars. The true continuation points are connected by a grey line (to guide the eye).

DATA SET D: COMPUTER-GENERATED DATA

In order to provide a relatively long series of known high-dimensional dynamics (between the extremes of Data Set A and Data Set C) with weak nonstationarity, we generated 100,000 points by numerically integrating the equations of motion for a damped, driven particle

$$\frac{d^2\mathbf{x}}{dt^2} + \gamma\frac{d\mathbf{x}}{dt} + \nabla V(\mathbf{x}) = \mathbf{F}(t) \tag{48}$$

in an asymmetrical four-dimensional four-well potential (Figure 11)

$$V(\mathbf{x}) = A_4\left(x_1^2 + x_2^2 + x_3^2 + x_4^2\right)^2 - A_2\,|x_1 x_2| - A_1 x_1 \tag{49}$$

with periodic forcing

$$\mathbf{F}(t) = F\sin(\omega t)\,\widehat{\mathbf{x}}_3. \tag{50}$$

$\mathbf{x} = (x_1, x_2, x_3, x_4)$ denotes the location of the particle. This system has nine degrees of freedom (four position, four velocity, and one forcing time). The equations were integrated with 64-bit real numbers using a fixed-step fourth-order Runge-Kutta algorithm (to eliminate possible coupling to an active stepper). The potential has four wells that are tilted by the parameter A_1. This parameter slowly drifted during the generation of the data according to a small-biased random walk from 0.02 to 0.06. As a scalar observable we chose

$$\sqrt{(x_1 - 0.3)^2 + (x_2 - 0.3)^2 + x_3^2 + x_4^2}. \tag{51}$$

This observable projects the four wells onto three distinguishable states. These three states, and the effect of the drift of A_1 can be seen in the probability distributions at the beginning and the end of the data set (Figure 12). The magnitude of the drift is chosen to be small enough such that the nature of the dynamics does not change, but large enough such that the relative probabilities are different at the beginning and end of the data set. Data Set D was generated with $A_4 = 1$, $A_2 = 1$, A_1 drifting from 0.02 to 0.06, $\gamma = 0.01$, $F = 0.135$, and with $\omega = 0.6$ (for these parameters, in the absence of the weak forcing by the drift, this system is not chaotic).

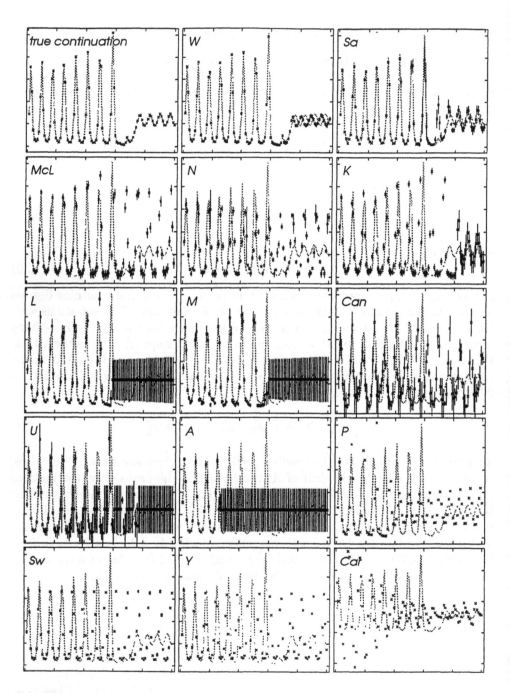

FIGURE 10 Continuations of Data Set A (laser). The letters correspond to the code of the entrant. Grey lines indicate the true continuation, × the predicted values, and vertical bars the predicted uncertainty.

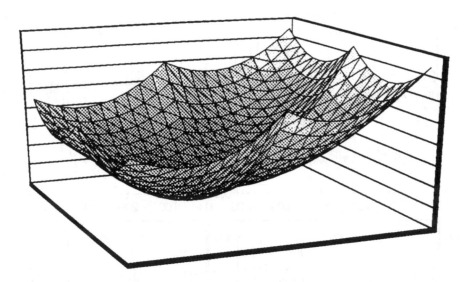

FIGURE 11   The potential $V(\mathbf{x})$ for Data Set D, plotted above the (x_1, x_2) plane.

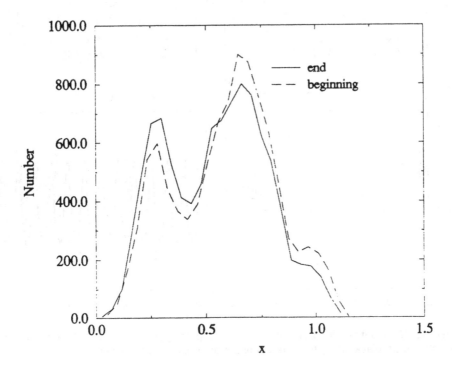

FIGURE 12   Histogram of the probability distribution at the beginning and end of Data Set D, indicating three observable states and the small drift in the relative probabilities.

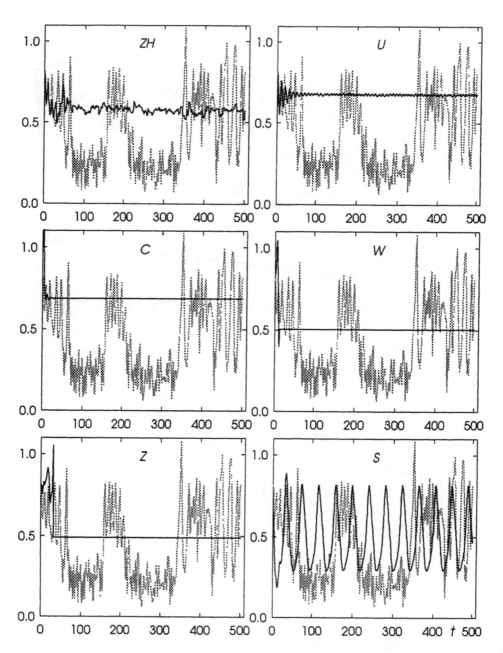

FIGURE 13 Continuations of Data Set D (computer-generated data). The predictions are shown as black lines, the true continuation is indicated in grey.

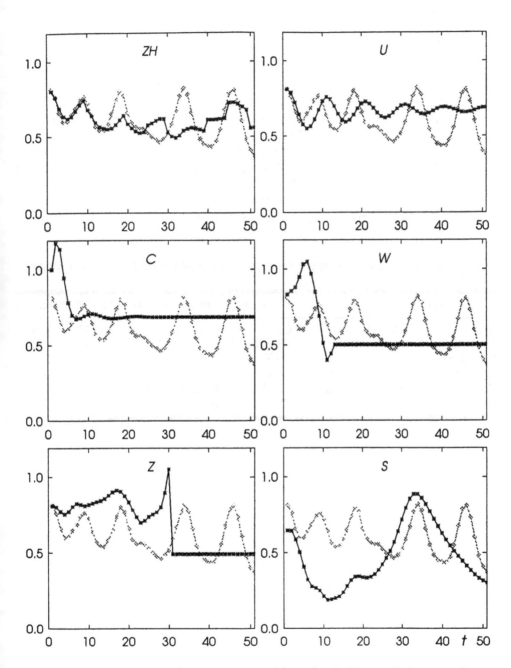

FIGURE 14   Continuations of first 50 points of Data Set D. The predictions are shown as black lines, the true continuation is indicated in grey.

The contributions received before the deadline are listed in Table 3; they are evaluated by normalized mean squared error, NMSE, averaged over the first 15, 30, and 50 predictions. Figure 13 shows the predictions received for this data set over the entire prediction range of 500 points. The first 50 points of the prediction range are plotted again in Figure 14. The first 15 steps of the best entry (ZH = Zhang and Hutchinson) is also shown in Figure 7 in the main text, in addition to the prediction of the evolution of the probability density function, submitted by Fraser and Dimitriadis after the deadline of the competition. On this time scale, the spread of an ensemble of continuations due to the stochasticity of the algorithm used to generate Dat Set D is small ($\sim 1\%$).

The program and parameter file that we used to generate Data Set D are available through anonymous ftp to `ftp.santafe.edu` .

TABLE 3 Entries received before the deadline for the prediction of Data Set D (computer-generated data).

index	method	type	computer	time	NMSE(15)	NMSE(30)	NMSE(50)
ZH	conn	...-30-30-1 & 30-100-5	CM-2 (16k)	8 days	0.086	0.57	0.87
U	tree	k-d tree; AIC	VAX 6420	30 min	1.3	1.4	1.4
C	conn	recurrent, 4-4c-1	VAX 8530	n/a	6.4	3.2	2.2
W	conn	1-30-30-1; lags 20,5,5	SPARC 2	1 day	7.1	3.4	2.4
Z	linear	36 AR(8), last 4k pts.	SPARC	10 min	4.8	5.0	3.2
S	conn	feedforward	SPARC 2	20 hrs	17.	9.5	5.5

REFERENCES

1. Abelson, H. "The Bifurcation Interpreter: A Step Towards the Automatic Analysis of Dynamical Systems." *Intl. J. Comp. & Math. Appl.* **20** (1990): 13.
2. Afraimovich, V. S., M. I. Rabinovich, and A. L. Zheleznyak. "Finite-Dimensional Spatial Disorder: Description and Analysis." In *Time Series Prediction: Forecasting the Future and Understanding the Past*, edited by A. S. Weigend and N. A. Gershenfeld, 539–557. Santa Fe Institute Studies in the Sciences of Complexity, Proc. Vol. XV. Reading, MA: Addison-Wesley, 1993.
3. Akaike, H. "Statistical Predictor Identification." *Ann. Inst. Stat. Math.* **22** (1970): 203–217.
4. Alesić, Z. "Estimating the Embedding Dimension." *Physica D* **52** (1991): 362–368.
5. Barron, A. R. "Universal Approximation Bounds for Superpositions of a Sigmoidal Function." *IEEE Trans. Info. Theory* **39(3)** (1993): 930–945.
6. Beck, C. "Upper and Lower Bounds on the Renyi Dimensions and the Uniformity of Multifractals." *Physica D* **41** (1990): 67–78.
7. Bourgoin, M., K. Sims, S. J. Smith, and H. Voorhees. "Learning Image Classification with Simple Systems and Large Databases." *IEEE Trans. Pat. Anal. & Mach. Intel.*, 1993.
8. Box, G. E. P., and F. M. Jenkins. *Time Series Analysis: Forecasting and Control*, 2nd ed. Oakland, CA: Holden-Day, 1976.
9. Bradley, E. 1992. "Taming Chaotic Circuits." Ph.D. Thesis, Massachusetts Institute of Technology, September 1992.
10. Brock, W. A., W. D. Dechert, J. A. Scheinkman, and B. LeBaron. "A Test For Independence Based on the Correlation Dimension." Madison, WI: University of Wisconsin Press, 1988.
11. Broomhead, D. S., and G. P. King. "Extracting Qualitative Dynamics from Experimental Data." *Physica D* **20** (1986): 217–236.
12. Broomhead, D. S., and D. Lowe. "Multivariable Functional Interpolation and Adaptive Networks." *Complex Systems* **2** (1988): 321–355.
13. Brown, R., P. Bryant, and H. D. I. Abarbanel. "Computing the Lyapunov Spectrum of a Dynamical System from an Observed Time Series." *Phys. Rev. A* **43** (1991): 2787–806.
14. Buntine, W. L., and A. S. Weigend. "Bayesian Backpropagation." *Complex Systems* **5** (1991): 603–643.
15. Casdagli, M. 1989. "Nonlinear Prediction of Chaotic Time Series." *Physica D* **35** (1989): 335–356.
16. Casdagli, M. "Chaos and Deterministic versus Stochastic Nonlinear Modeling." *J. Roy. Stat. Soc. B* **54** (1991): 303–328.
17. Casdagli, M., S. Eubank, J. D. Farmer, and J. Gibson. "State Space Reconstruction in the Presence of Noise." *Physica D* **51D** (1991): 52–98.

18. Casdagli, M. C., and A. S. Weigend. "Exploring the Continuum Between Deterministic and Stochastic Modeling." In *Time Series Prediction: Forecasting the Future and Understanding the Past,* edited by A. S. Weigend and N. A. Gershenfeld, 347–366. Reading, MA: Addison-Wesley, 1993.

19. Casdagli, M., and A. S. Weigend. "Exploring the Continuum Between Deterministic and Stochastic Modeling." In *Time Series Prediction: Forecasting the Future and Understanding the Past,* edited by A. S. Weigend and N. A. Gershenfeld, 347–366. Santa Fe Institute Studies in the Sciences of Complexity, Proc. Vol. XV. Reading, MA: Addison-Wesley, 1993.

20. Catlin, D. E. *Estimation, Control, and the Discrete Kalman Filter.* Applied Mathematical Sciences, Vol. 71. New York: Springer-Verlag, 1989.

21. Chaitin, G. J. "On the Length of Programs for Computing Finite Binary Sequences." *J. Assoc. Comp. Mach* **13** (1966): 547–569.

22. Chaitin, G. J. *Information, Randomness & Incompleteness.* Series in Computer Science, Vol. 8, 2nd ed. Singapore: World-Scientific, 1990.

23. Chatfield, C. "What is the Best Method in Forecasting?" *J. Appl. Stat.* **15** (1988): 19–38.

24. Chatfield, C. *The Analysis of Time Series,* 4th ed. London: Chapman and Hall, 1989.

25. Clemens, J. C. "Whole Earth Telescope Observations of the White Dwarf Star (PG1159-035)." In *Time Series Prediction: Forecasting the Future and Understanding the Past,* edited by A. S. Weigend and N. A. Gershenfeld, 239–150. Santa Fe Institute Studies in the Sciences of Complexity, Proc. Vol. XV. Reading, MA: Addison-Wesley, 1993.

26. Collet, P., and J.-P. Eckmann. *Iterated Maps on the Interval as Dynamical Systems.* Boston: Birkhäuser, 1980.

27. Cover, T. M. "Geometrical and Statistical Properties of Systems of Linear Inequalities with Applications in Pattern Recognition." *IEEE Trans. Elec. Comp.* **14** (1965): 326–334.

28. Cover, T. M., and J. A. Thomas. *Elements of Information Theory.* New York: John Wiley, 1991.

29. Cremers, J., and A. Hübler. "Construction of Differential Equations from Experimental Data." *Z. Naturforsch.* **42(a)** (1987): 797–802.

30. Crutchfield, J. P., and B. S. McNamara. "Equations of Motion from a Data Series." *Complex Systems* **1** (1987): 417–452.

31. Crutchfield, J. P., and K. Young. "Inferring Statistical Complexity." *Phys. Rev. Lett.* **63** (1989): 105–108.

32. Cybenko, G. "Approximation by Superpositions of a Sigmoidal Function." *Math. Control, Signals, & Sys.* **2(4)** (1989).

33. Diebold, F. X., and J. M. Nason. "Nonparametric Exchange Rate Prediction?" *J. Intl. Econ.* **28** (1990): 315–332.

34. Dirst, M., and A. S. Weigend. "Baroque Forecastsing: On Completing J. S. Bach's Last Fugue." In *Time Series Prediction: Forecasting the Future and Understanding the Past,* edited by A. S. Weigend and N. A. Gershenfeld,

151–172. Santa Fe Institute Studies in the Sciences of Complexity, Proc. Vol. XV. Reading, MA: Addison-Wesley, 1993.

35. Duda, R. O., and P. E. Hart. *Pattern Classification and Scene Analysis*. New York: Wiley, 1973.

36. Dutta, P., and P. M. Horn. "Low-Frequency Fluctuations in solids—1/f Noise." *Rev. Mod. Phys.* **53** (1981): 497–516.

37. Farmer, J. D., and J. J. Sidorowich. "Predicting Chaotic Time Series." *Phys. Rev. Lett.* **59(8)** (1987): 845–848.

38. Farmer, J. D., and J. J. Sidorowich. "Exploiting Chaos to Predict the Future and Reduce Noise." In *Evolution, Learning, and Cognition*, edited by Y. C. Lee. Singapore: World Scientific, 1988.

39. Fraser, A. M., and A. Dimitriadis. "Forecasting Probability Densities by Using Hidden Markov Models with Mixed States." In *Time Series Prediction: Forecasting the Future and Understanding the Past*, edited by A. S. Weigend and N. A. Gershenfeld, 265–282. Santa Fe Institute Studies in the Sciences of Complexity, Proc. Vol. XV. Reading, MA: Addison-Wesley, 1993.

40. Fraser, A. M., and H. L. Swinney. "Independent Coordinates for Strange Attractors from Mutual Information." *Phys. Rev. A* **33** (1986): 1134–1140.

41. Fraser, A. M. "Reconstructing Attractors from Scalar Time Series: A Comparison of Singular System and Redundancy Criteria." *Physica D* **34** (1989): 391–404.

42. Fraser, A. M. "Information and Entropy in Strange Attractors." *IEEE Trans. Info. Theory* **IT-35** (1989): 245–262.

43. Fraser, A. M. "Reconstructing Attractors from Scalar Time Series: A Comparison of Singular System and Redundancy Criteria." *Physica D* **34** (1989): 391–404.

44. Fraser, A. M. Personal communication, 1993.

45. Fredkin, E., and T. Toffoli. "Conservative Logic." *Intl. J. Theor. Phys.* **21** (1982): 219–253.

46. Friedman, J. H. "Multivariate Adaptive Regression Splines." *Ann. Stat.* **19** (1991): 1–142.

47. Funahashi, K.-I. "On the Approximate Realization of Continuous Mappings by Neural Networks." *Neur. Net.* **2** (1989): 183–192.

48. Geman, S., E. Bienenstock, and R. Doursat. "Neural Networks and the Bias/Variance Dilemma." *Neur. Comp.* **5** (1992): 1–58.

49. Gencay, R., and W. D. Dechert. "An Algorithm for the n Lyapunov Exponents of an n-Dimensional Unknown Dynamical System." *Physics D* **59** (1992): 142–157.

50. Gershenfeld, N. A. "An Experimentalist's Introduction to the Observation of Dynamical Systems." In *Directions in Chaos*, edited by B.-L. Hao, Vol. 2, 310–384. Singapore: World Scientific, 1989.

51. Gershenfeld, N. A. "Dimension Measurement on High-Dimensional Systems." *Physica D* **55** (1992): 135–154.

52. Gershenfeld, N. A. "Embedding, Expectations, and Noise." Preprint, 1993.

53. Gershenfeld, N. A. "Information in Dynamics." In *Proceedings of the Workshop on Physics of Computation*, edited by D. Matzke, 276–280. Los Alamitos, CA: IEEE Press, 1993.

54. Gingerich, O. *The Great Copernicus Chase and Other Adventures in Astronomical History*. Cambridge, MA: Sky, 1992.

55. Giona, M., F. Lentini, and V. Cimagalli. "Functional Reconstruction and Local Prediction of Chaotic Time Series." *Phys. Rev. A* **44 (1991)**: 3496–3502.

56. Gödel, K. "Über formal unentscheibare Sätze der *Principia Mathematica* und verwandter Systeme, I." *Monatshefte für Mathematik und Physik* **38** (1931): 173–198. An English translation of this paper is found in *On Formally Undecidable Propositions* by K. Gödel (New York: Basic Books, 1962).

57. Granger, C. W. J., and A. P. Andersen. *An Introduction to Bilinear Time Series Models*. Gottingen: Vandenhoek and Ruprecht, 1978.

58. Grassberger, P., and I. Procaccia. "Characterization of Strange Attractors." *Phys. Rev. Lett.* **50** (1983): 346–349.

59. Grassberger, P. "Finite Sample Corrections to Entropy and Dimension Estimates." *Phys. Lett A* **128** (1988): 369–373.

60. Grebogi, C., S. M. Hammel, J. A. Yorke, and T. Sauer. "Shadowing of Physical Trajectories in Chaotic Dynamics: Containment and Refinement." *Phys. Rev. Lett.* **65** (1990): 1527.

61. Green, M. L., and R. Savit. "Dependent Variables in Broadband Continuous Time Series." *Physica D* **50** (1991): 521–544.

62. Guckenheimer, J. "Noise in Chaotic Systems." *Nature* **298** (1982): 358–361.

63. Guillemin, V., and A. Pollack. *Differential Topology*. Englewood Cliffs, NJ: Prentice-Hall, 1974.

64. Gutowitz, H., ed. *Cellular Automata, Theory and Experiment*. Cambridge, MA: MIT Press, 1991.

65. Hammerstrom, D. "Neural Networks at Work." *IEEE Spectrum* **June** (1993): 26–32.

66. Haussler, Personal communication, 1993.

67. Hentschel, H. G. E., and I. Procaccia. "The Infinite Number of Generalized Dimensions of Fractals and Strange Attractors." *Physica D* **8** (1983): 435–444.

68. Hertz, J. A., A. S. Krogh, and R. G. Palmer. *Introduction to the Theory of Neural Computation*. Santa Fe Institute Studies in the Sciences of Complexity, Lect. Notes Vol. I. Redwood City, CA: Addison-Wesley, 1991.

69. Hinton, G. E., and T. J. Sejnowski. "Learning and Relearning in Boltzmann Machines." In *Parallel Distributed Processing*, edited by D. E. Rumelhart and J. L. McClelland, Vol. 1. Cambridge, MA: MIT Press, 1986.

70. Hinton, G. E., and D. van Camp. "Keeping Neural Networks Simple by Minimizing the Description Length of the Weights." Preprint, Computer Science Department, University of Toronto, June 1993.

71. Hofstadter, D. R. *Gödel, Escher, Bach: An Eternal Golden Braid*. New York: Basic Books, 1979.

72. Hübler, A. "Adaptive Control of Chaotic Systems." *Helv. Phys. Acta* **62** (1989): 343–346.
73. Hübner, U., C. O. Weiss, N. B. Abraham, and D. Tang. "Lorenz-Like Chaos in NH_3-FIR Lasers." In *Time Series Prediction: Forecasting the Future and Understanding the Past*, edited by A. S. Weigend and N. A. Gershenfeld, 73–104. Santa Fe Institute Studies in the Sciences of Complexity, Proc. Vol. XV. Reading, MA: Addison-Wesley, 1993.
74. Kantz, H. "Noise Reduction by Local Reconstruction of the Dynamics." In *Time Series Prediction: Forecasting the Future and Understanding the Past*, edited by A. S. Weigend and N. A. Gershenfeld, 475–490. Santa Fe Institute Studies in the Sciences of Complexity, Proc. Vol. XV. Reading, MA: Addison-Wesley, 1993.
75. Kaplan, D. T. "A Geometrical Statistic for Detecting Deterministic Dynamics." In *Time Series Prediction: Forecasting the Future and Understanding the Past*, edited by A. S. Weigend and N. A. Gershenfeld, 415–428. Santa Fe Institute Studies in the Sciences of Complexity, Proc. Vol. XV. Reading, MA: Addison-Wesley, 1993.
76. Kennel, M. B., R. Brown, and H. D. I. Abarbanel. "Determining Minimum Embedding Dimension Using a Geometrical Construction." *Phys. Rev. A* **45** (1992): 3403–3411.
77. Knuth, D. E. *Semi-Numerical Algorithms. Art of Computer Programming*, Vol. 2, 2nd ed. Reading, MA: Addison-Wesley, 1981.
78. Kolmogorov, A. "Interpolation und Extrapolation von stationären zufälligen Folgen." *Bull. Acad. Sci. (Nauk)* **5** (1941): 3–14.
79. Kolmogorov, A. N. "Three Approaches to the Quantitative Definition of Information." *Prob. Infor. Trans.* **1** (1965): 4–7.
80. Koza, J. R. *Genetic Programming*. Cambridge, MA: MIT Press, 1993.
81. Kung, S. Y. *Digital Neural Networks*. Englewood Cliffs, NJ: Prentice Hall, 1993.
82. Laird, P., and R. Saul. "Discrete Sequence Prediction and Its Applications." *Machine Learning*, 1993.
83. Landauer, R. "Information is Physical." *Physics Today* **44** (1991): 23.
84. Lang, K. J., A. H. Waibel, and G. E. Hinton. "A Time-Delay Neural Network Architecture for Isolated Word Recognition." *Neur. Net.* **3** (1990): 23–43.
85. Lapedes, A., and R. Farber. "Nonlinear Signal Processing Using Neural Networks." Technical Report No. LA-UR-87-2662, Los Alamos National Laboratory, Los Alamos, NM, 1987.
86. LeBaron, B. "Nonlinear Diagnostics and Simple Trading Rules for High-Frequency Foreign Exchange Rates." In *Time Series Prediction: Forecasting the Future and Understanding the Past*, edited by A. S. Weigend and N. A. Gershenfeld, 457–474. Santa Fe Institute Studies in the Sciences of Complexity, Proc. Vol. XV. Reading, MA: Addison-Wesley, 1993.

87. le Cun, Y. "Generalization and Network Design Strategies." In *Connectionism in Perspective*, edited by R. Pfeifer, Z. Schreter, F. Fogelman, and L. Steels. Amsterdam: North Holland, 1989.

88. le Cun, Y., J. S. Denker, and S. A. Solla. "Optimal Brain Damage." In *Advances in Neural Information Processing Systems 2 (NIPS*89)*, edited by D. S. Touretzky, 598–605. San Mateo, CA: Morgan Kaufmann, 1990.

89. Lequarré, J. Y. "Foreign Currency Dealing: A Brief Introduction." In *Time Series Prediction: Forecasting the Future and Understanding the Past*, edited by A. S. Weigend and N. A. Gershenfeld, 131–138. Santa Fe Institute Studies in the Sciences of Complexity, Proc. Vol. XV. Reading, MA: Addison-Wesley, 1993.

90. Lewis, P. A. W., B. K. Ray, and J. G. Stevens. "Modeling Time Series by Using Multivariate Adaptive Regression Splines (MARS)." In *Time Series Prediction: Forecasting the Future and Understanding the Past*, edited by A. S. Weigend and N. A. Gershenfeld, 297–318. Santa Fe Institute Studies in the Sciences of Complexity, Proc. Vol. XV. Reading, MA: Addison-Wesley, 1993.

91. Liebert, W., and H. G. Schuster. "Proper Choice of the Time Delay for the Analysis of Chaotic Time Series." *Phys. Lett. A* **142** (1989): 107–111.

92. Lorenz, E. N. "Deterministic Non-periodic Flow." *J. Atmos. Sci.* **20** (1963): 130–141.

93. Lorenz, E. N. "Computational Chaos—A Prelude to Computational Instability." *Physica D* **35** (1989): 299–317.

94. Makridakis, S., A. Andersen, R. Carbone, R. Fildes, M. Hibon, R. Lewandowski, J. Newton, E. Parzen, and R. Winkler. *The Forecasting Accuracy of Major Time Series Methods*. New York: Wiley, 1984.

95. Makridakis, S., and M. Hibon. "Accuracy of Forecasting: An Empirical Investigation." *J. Roy. Stat. Soc. A* **142** (1979): 97–145.

96. Marteau, P. F., and H. D. I. Abarbanel. "Noise Reduction in Chaotic Time Series Using Scaled Probabilistic Methods." *J. Nonlinear Sci.* **1** (1991): 313.

97. May, R. M. "Simple Mathematical Models with Very Complicated Dynamics." *Nature* **261** (1976): 459.

98. Melvin, P., and N. B. Tufillaro. "Templates and Framed Braids." *Phys. Rev. A* **44** (1991): R3419–R3422.

99. Meyer, T. P., and N. H. Packard. 1992. "Local Forecasting of High-Dimensional Chaotic Dynamics." In *Nonlinear Modeling and Forecasting*, edited by M. Casdagli and S. Eubank. Santa Fe Institute Studies in the Sciences of Complexity, Proc. Vol. XII, 249–263. Reading, MA: Addison-Wesley.

100. Mitchison, G. J., and R. M. Durbin. "Bounds on the Learning Capacity of Some Multi-Layer Networks." *Biol. Cyber.* **60** (1989): 345–356.

101. Miller, W. T., R. S. Sutton, and P. J. Werbos. *Neural Networks for Control*. Cambridge, MA: MIT Press, 1990.

102. Moody, J. "The *Effective* Number of Parameters: An Analysis of Generalization and Regularization in Nonlinear Systems." In *Advances in Neural Information Processing Systems 4*, edited by J. E. Moody, S. J. Hanson, and R. P. Lippmann, 847–854. San Mateo, CA: Morgan Kaufmann, 1992.

103. Moore, C. "Generalized Shifts: Unpredictability and Undecidability in Dynamical Systems." *Nonlinearity* 4 (1991): 199–230.

104. Mozer, M. X. "Neural Net Architectures for Temporal Sequence Processing." In *Time Series Prediction: Forecasting the Future and Understanding the Past*, edited by A. S. Weigend and N. A. Gershenfeld, 243–264. Santa Fe Institute Studies in the Sciences of Complexity, Proc. Vol. XV. Reading, MA: Addison-Wesley, 1993.

105. Nowlan, S. J., and G. E. Hinton. "Simplifying Neural Networks by Soft Weight-Sharing." *Neur. Comp.* 4 (1992): 473–493.

106. Nychka, D., S. Ellner, D. McCaffrey, and A. R. Gallant. "Finding Chaos in Noisy Systems." *J. Roy. Stat. Soc. B* **54(2)** (1992): 399–426.

107. Oppenheim, A. V., and R. W. Schafer. *Discrete-Time Signal Processing.* Englewood Cliffs, NJ: Prentice Hall, 1989.

108. Ott, E., C. Grebogi, and J. A. Yorke. "Controlling Chaos." *Phys. Rev. Lett.* **64** (1990): 1196.

109. Packard, N. H., J. P. Crutchfield, J. D. Farmer, and R. S. Shaw. "Geometry from a Time Series." *Phys. Rev. Lett.* **45(9)** (1980): 712–716.

110. Paluš, M. "Identifying and Quantifying Chaos by Using Information-Theoretic Functionals." In *Time Series Prediction: Forecasting the Future and Understanding the Past*, edited by A. S. Weigend and N. A. Gershenfeld, 387–414. Santa Fe Institute Studies in the Sciences of Complexity, Proc. Vol. XV. Reading, MA: Addison-Wesley, 1993.

111. Parlitz, U. "Identification of True and Spurious Lyapunov Exponents from Time Series." *Intl. J. Bif. & Chaos* 2 (1992): 155–165.

112. Petersen, K. *Ergodic Theory*, 2nd ed. Cambridge Studies in Advanced Mathematics, Vol. 2. Cambridge, MA: Cambridge University Press, 1989.

113. Pettis, K. W., T. A. Bailey, A. K. Jain, and R. C. Dubes. "An Intrinsic Dimensionality Estimator from Near-Neighbor Information." *IEEE Trans. Patt. Anal. & Mach. Intel.* **PAMI-1** (1979): 25–37.

114. Pi, H., and C. Peterson. "Finding the Embedding Dimension and Variable Dependences in Time Series." Preprint LU TP 93-4, Department of Theoretical Physics, University of Lund, March 1993. Submitted to *Neural Computation*.

115. Pineda, F. J., and J. C. Sommerer. "Estimating Generalized Dimensions and Choosing Time Delays: A Fast Algorithm." In *Time Series Prediction: Forecasting the Future and Understanding the Past*, edited by A. S. Weigend and N. A. Gershenfeld, 367–386. Santa Fe Institute Studies in the Sciences of Complexity, Proc. Vol. XV. Reading, MA: Addison-Wesley, 1993.

116. Poggio, T., and F. Girosi. "Networks for Approximation and Learning." *Proc. IEEE* **78(9)** (1990): 1481–1497.

117. Powell, M. J. D. "Radial Basis Functions for Multivariate Interpolation: A Review." In *IMA Conference on "Algorithms for the Approximation of Functions and Data,"* edited by J. C. Mason and M. G. Cox. Shrivenham: RMCS, 1987.

118. Principe, J. C., B. de Vries, and P. Oliveira. "The Gamma Filter—A New Class of Adaptive IIR Filters with Restricted Feedback." *IEEE Trans. Sig. Proc.* **41** (1993): 649–656.

119. Press, W. H., B. P. Flannery, S. A. Teukolsky, and W. T. Vetterling. *Numerical Recipes in C: The Art of Scientific Computing*, 2nd ed. Cambridge: Cambridge University Press, 1992.

120. Priestley, M. *Spectral Analysis and Time Series.* London: Academic Press. 1981.

121. Rabiner, L. R., and B. H. Juang. "An Introduction to Hidden Markov Models." *IEEE ASSP Magazine* **January** (1986): 4–16.

122. Rico-Martinez, R., I. G. Kevrekidis, and R. A. Adomaitis. 1993. "Noninvertibility in Neural Networks." In *Proceedings of ICNN, San Francisco, 1993*, 382–386. Piscataway, NJ: IEEE Press, 1993.

123. Rigney, D. R., A. L. Goldberger, W. C. Ocasio, Y. Ichimaru, G. B. Moody, and R. G. Mark. "Multi-Channel Physiological Data: Description and Analysis." In *Time Series Prediction: Forecasting the Future and Understanding the Past*, edited by A. S. Weigend and N. A. Gershenfeld, 105–130. Santa Fe Institute Studies in the Sciences of Complexity, Proc. Vol. XV. Reading, MA: Addison-Wesley, 1993.

124. Rissanen, J., and G. G. Langdon. "Universal Modeling and Coding." *IEEE Trans. Info. Theory.* **IT-27** (1981): 12–23.

125. Rissanen, J. "Stochastic Complexity and Modeling." *Ann. Stat.* **14** (1986): 1080–1100.

126. Rissanen, J. "Stochastic Complexity." *J. Roy. Stat. Soc. B* **49** (1987): 223–239.

127. Ruelle, D., and J. P. Eckmann. "Ergodic Theory of Chaos and Strange Attractors." *Rev. Mod. Phys.* **57** (1985): 617–656.

128. Rumelhart, D. E., G. E. Hinton, and R. J. Williams. "Learning Internal Representations by Error Propagation." In *Parallel Distributed Processing: Explorations in the Microstructure of Cognition. Volume I: Foundations*, edited by D. E. Rumelhart and J. L. McClelland, 318–362. Cambridge, MA: MIT Press/Bradford Books, 1986.

129. Rumelhart, D. E., G. E. Hinton, and R. J. Williams. "Learning Representations by Back-Propagating Errors." *Nature* **323** (1986): 533–536.

130. Rumelhart, D. E., R. Durbin, R. Golden, and Y. Chauvin. "Backpropagation: The Basic Theory." In *Backpropagation: Theory, Architectures and Applications*, edited by Y. Chauvin and D. E. Rumelhart. Hillsdale, NJ: Lawrence Erlbaum, 1993.

131. Russell, D. A., J. D. Hanson, and E. Ott. "Dimension of Strange Attractors." *Phys. Rev. Lett.* **45** (1980): 1175.

132. Sakamoto, Y., M. Ishiguro, and G. Kitagawa. *Akaike Information Criterion Statistics.* Dordrecht: D. Reidel, 1986.

133. Sauer, T. "A Noise Reduction Method for Signals from Nonlinear Systems." *Physica D* **58** (1992): 193–201.

134. Sauer, T., J. A. Yorke, and M. Casdagli. "Embedology." *J. Stat. Phys.* **65(3/4)** (1991): 579–616.

135. Sauer, T. "Time Series Prediction by Using Delay Coordinate Embedding." In *Time Series Prediction: Forecasting the Future and Understanding the Past*, edited by A. S. Weigend and N. A. Gershenfeld, 175–194. Santa Fe Institute Studies in the Sciences of Complexity, Proc. Vol. XV. Reading, MA: Addison-Wesley, 1993.

136. Saund, E. "Dimensionality-Reduction Using Connectionist Networks." *IEEE Transactions on Pattern Analysis and Machine Intelligence (T-PAMI)* **11** (1989): 304–314.

137. Schuster, A. "On the Investigation of Hidden Periodicities with Applications to a Supposed 26-Day Period of Meteorological Phenomena." *Terr. Mag.* **3** (1898): 13–41.

138. Schwartz, E. I. "Where Neural Networks are Already at Work: Putting AI to Work in the Markets." *Bus. Week* **November 2** (1992): 136–137.

139. Shannon, C. E. "A Mathematical Theory of Communication." *Bell Syst. Tech. J.* **27** (1948): 379–423, 623–656. Reprinted in *Key Papers in the Development of Information Theory,* edited by D. Slepian, 5–18. New York: IEEE Press.

140. Shaw, R. S. "Strange Attractors, Chaotic Behavior and Information Flow." *Z. Naturforsch.* **36A** (1981): 80–112.

141. Smith, L. A. "Does a Meeting in Santa Fe Imply Chaos?" In *Time Series Prediction: Forecasting the Future and Understanding the Past*, edited by A. S. Weigend and N. A. Gershenfeld, 323–344. Santa Fe Institute Studies in the Sciences of Complexity, Proc. Vol. XV. Reading, MA: Addison-Wesley, 1993.

142. Smolensky, P., M. C. Mozer, and D. E. Rumelhart, eds. *Mathematical Perspectives on Neural Networks.* Hillsdale, NJ: Lawrence Erlbaum, 1994.

143. Solomonoff, R. J. "A Formal Theory of Induction Inference, Parts I and II." *Information & Control* **7** (1964): 1–22, 221–254.

144. Subba Rao, T. "Analysis of Nonlinear Time Series (and Chaos) by Bispectral Methods." In *Nonlinear Modeling and Forecasting*, edited by M. Casdagli and S. Eubank. Santa Fe Institute Studies in the Sciences of Complexity, Proc. Vol. XII, 199–226. Reading, MA: Addison-Wesley, 1992.

145. Sussman, G. J., and J. Wisdom. "Numerical Evidence that the Motion of Pluto is Chaotic." *Science* **241** (1988): 433–437.

146. Sussman, G. J., and J. Wisdom. "Chaotic Evolution of the Solar System." *Science* **257** (1992): 56–62.

147. Svarer, C., L. K. Hansen, and J. Larsen. "On Design and Evaluation of Tapped-Delay Neural Network Architectures." In *IEEE International Conference on Neural Networks, San Francisco (March 1993)*, 46–51. Piscataway, NJ: IEEE Service Center, 1993.

148. Swinney, H. "Spatio-Temporal Patterns: Observations and Analysis." In *Time Series Prediction: Forecasting the Future and Understanding the Past*, edited by A. S. Weigend and N. A. Gershenfeld, 557–568. Santa Fe Institute Studies in the Sciences of Complexity, Proc. Vol. XV. Reading, MA: Addison-Wesley, 1993.

149. Takens, F. "Detecting Strange Attractors in Turbulence." In *Dynamical Systems and Turbulence*, edited by D. A. Rand and L.-S. Young. Lecture Notes in Mathematics, Vol. 898, 336–381. Warwick, 1980. Berlin: Springer-Verlag, 1981.

150. Temam, R. *Infinite-Dimensional Dynamical Systems in Mechanics and Physics*. Applied Mathematical Sciences, Vol. 68. Berlin: Springer-Verlag, 1988.

151. Theiler, J. "Estimating Fractal Dimension." *J. Opt. Soc. Am. A* **7(6)** (1990): 1055–1073.

152. Theiler, J. "Some Comments on the Correlation Dimension of $1/f^\alpha$ Noise." *Phys. Lett. A* **155** (1991): 480–493.

153. Theiler, J., P. S. Linsay, and D. M. Rubin. "Detecting Nonlinearity in Data with Long Coherence Times." In *Time Series Prediction: Forecasting the Future and Understanding the Past*, edited by A. S. Weigend and N. A. Gershenfeld, 429–456. Santa Fe Institute Studies in the Sciences of Complexity, Proc. Vol. XV. Reading, MA: Addison-Wesley, 1993.

154. Thearling, K. Personal communication, 1992

155. Tong, H., and K. S. Lim. "Threshold Autoregression, Limit Cycles and Cyclical Data." *J. Roy. Stat. Soc. B* **42** (1980): 245–292.

156. Tong, H. *Nonlinear Time Series Analysis: A Dynamical Systems Approach*. Oxford: Oxford University Press, 1990.

157. Trunk, G. V. "Representation and Analysis of Signals: Statistical Estimation of Intrinsic Dimensionality and Parameter Identification." *General Systems* **13** (1968): 49–76.

158. Tukey, J. W. *Exploratory Data Analysis*. Reading, MA: Addison-Wesley, 1977.

159. Turing, A. M. "On Computable Numbers, with an Application to the Entscheidungsproblem." *Proc. London Math. Soc.* **42** (1936): 230–265.

160. Ulam, S. "The Scottish Book: A Collection of Mathematical Problems." Unpublished manuscript, 1957. See also the special issue on S. Ulam: *Los Alamos Science* **15** (1987).

161. Volterra, V. *Theory of Functionals and of Integral and Integro-Differential Equations*. New York: Dover, 1959.

162. Wallace, C. S., and D. M. Boulton. "An Information Measure for Classification." *Comp. J.* **11** (1968): 185–195.

163. Wan, E. A. "Time Series Prediction by Using a Connectionist Network with Internal Delay Lines." In *Time Series Prediction: Forecasting the Future and Understanding the Past*, edited by A. S. Weigend and N. A. Gershenfeld, 175–194. Santa Fe Institute Studies in the Sciences of Complexity, Proc. Vol. XV. Reading, MA: Addison-Wesley, 1993.

164. Weigend, A. S. "Connectionist Architectures for Time Series Prediction of Dynamic Systems." Ph.D. Thesis, Stanford University, 1991.
165. Weigend, A. S., B. A. Huberman, and D. E. Rumelhart. "Predicting the Future: A Connectionist Approach." *Intl. J. Neur. Sys.* **1** (1990): 193–209.
166. Weigend, A. S., and D. E. Rumelhart. "The Effective Dimension of the Space of Hidden Units." In *Proceedings of International Joint Conference on Neural Networks, Singapore*, 2069–2074. Piscataway, NY: IEEE Service Center, 1991.
167. Weigend, A. S., and D. E. Rumelhart. "Generalization Through Minimal Networks with Application to Forecasting." In *INTERFACE '91—23rd Symposium on the Interface: Computing Science and Statistics*, edited by E. M. Keramidas, 362–370. Conference held in Seattle, WA, in April 1991. Interface Foundation of North America, 1991.
168. Weigend, A., B. A. Huberman, and D. E. Rumelhart. "Predicting Sunspots and Exchange Rates with Connectionist Networks." In *Nonlinear Modeling and Forecasting*, edited by M. Casdagli and S. Eubank. Santa Fe Institute Studies in the Sciences of Complexity, Proc. Vol. XII, 395–432. Redwood City, CA: Addison-Wesley, 1992.
169. Weigend, A. S. "On Overfitting and the Effective Number of Hidden Units." In *Proceedings of the 1993 Connectionist Models Summer School*, edited by M. C. Mozer, P. Smolensky, D. S. Touretzky, J. L. Elman, and A. S. Weigend, 335–342. Hillsdale, NJ: Erlbaum Associates, 1994.
170. Werbos, P. "Beyond Regression: New Tools for Prediction and Analysis in the Behavioral Sciences." Ph.D. Thesis, Harvard University, Cambridge, MA, 1974.
171. White, H. "Connectionist Nonparametric Regression: Multilayer Feedforward Networks Can Learn Arbitrary Mappings." *Neur. Net.* **3** (1990): 535–549.
172. White, D. A., and D. A. Sofge, eds. *Handbook of Intelligent Control.* Van Nostrand Reinhold, 1992.
173. Widrow, B., and M. E. Hoff. "Adaptive Switching Circuits." In *1960 IRE WESCON Convention Record*, Vol. 4, 96–104. New York: IRE, 1960.
174. Wiener, N. *The Extrapolation, Interpolation and Smoothing of Stationary Time Series with Engineering Applications.* New York: Wiley, 1949.
175. Yip, K. M.-K. "Understanding Complex Dynamics by Visual and Symbolic Reasoning." *Art. Intel.* **51** (1991): 179–221.
176. Yule, G. "On a Method of Investigating Periodicity in Disturbed Series with Special Reference to Wolfer's Sunspot Numbers." *Phil. Trans. Roy. Soc. London* **A 226** (1927): 267–298.
177. Zhang, X., and J. Hutchinson. "Simple Architectures on Fast Machines: Practical Issues in Nonlinear Time Series Prediction." In *Time Series Prediction: Forecasting the Future and Understanding the Past*, edited by A. S. Weigend and N. A. Gershenfeld, 219–242. Santa Fe Institute Studies in the Sciences of Complexity, Proc. Vol. XV. Reading, MA: Addison-Wesley, 1993.

Index

Milton Keynes UK
Ingram Content Group UK Ltd.
UKHW020010071024
449327UK00031B/2729